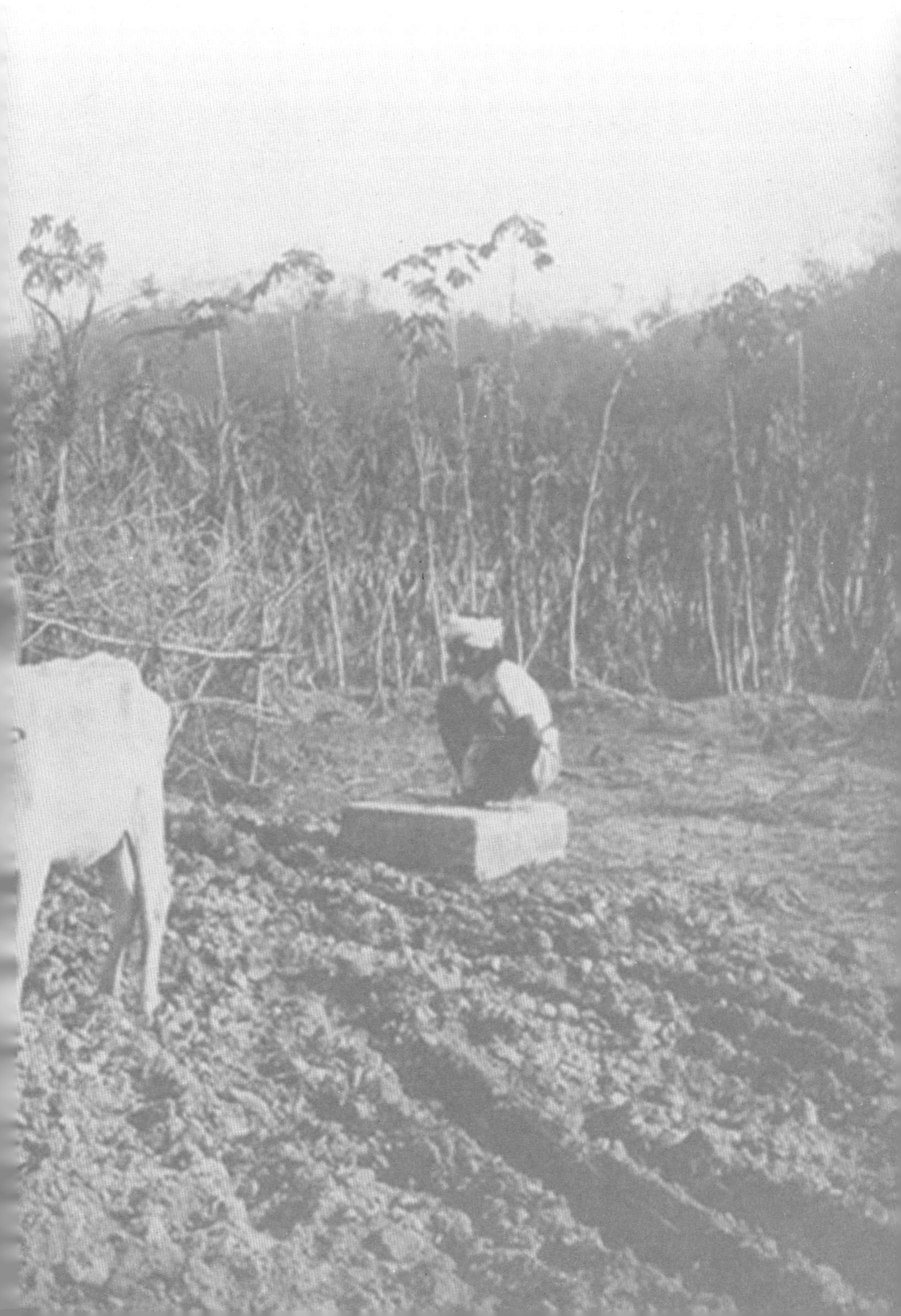

TILLING THE LAND

TILLING THE LAND
Agricultural Knowledge and
Practices in Colonial India

edited by
DEEPAK KUMAR
BIPASHA RAHA

PRIMUS
BOOKS

PRIMUS BOOKS
An imprint of Ratna Sagar P. Ltd.
Virat Bhavan
Mukherjee Nagar Commercial Complex
Delhi 110 009

Offices at CHENNAI LUCKNOW
AGRA AHMEDABAD BENGALURU COIMBATORE DEHRADUN GUWAHATI HYDERABAD
JAIPUR JALANDHAR KANPUR KOCHI KOLKATA MADURAI MUMBAI PATNA
RANCHI VARANASI

© *Deepak Kumar and Bipasha Raha for Introducduction and editorial selection 2016*
© *Individual contributors for their respective essays 2016*

All rights reserved. No part of this publication may be reproduced,
stored in a retrieval system or transmitted, in any form or by any means,
without the prior permission in writing from Primus Books, or as
expressly permitted by law, by licence, or under terms agreed with
the appropriate reproduction rights organization.
Enquiries concerning reproduction outside the scope of the above should
be sent to Primus Books at the address above.

First published 2016

ISBN: 978-93-84092-80-1 (hardback)
ISBN: 978-93-84092-81-8 (POD)
ISBN: 978-93-84092-82-5 (e-book)

Published by Primus Books

Lasertypeset by Sai Graphic Design
Arakashan Road, Paharganj, New Delhi 110 055

Printed and bound by Sanat Printers, Kundli, Haryana

This book is meant for educational and learning purposes. The author(s) of the book
has/have taken all reasonable care to ensure that the contents of the book do not
violate any existing copyright or other intellectual property rights of any person in
any manner whatsoever. In the event the author(s) has/have been unable to track
any source and if any copyright has been inadvertently infringed, please notify
the publisher in writing for corrective action.

In the memory of
PROFESSOR BARUN DE

Contents

List of Tables		ix
Preface		xi
1.	Introduction	1
2.	Science in Agriculture: A Study in Victorian India DEEPAK KUMAR	20
3.	Contextualizing Modern Science in Agriculture in Colonial Bengal, 1876–1928: A Case of Productionist Discourse ARNAB ROY	49
4.	Modernizing Agriculture in the Colonial Era: A View from Some Hindi Periodicals, 1880–1940 SANDIPAN BAKSI	71
5.	Missionaries as Agricultural Pioneers: Protestant Missionaries and Agricultural Improvements in Twentieth-Century India RAJSEKHAR BASU	99
6.	Transformation of Agricultural Practices: An Indigenous Experiment in Colonial Bengal BIPASHA RAHA	122
7.	Barbaric Hoe and Civilized Plough: Tribes, Civilizational Discourse and Colonial Agriculture in the Khasi-Jaintia Hills of North-Eastern India SAJAL NAG	145
8.	Agricultural Knowledge and Practices in a Bengal District: Burdwan under Colonial Rule ACHINTYA KUMAR DUTTA	168
9.	Strains of Settlement: Reclamation and Cultivation in the Sundarbans—Myth and History SUTAPA CHATTERJEE	192

10. Cattle, Cruelty, Cow Doctors: Examining Animal Health in Rural Bengal, 1850–1920
   SAMIPARNA SAMANTA   214

11. Cattle Breeding Policies in Colonial India
    HIMANSHU UPADHYAY   238

12. Agrarian Distress: The Political Economy of British India in the 1930s
    S.M. MISHRA   260

13. In Quest of Plenty: Hunger and Agricultural Technology in India, 1955–1967
    MADHUMITA SAHA   282

    Select Bibliography   303

    Notes on Editors and Contributors   315

    Index   317

Tables

6.1	The Normal Acreage of Principal Crops in the District (1910)	133
7.1	Number of Wasteland Applications	158
8.1	Output, Consumption and Surplus of Rice (in tons) in Burdwan District (Decennial Statistics)	170
8.2	Area (in acres) Cropped in Burdwan District (Decennial Statistics)	176
8.3	Area (in acres) Irrigated in Burdwan District (Decennial Statistics)	181
10.1	Animal Mortality Rates in Assam from 16 to 31 May 1869	216

Preface

Over the years, much attention has been devoted to the colonial government's unparalleled attempt to maximize land revenue which initiated devastating alterations in the social framework of the rural countryside and prompted a series of experiments with land revenue settlements that heralded a significant transformation of the economy in different parts of the country. The impact was such that even during colonial rule, people found it difficult to remain oblivious to the often devastating repercussions, like famines, pauperization, rural indebtedness, etc. Writers like Romesh Chandra Dutt articulated their concern in their work emphasizing the manner in which the frailty of the narrow base of the Indian economy was exposed by recurrent famines.

Interest in agrarian history can be traced back to the 1950s and considerable serious research has been undertaken on colonial land policies and their implications in different regions. Any attempt at tracing the historiography of agrarian history of the colonial period should rightfully start with a discussion on the seminal contribution made by B.B. Chaudhuri. His entry into this domain of research began with a study of the problems of the Bengal peasantry before and after the introduction of the Permanent Settlement in Bengal. He then focussed on eastern India. Among the issues he addressed were the official policy of land revenue management, system of agricultural production, agrarian relations, commercialization of agriculture and the rural credit scenario. He began with the assumption that colonial policy of maximization of revenue set the tone for land and tenurial legislations. This led to the study of the impact of the commercialization of agriculture on the peasantry and agrarian relations that brought into historical discourse the issue of peasant resistance and the peasant acquired an independent identity. Suprakash Roy's probe into the struggles of the peasants against zamindari, sahukari and sarkari oppressions is one of the earliest such studies. Kathleen Gough offered an entirely new perspective by classifying peasant rebellions in colonial India using different yardsticks, like their goals, ideological compulsions and methods.

Daniel and Alice Thorner opened up new areas of research by studying the agrarian structure that evolved consequent to the formulation of the British land tenure systems. Soon, agriculture, agrarian relations, agricultural organizations, agricultural labour, rural credit, the emergence of a dominant peasantry and peasant responses to these changes attracted new researches. Nilmani Mukherjee's elaborate examination of the ryotwari system in Madras was one of the earliest. Beginning with the formulation of the tenurial system, he went on to study its impact on the regional economy. Dharma Kumar's study of land and caste and the practice of bonded labour, highlighting the social roots of bondage, was a significant pioneering work. The publication of Bipan Chandra's seminal work kindled greater interest in the economic impact of colonial rule. Ravinder Kumar followed this up with his study of the impact of colonial rule on western India in the nineteenth century. For an understanding of the Deccan Riots, it becomes necessary to be acquainted with the credit system that existed in the region and the consequent credit relations.

'Agrarian' refers to both agriculture as a sector of the economy as well as the broad social structure that underlies it. The former includes issues pertaining specifically to agriculture, viz., techniques of production, land productivity, organization of production and marketing, as also allied activities such as dairy, animal husbandry, etc. The latter, on the contrary, refers to land settlements, agrarian relations, and the rights of the peasantry—their condition, credit relations, tenancy legislations and rent questions. In most agrarian studies of the colonial period conducted in the 1970s and 1980s, it is the latter that has been primarily focused upon.

Gautam Bhadra, who studies the various nuances of the nineteenth century, added to the existing knowledge on the evolution of land relations in eastern India. Sirajul Islam began with a study of the attempts of Lord Cornwallis to put land revenue extraction on a more secure footing, which evolved into the Permanent Settlement, that went on to probe how the subsequent legislations, till 1819, attempted to enable zamindars to extract peasant surplus. Chittabrata Palit contributed to the existing historiography on eastern India with his work on agrarian relations in the rural countryside of Bengal. He focused on the increasing tension in the period that saw peasant resistance in the forms of Wahabi, Farazi, Santal and the Indigo rebellions. In the same year was published Nilmani Mukherjee's biography of his forefather, Joykrishna Mukherjee, the zamindar of Uttarpara, in which he attempts to examine the zamindari system in Bengal through the life and times of one of the leading zamindars of Bengal. In the meanwhile, considerable work continued to be undertaken on peasant resistance in Bengal. Blair B. Kling had shown how all sections of rural society had participated in the Indigo Rebellion in 1860. Narahari Kaviraj focused on the

Wahabi and Farazi movements that convulsed large areas in eastern Bengal, while Sunil Sen worked on the Tebhaga Movement which saw the participation of a highly politicized peasantry. The fact that he had himself participated in the Movement gave greater authenticity to his account. This was followed by Ranajit Guha's famous study on the Indigo Rebellion and added greatly to our understanding of peasant resistance to exploitation and its handling by the middle class. The agrarian disturbances that swept across wide tracts in eastern Bengal in 1873, referred to as the Pabna Revolt, caused by high rental exactions of the zamindars, was the subject matter of Kalyan Kumar Sengupta's study while B.B. Chaudhuri looked at these events from an entirely different angle. He continued to ceaselessly study the agrarian history of colonial Bengal. No study of the historiography of agrarian Bengal would be complete without mention of Ratnalekha Ray's seminal work on changes in the agrarian society of Bengal caused by British land legislations, over a period spanning almost a century.

I.J. Catanach's study on rural credit in western India traces the background to peasant indebtedness, the stranglehold of the moneylenders on the rural economy, that led to the resistance of the peasants and the beginning of the cooperative movement in the Bombay presidency. Bonded labour, the *hali* system, the nature of patronage and the element of exploitation in agrarian relations, prevalent in southern Gujarat, were studied by Jan Bremen, who also examined the role of the caste system. Aditya Mukherjee, on the other hand, made a case study of five districts of the Brahmaputra valley to provide a glimpse of the agrarian conditions in a region on the diametrically opposite side, i.e. Assam. His research opened up a new area of study. Meanwhile, our understanding of agrarian north India was enriched by two significant studies, one by Thomas R. Metcalf and the other by Asya Siddiqi. The former focused on land relations in the nineteenth century and the latter on Uttar Pradesh.

The 1980s opened with the publication of the *Subaltern Studies* series. Historians associated with this project, led by Ranajit Guha, took up diverse issues pertaining to changing agrarian relations and the peasantry, primarily through case studies of different regions ranging from Gorakhpur to the Gudem Rampa hills. The study of the peasantry, thus acquired a new dimension. Ranajit Guha supplemented the growing historiography with an eloquent work on the elements of consciousness of the insurgent peasants. The publication of the *Cambridge History of India*, in two volumes, added considerably to the expanding knowledge of agrarian changes in India from the Mughal to the colonial era.

In the next two decades, more Subaltern Studies followed. The changing agrarian landscape in Punjab in the colonial context attracted other voices and other perspectives through Himadri Banerjee, Imran Ali, Mridula

Mukherjee, Neeladri Bhattacharya, Indu Banga and Chhanda Chatterjee among others. S. Ambirajan, Maijd Siddiqui and Shahid Amin worked on northern India, Uttar Pradesh in particular. Crispin Bates and Prabhu Prasad Mohapatra worked on central India, Neil Charlesworth on Bombay presidency and David Hardiman on Gujarat. K. Saradamoni's study of the Pulayas in Kerala continued the trend set by Dharma Kumar's study on bonded labour. David Ludden and later Arun Bandyopadhyay did significant work on the agrarian economy of Tamil Nadu. However, the historiography on agrarian Bengal in the colonial era has continued to hold the interest of academics. B.B. Chaudhuri has contributed immensely through the 1980s and the 1990s. Partha Chatterjee, Sugata Bose and Saugata Mukherjee have studied agrarian structures, tenancy legislations, peasant issues and agrarian politics.

In recent years, since the beginning of the twenty-first century, the scope of historical research has broadened considerably. In an attempt to unravel various aspects of the lived experiences of the colonial era, the historians have added different dimensions to the nature of historical research. With growing emphasis on social and cultural history, agrarian history has somehow taken a back seat. While it had once been fashionable to study agrarian issues so vital to life in the past, historians have now moved on to the histories of science, health, medicine, environment, gender, sports, etc. B.B. Chaudhuri on Bengal, Mridula Mukherjee on Punjab, and Arupjyoti Saikia on Assam are among those who have continued to persist in the quest to add to the existing knowledge on agrarian history. Lately, some light is also beginning to be thrown on the area of agrarian thought.

This volume attempts to rekindle interest in the history of agriculture, its knowledge and practices. The emphasis is not so much on the changing agrarian social or revenue structure since the colonial period, as was the preoccupation of scholars over the years, but on issues pertaining specifically to agricultural knowledge and practices.

It is now time to acknowledge the intellectual and moral debt that we have incurred as we endeavour to look at India's agrarian history a little differently. The idea of this book was first conceptualized during the conference of the International Association of the Historians of Asia held in Jawaharlal Nehru University in 2008. We must first thank all the contributors who have helped to enrich this volume in spite of their busy schedules. Arnab Roy helped prepare a select bibliography. The editors acknowledge gratefully the labour and cooperation of the contributors and the publisher. An edited volume is always a joint enterprise and we hope that readers will find this work useful.

DEEPAK KUMAR
BIPASHA RAHA

1

Introduction

Deepak Kumar and *Bipasha Raha*

Britain has enslaved India, it is time she should enfranchise her; she has plundered her, it is time she should invest her with a portion of wealth . . . Turning to the capitalists of Britain, we say, embark some portion of your redundant riches, in speculations of Indian agriculture. Grow sugar, rice, cotton, indigo, even wheat, even tea, encourage the silkworm, culture of drugs (tobacco and opium) and abolish the monopolies of those drugs and that of salt.
— *Alexander's East India and Colonial Magazine,* November 1834

Does the problem of Indian agriculture begin and end only with the knowledge and use of scientific methods of agriculture? Are there any more vital reasons which adversely affect Indian agriculture?
—*The Mahratta,* October 1926

The crucial word in the first quote is 'speculation'. One wonders whether agriculture in pre-colonial India was speculative. Cotton, sugar, indigo and opium were grown earlier also but under the benign Raj, they became 'staple'.[1] With the ever growing preference for cash crops, the traditional agricultural knowledge and practices came under new scrutiny and pressure. Colonial capital was now the prime mover. Even then the number of famines kept mounting—the famines of 1830s, 1860s and 1870s being utterly severe.[2] The second quote from a 'native' newspaper hits the nail. Was scientific knowledge alone the solution? Alluding to the extreme poverty and degradation in 1835, the then Governor General, William Bentinck roared,

For all these evils, knowledge, knowledge, knowledge, is the universal cure. We must not forget that the Government is the landlord of the country, possessing both the means and knowledge of improvement, and, putting all obligations of public duty aside, is most interested in the advancement of the wealth and comfort of its numerous tenantry.[3]

But these high-level pronouncements were not meant for implementation. The ground realities were different. Of course, the Raj generated a lot of data and produced many texts on agriculture, but the improvements depended on many socio-economic internalities as well as externalities. The rulers were quick to diagnose the ills—that the agricultural fields were scattered and fragmented, holdings were small, irrigation deficient and the tools primitive. Buchanan wailed over the 'want of skill' while some like Wight were satisfied with the local plough. The dominant discourse nevertheless was of decadence and misery.[4] But this does not explain why the then available capital could not be used for improvement and why technical knowledge was not employed for this purpose. Instead, the peasants were burdened with a revenue system that crippled them. The pre-colonial economy had some occupational diversity; the artisans and the farmers supported and complemented each other to a large extent. But the colonial economy destroyed the artisans and this led to an unprecedented increase in the number of landless labourers. The balance was gone. To add to woes, an extensive cash crop economy, geared to serve the external market, was cultivated. The result was an 'external integration' with the metropolis, without an accompanying growth of an 'integrated' internal market. Hamza Alavi rightly calls it 'internal disarticulation'.[5] All push on the cash crops could not satisfy even the colonial planters. As a planter jingle goes:

The Lion King stretched out his hand,
Talked of the cheapness of labour and richness of land,
Of twenty maunds a bigha
Take the cipher from the aught, divide the ten by two,
The result will be the produce exceeded but by few,
Then things went on right jolly,
Till the district was dotted o'er with monuments of folly.[6]

It is not without reason that the colonies were also called plantations. Naturally the planters who came from distant lands to invest and work on plantations would have profit as the sole motivation. This was a new enterprise. It required good knowledge of seeds, plants, soil, manure, irrigation, etc. Modern science took root in the colonies, inter alia with this knowledge. The early colonizers traded almost exclusively in plant products and this had a lot to do with botanical knowledge.

The rule of the East India Company had initiated an era of exploration that was unprecedented in Indian history. This exploratory phase had certain interesting characteristics. First, the explorers always looked for things and materials that would be 'useful', commercially or otherwise. Second, they never lost sight of the knowledge value of their enterprise. They worked under some sort of a dual mandate under which both the commercial and the knowledge components had to be taken care of. Third, unlike Victorian

Africa, where the explorers arrived much before the flag and trade, in India, these three moved together in tandem. There was not much planning and coordination, but their interests did converge and there did exist a 'method' in the colonial 'madness'. Fourth, they could not only claim conceptual and technological superiority but also dismiss the 'other' or local knowledge, tools and techniques as useless and antiquated.

Ancient Inheritance

The archaeological explorations from the sites of the Indus Valley Civilization have thrown good evidences of botanical knowledge. Wheat, barley, rice, cotton and different kinds of goods had been found, and in the Vedic period, there are numerous references to plants and cereals in the *Samhitas* and the *Brahmanas*. Of course, botanical knowledge had not been systematized, but people knew about the morphology and distribution of plants. Even some categorization was attempted. For example, some plants were described as those of the spreading variety (*prastranati*), some bushy (*stambini*), some jointed (*kandini*), some with large branches (*vishakha*), and so on. They were broadly divided as trees (*vrksha*), herbs (*osadhi*) and creepers (*virudh*). Plants appear extensively in the medical literature. For Charaka, an expert physician is one who knows the herbs botanically, pharmacologically, and in every other respect. Without plant knowledge, *Bhesajavidya* would be utterly incomplete. In fact, a distinct knowledge (*veda*) appeared known as *Vrksayurveda* and Kautilya's *Arthasastra* devoted separate sections to it, dealing with collection and selection of seeds, germination, grafting, cutting, manuring and also the meteorological conditions. Later in the first century AD, Parasara composed *Vrksayurveda* which gives, in a rudimentary way, an outline of plant morphology, nature and properties of soil, description and distribution of forests, functions and classification of flowers and fruits, and details on seeds (*Bija*). The text by Parasara is really unique and truly reflects the state of botanical knowledge. Arboriculture and horticulture are also referred to in many texts. The *Upavanvinoda* (pleasures of gardening) are referred to in great detail. However, agriculture remained central to plant knowledge and vice versa. The Vedic farmers knew the seeds, the methods to sow and also the use of plough. It is said that the system of sowing began in India. They also knew the method of improving the fertility of the soil through crop rotation. They knew the significance of irrigation and constructed channels for carrying water from different sources. It is possible that they used wheels and earthen pots (*ghada*) to lift water. The Mauryans are known to have patronized irrigation schemes and constructed large tanks. In south India, huge tanks were constructed and some of them survived for centuries. The *Arthasastra* also talks of meteorological knowledge. Rainfall was monitored and recorded. Soil classification and land use was discussed in great detail. Land was classified in terms of *karsa* (cultivated), *usara* (wasteland) and *gocara* (pasture). The *Amarkosa* describes twelve types of lands in terms of their fertility and physical characteristics and it also describes a number of agricultural implements. The treatment of seeds and manuring the soil to ensure a good crop are discussed in great detail. All this was empirical knowledge which was found useful through long periods of observation and understanding.

Was there anything like a science of plants in pre-colonial India? Probably not; of course, there are references to *Krishitantra*, *Vrksayurveda* and *Bhesajvidya* in the works of Kasyapa, Parasar, Saraswata and Kautilya, but this knowledge was definitely made subservient to philosophy, the science of medicine or the knowledge and techniques of agriculture. Majumdar attributes this neglect to the fact that scientific cognition and results of observations were not kept sufficiently distinct from the popular notions, guesses and superstitions. He adheres to a familiar trajectory: a brilliant beginning, marked progress to a certain stage, and a tragic stagnation, thanks to the Muslim invaders.[7] It is difficult to believe in the stagnation theory but the area is certainly under-researched. The Mughals are known for their keen interest in plants and gardens. A seventeenth-century text, *Dar Fann-i-Falahat*, describes the various methods of grafting, preparation of soil, harvesting techniques, water and manure requirements, etc. Interestingly, it also refers to male and female plants! Later, several tracts were written in Persian on various useful plants, for example, *Nakhl-bandiya* by Ahmed Ali in 1790 and *Nuskha-i-Kukh-bad* by Amanullah Husain.[8] Yet it may be correct to infer that the 'oriental' learning had no 'state of the art' knowledge. But here certain questions do emerge. What were the 'European' and 'non-European' precepts of plant or other kinds of knowledge in the early modern world? Could they interact; could they change? Was a synthesis or co-production possible? How did Europeans and South Asians develop working relationships in 'knowledge making enterprises'? Interaction did take place but a synthesis or co-production remained a far cry.

Colonial Experiments

During the second half of the eighteenth century, the East India Company had secured a firm grip at least over Bengal and Madras and its rivals were on the wane. The Company appreciated the significance of botanical and geographical investigations and encouraged its interested employees to undertake such activities. An early example is of James Anderson who joined the East India Company in 1759. In 1778, he obtained a large piece of land at Nungambakkam near Fort St. George, from the Madras Government, which he developed as a botanical garden where he experimented with introduction of cochineal insects, silkworms and plants of commercial value such as sugar cane, coffee, American cotton and also European apples. A little later, a botanical garden was established at Shibpur near Calcutta by Robert Kyd. This garden was nursed and developed by William Roxburgh. This botanist was one of the first to reflect on the plight of the ordinary cultivators and think of agrarian solutions. While working at Samalcotta near Madras, Roxburgh was moved by the poverty of the people and called for the introduction of plants (like jackfruit) that would furnish sustenance to the poor in times of scarcity. For scarcities, he frankly blamed the system and administration.[9] Almost a century later, a visiting agricultural scientist J.A.Voelcker made similar remarks.[10] But these were individual voices and the government of the day was not obliged to listen to them.

The colonizers looked at farming as an enterprise 'defined by input-output accounting'.[11] Surveys by Buchanan, Munro and others brought forth certain statistical details on agricultural resources. The revenue settlement done by Cornwallis was aimed at maximum appropriation of agricultural wealth as directly as possible. But the state input in agriculture was to be minimal and indirect. The improvement was sought through private societies and efforts. In September 1820, the Agricultural and Horticultural Society of India (AHSI) was established and its founder, William Carey, hoped that 'it will ultimately be of great benefit to the country and contribute to prepare its inhabitants for the time when they shall beat their swords into ploughshares, and their spears into pruning hooks'.[12] A hope largely belied as rural India continued to groan and bleed under an unprecedented pressure to grow more and more cash crops.

It is from this point that the first paper in this volume begins and moves chronologically. For almost a century, there was no official agricultural policy but the private societies and planters were bestowed the responsibility of 'improvement'. The AHSI imported seeds for distribution and organized agricultural fairs. Experimental farms were established in different parts of the country. They were expected to experiment on exotic seeds, implements, manures and acclimatize the same to local conditions. Even planters from America descended upon Dharwar and Gujarat to grow cotton.[13] But all these experiments could not bring the desired result, they remained 'exotic'. The author finds three distinct phases in the attitude of the colonial government. The first was one of 'masterly inactivity' on the part of the latter and 'superactivity' on that of the private societies and the planters, who saw immense possibilities of plantation agriculture and a cash crop boom in the luxuriance of the tropical virgin forests. In the next phase, there were hectic but confused activities, when several 'well intentioned' scientific men tried to impress upon the government, without much success, to introduce science into agriculture. Private bodies like the Indian Tea Association showed more interest in entomological researches than the Government of India. In the third phase, the government recognized the importance of science in agriculture, but was reluctant to make financial investments in projects that would not be immediately remunerative. They were willing to invest in railways but not in irrigation and agriculture.[14] Attempts at dissemination of scientific agriculture faced several obstacles. The author attempts to probe some of them in terms of official policies and local response. The pragmatic, though illiterate, cultivator would refuse to improve the yield as the profit would be taken away by the zamindar. But in the wake of the most severe famines in the 1860s and 1870s, the government was forced to establish model farms, agricultural departments, agricultural schools, etc. Richard Temple, Lt. Governor of Bengal, wanted instruction in agriculture 'to form an essential branch of our State Education in India'.[15] At the Agricultural Conference in October 1890, two significant questions were raised: (1) Is special teaching in agriculture desirable? (2) Should it proceed from above downwards, or from below upwards?[16] The dilemma of 1820s–30s was still there. Voelcker, the Royal Agricultural Chemist who was then visiting India, tried to resolve the issue and produced an excellent report which the government thought more prudent to shelve in the name of financial constraints. But during the late Victorian era, a good deal of public awareness and pressure had built up. The rising Indian bourgeoisie was not oblivious to the importance of science in agriculture and raised related issues on several forums. The major problem, concludes Kumar, was not cultural stagnation or social conservatism of the Indians; it was rather finding economically viable, appropriate technological solutions. The government first ignored it and later, 'tried to use science as a crutch which could at best provide a halting gait'.

People's Response

More than the government of the day, some people knew that novelty or 'exotic' did not really mean improvement. Towards the last decade of the nineteenth century, one finds a large number of journals, articles, resolutions, pamphlets, etc., asking for agricultural education. These also dealt with questions like who was to benefit from the agricultural schools, what should be the curriculum, and of what use were the bulletins issued by the agricultural department and their model farms? There were farmers who promptly dismissed any use of theoretical knowledge or science for an agriculturist. Pointing to the experimental stations maintained at great cost by various agriculture societies and governments, they asked what good had resulted to the agricultural community from these costly experiments.[17] When an Agriculture Research Institute was opened at Pusa, thanks to a magnificent grant in 1903 of $100,000 from an American philanthropist, similar questions were asked. How was it going to serve the illiterate cultivators with reports and bulletins in a highly technical language?[18] Pusa was intended to be 'scientific' and the practical details of agricultural improvement were left to the provincial agricultural departments.[19] The *Amrita Bazar Patrika* asked for vernacular leaflets, brochures and directions clothed in as popular a garb as possible. Along with the use of the vernacular press for the dissemination of useful information, the agency of pathsalas and primary schools, post offices and panchayats needed to be utilized. It argued that 'if one has to wait till the masses educate themselves so far as to understand the dry, obscure and scientific language of the agricultural ledgers and memoirs, one will have to wait till, say, doomsday!'[20] So the medium had to be the local language and conveyed in a manner which could be easily understood. But the officials had another dilemma; they were convinced that the Indian languages were not competent to convey the 'new' knowledge. As Moreland wrote, 'if the sciences are to be taught, either the English scientific terminology must be borrowed or a new terminology invented. It must be remembered that the Indian classical languages cannot fill the gap as they do in the case of philosophy . . .'.[21] One may argue that by the turn of the century an agrarian bourgeoisie had appeared and its interactions and altercations with the colonial government had created what Arnab Roy calls a 'productionist' discourse. His paper examines the attitude, role and Indian engagement in internalizing and critically assimilating a modern scientific tradition of agriculture in colonial Bengal. The reasons that prompted the colonial government's interest in promoting modern agriculture and the engagement of the Bengalis are questions that the author attempts to answer. Roy uses vernacular source materials to understand the Indian literati's engagement in furthering scientific agriculture. He argues that the colonial authority had to take up a productionist discourse on agriculture, either to safeguard profit from land revenue or to divert peasant discontent or to consolidate its own cultural and social supremacy. Introduction of modern science into agricultural practices was possible through a proper dissemination of knowledge to the cultivating classes and through the formation of an educational curriculum in agricultural science. It was the former that was given maximum emphasis, not the latter. Roy's paper highlights some hitherto unknown facts on the agricultural farms that emerged in Bengal, particularly in the princely states. He goes on to discuss the paradox of the Bengali intelligentsia who, on the one hand, resorted to Hindu revivalism and, on the other, harped on their penchant for modern scientific knowledge. The government had its own financial excuses. Moreover, the country was too big and its population so varied in cultures and attitudes. Many Indians recognized this. As a Tamil journal from Cuddalore reflected,

The ryots of the country have cooperated with the government and put to practice the knowledge imparted to them by scientific men. The cause of this failure is due to in a large measure to the following circumstance: the country is very vast in extent and the people are very large in number. The efforts of government to impart scientific instruction or to evoke competition or to induce emulation by mean of associations, cattle-shows, experimental farms and the like being confined to small areas, have naturally failed to make any impression on the matter. In our district the Palur Farm has been in existence for the last so many years and yet it is doubtful whether its objects are known to many villagers in the neighborhood. The officials entrusted with the duty of disseminating scientific knowledge are men having manifold duties and their number is also inadequate for the immense population. The result therefore is naturally poor and sometimes even discouraging.[22]

Sandipan Baksi focuses on some leading Hindi periodicals. He rightly argues that the response to agricultural science could be more interestingly traced in the vernacular press as it had greater reach outside the major urban centres. There indeed was a close relationship between the rise of the vernacular press and the development of engagement with modern science and technology in different sections of society. He shows how the Hindi periodicals reflected an acute and persistent sense of dissatisfaction with the existing state of affairs, the dismal levels of agricultural production and productivity, the acute distress in agriculture, and the difficult conditions of subsistence of the cultivating classes. Hindi press echoed the same sentiments which one finds in Bengali or any other Indian language publications.[23] At times, the development of agriculture was argued as the path to revive lost glory. Interestingly, while the need to develop agricultural production and productivity was couched in revivalist terms, the perceived means for attainment of that goal were modern. The persistent disaffection, bordering almost on outrage, with the state of agriculture in the late nineteenth and early twentieth century was clearly associated with an urge for modernization of agriculture. Baksi argues that the discourse on modernization of agricultural science and technology originated from those who represented the interests of a section of the cultivator class that saw a conflict of interests with the zamindars but was not disposed towards sustained opposition to them. Nor did these voices emerge from any radical articulation of the interests of the bulk of the peasantry of the region, given the absence of any reference to social or agrarian movements. Instead, in later years, they tended to incline towards the idea of *gram sudhar* or village improvement but without any focus on land relations or even radical improvement in agricultural productivity.

Rajsekhar Basu's essay examines missionary involvement with agricultural productivity and the introduction of better breeds of cattle and poultry in different rural localities. He argues that by the end of the nineteenth century, Christian overseas missions supported a shift from an all out emphasis on proselytization to social service, which was directly linked to the development of humanity. So along with medicine and health, they turned to agriculture as well and agricultural missions were established to improve conditions of the rural population in different parts of Asia, Africa and Latin America. By this time, research institutes for cereal crops like rice and wheat had been set up and teaching curricula in Indian universities had tried to integrate knowledge systems drawn in from different Western scientific disciplines, viz., entomology, pathology, plant genetics and soil chemistry. So the missionaries decided to focus more on rural agricultural labour and directed their conversion efforts to the low caste among them. It was considered logical that these newly converted Christian communities had to be provided with some degree of economic autonomy and education. This explains setting up of Christian villages which could provide social security to their new converts, and also solve their problems related to poverty. But Basu shows how missionary involvement with the Indian rural world included not only vocational and industrial training, but other activities like agricultural education, crop activity, cattle improvement, cooperatives, village-based agro-industries and land reclamation. The author also examines government attempts to popularize scientific agriculture and the main influences behind such attempts. He says that by the early 1920s, the debate on agricultural improvements became intense. He also discusses the role played by some incredible individuals in initiating agricultural improvements in India, viz., Sam Higginbottom who attempted transformation in the agricultural educational curriculum and John Goheen who introduced an agricultural programme in Sangli that was directed towards meeting the demands of the farmers of the locality, and so forth.[24] By the 1930s, in spite of the Depression, agricultural missionaries continued to be an important part of the missionary-sponsored welfare activities. They were involved with institutions which tried to train agricultural leaders who would be able to deal with the country's rural problems. They also tried to initiate some steps for the socio-economic uplift of the villages themselves.

In the next paper, Bipasha Raha weaves the discourse around one of the greatest cultural icons modern India had produced, i.e. Rabindranath Tagore. Having diagnosed some of the ills besetting Indian agriculture, Tagore evolved a programme of rural revitalization that laid particular emphasis on agrarian development. The poet first familiarized himself with agrarian developments in the Western world. His own agrarian experiments were carried out in three phases where he tried out his ideas: Shilaidaha, Patisar and Sriniketan. Emphasis was on scientific agriculture, mechanization and removal of poverty of the peasantry. Raha shows that the poet advocated the adoption of the principle of cooperation at a time when there was not much government initiative in agrarian development. It was a daunting task that the poet undertook as the ryots were not amenable to the adoption of new methods. When Tagore began his work, he had no models before him to emulate. While the projects at Shilaidaha and Patisar had to be abandoned midway for various reasons, Sriniketan, during the poet's lifetime, attempted something more stable and enduring. It contributed significantly towards a comprehensive development of the life and welfare of the peasantry and the rural folk in general. It was the poet's hope that the results achieved there would ensure widespread application of his programme. The role of Sriniketan was that of an educator. Analysing the nature of the work undertaken in Sriniketan, Raha argues that the role of Sriniketan in experimenting with more advanced programmes of welfare, development, training and research, which could be used as a model by contemporary educators and rural rejuvenators, was unique from the point of view of the demands of the time.

Case Studies

This volume now moves to three specific case studies, one located at the hilly terrains of the North-East, second on the plains of *rarh* Bengal, and the third to the enchanting but neglected mangroves of the Sundarbans. Sajal Nag studies agricultural practices in the North-East. He asserts that the British colonists considered the phenomena of subsistence agriculture and shifting cultivation as primitive agriculture. The British could not believe that the vast agricultural pursuits in India did not generate the same kind of profit as it did in Europe. Hence, Nag argues, the integrated discourse of backward agriculture, lazy natives and migration of stout farmers from neighbouring provinces were put forward by the British throughout colonial India. The tribals were not spared from this discourse as the hill tribes who practised swidden cultivation were also seen as lazy natives who performed less than their potential and produced so little that they could not meet the revenue demands of the state. Nag examines the colonial attempts to transform the tribal agriculture of the Khasis of north-eastern India from a subsistence level of commodity production, in an effort to save the expensive forests and even import and settle agriculturists from Naga hills who were skilled in terrace cultivation in the area, so that they could teach new technology to the Khasi farmers. The author analyses the implications of shifting cultivation in the perception of the British colonists and their attempts to pursuade the Khasi-Jaintias to give up the practice and adopt settled rice culture. He then discusses the ensuing debate in official circles on the subject of *jhumming* in the Jaintia hills and the viability of terrace cultivation. It was generally agreed that shifting cultivation was a destroyer of forest cover. Nag shows that, after some initial efforts, although the state administration gave up the project of shifting to terrace cultivation temporarily, the investigation reports reveal interesting details regarding the existing state of subsistence endeavours of the people, the arguments in favour and against shifting cultivation by the people themselves as well as the administration, the state of forests and the transforming landscapes as well as property relations of the Khasi and Jaintia people. This paper depicts the stress the tribal system of agriculture was facing from the colonial commercial interests and how all this was transforming the economy and society of the people. In the colonial discourse, slash and burn cultivation was seen as the lowest form of agriculture pursued by communities who were tribal and in a savage and barbaric stage of civilization and one way of introducing them to civilization was to transform their agriculture.

Taking Burdwan in Bengal as a case study, Achintya Datta throws some light on how the peasants of a fertile region maintained techniques of cultivation and production and also responded to the changes introduced in this field and to the extension of knowledge. He argues that an appraisal of the agrarian history of Burdwan, agrarian changes, viz., the means and methods of cultivation and production, use of implements, seeds and manures, irrigation system, etc., help to recognize the diversities and distinct features of agrarian practices and knowledge at a local level. The essay begins with a rejoinder to the debate on the growth rate of Indian food crop output and efficiency of the method and means of production. Datta describes how Burdwan as a surplus district had helped Bengal over decades 'to keep her flow of rice supply unabated to the areas where it was in great demand'. He discusses the resource endowments for agricultural production in the district. Particular attention is paid to the existing irrigation facilities and irrigational knowledge and practices. The author devotes a separate section to the method of cultivation and agrarian practices. He concludes with a critique of the colonial government's role and observes that it did not demonstrate any commendable gesture for irrigational works in Bengal. Public investment in irrigation was insignificant. Yet it was hard to deny scientificity in indigenous system of irrigation. The traditional indigenous knowledge about irrigation helped produce crops. But this system was not given due care and patronage for further improvement. The indigenous agricultural knowledge and practices of the people of Burdwan had potential, though condemned as backward by the colonizers, and this was proved by its ability to produce more than what was required for consumption and accumulation of surplus of food crop which was supplied regularly to the deficit region.

Sutapa Chatterjee takes us to the long antiquity of the Sunderbans characterized by marshy tracts, saline creeks, occasional cyclones, large trees and dense undergrowth. Here the jungle and wildlife reign supreme. This vast mangrove forest is the haunt of the Royal Bengal Tiger; snakes, crocodiles and other animals abound; and in these environs live a community of brave humans, whose courage is manifest in their daily battle against nature, seeking as they do to eke out a living by cultivation, some by venturing deep into the forest to collect honey and at times ending up as a victim to the lord of the jungle. Faced by such challenges of nature, the author argues, collective life in the region was marked by an unusual degree of fluidity, and often ephemerality. Those who made this region their home, cleared forest and brought land under the plough, hailed almost entirely from the margins in more senses than one—something that endowed the social fabric of a hostile geographical category with a historical unity and an element of popular spontaneity. This essay, thus, is about the formation of settlements and the spread of habitation; it is also crucially about the natural and human strains they were subjected to. In other words, it is both about the imprint of recurrent hazards on settled human existence and the chain of implosions endemic in a frontier agrarian society.

Animal Wealth

It is difficult to think of an agrarian society without animal wealth (*pashu dhan*) that it cares for and is justifiably proud of. From time immemorial, cattle have constituted the backbone of our agriculture, its main source of sustenance. This issue of animal wealth was frequently discussed in the columns of contemporary newspapers and journals in colonial Bengal. Agrarian enthusiasts put emphasis on its importance in society. When Rabindranath Tagore laid equal emphasis on animal husbandry and dairy farming in his experiments with rural rejuvenation that caught the interest of the literati in late nineteenth- and early twentieth-century Bengal, he was actually echoing a common concern. He was one of the earliest to draw attention to the question of fodder. He even encouraged the cultivation of *napier* grass and breeding of hardy animals for the purpose. Samiparna Samanta's essay addresses this important issue that was inextricably linked up with the rural world. She examines how in the late nineteenth and early twentieth centuries, domesticated animals were at the centre of a heated debate in Bengal as farmers, veterinarians, humane societies, and the Hindu middle class viewed cattle health and animal cruelty with varying degree of keenness and intensity. She examines attitudes towards domesticated animals and the even larger battle to improve cattle health in colonial Bengal. The colonial government panicked at the loss of agricultural cattle and draught animals to a lethal disease like rinderpest. It initiated legislations and veterinary education and restructured the city and its slaughterhouses to tackle the disease. The Bengali *bhadralok,* on the other hand, seemed caught between its reliance on science, Hindu past and local knowledge in their understanding of cattle improvement. The Bengali *bhadralok* defended their ancient knowledge systems, and at the same time, appropriated new knowledge. The author has attempted to demonstrate that there were also moments when the *bhadralok* translated Western notions of science and health into their own mental worlds. They were not passive recipients. By looking at late nineteenth- and early twentieth-century notions of cattle health through the lens of a colonial government and the voices of Bengali writers who campaigned for cattle improvement, the author notices an uneasy and subliminal cross-fertilization of tropes between the ethical/religious and the scientific, between *ahimsa* and *bigyan.*

The next essay by Himanshu Upadhyay concentrates only on the role played by scientific institutions with reference to cattle breeding policies during the period from 1905 to 1940s. He argues that cattle breeding policies that evolved in the first half of the twentieth century in colonial India gradually shifted focus and aimed at improving milking capabilities of the Indian cattle, almost to an extent of ignoring the draught qualities. The peasants were even expected to tolerate deficiency on draught qualities, in the interest to see increase in milk yields. In the years prior to the introduction of artificial insemination technology on a large scale, cattle breeding was perceived as a long-range work. There were frequent changes in cattle breeding policies. However, these changes were deprecated by experts, who felt that such changes which accompanied change in leadership, or faced with financial stringency, had potential to undo the advances or improvements already made. Upadhyay also deals with the impact of introduction of perennial irrigation in regions known for excellent cattle breeding traditions. Few agricultural scientists felt that irrigation facilities no doubt improved agriculture but made the peasants more sedentary and slow on cattle improvement. The author also reviews the issue of cross-breeding of Indian cattle with exotic bulls and argues that gradually some colonial experts felt that large-scale extension of cross-breeding with exotics led to an adverse impact on agrarian balance. On the contrary, even while keeping the focus on aiming at improving milk yields, the official discussions around cattle breeding policy repeatedly revolved around the idea of 'dual purpose' breed. The concept of dual purpose breed was never more intense than at the time of Royal Commission on Agriculture Proceedings. While the Commission pronounced that draught and milk characteristics were physiologically incompatible and hence it would be desirable to attempt one task at a time, the Imperial Animal Husbandry Expert, Arthur Olver (who served in that capacity during the period 1930–8) referred to the Indian peasant's preoccupation with 'general utility' cattle. The author reviewed the changing contours of the meanings assigned to cattle by both the native cattle breeder or cultivator and the colonial experts. Along with the emerging milk markets, increasing demand for milk and milk products by British rulers and the elites in the city, an altogether different form of dairying emerged in the form of city-based milch stables. While colonial experts voiced their caution against extending indiscriminate cross-breeding and rather laid stress on increasing milk yields through selective breeding, the cause of draught bullocks at best received only lip service. While warning bells were sounded by colonial experts such as Albert Howard, Arthur Olver, N.C. Wright, etc., the question of retaining the agriculture and livestock 'symbiosis' did not receive the attention that it deserved.

Discourse on Distress

Agrarian distress thus continued, was in fact compounded by the Great Depression of the late 1920s. By the turn of the century, colonial India had become an integral part of the world economy with increased demand for her cereals, jute and cotton. Her primary products had become essential for the feeding of men and machines in the West, and she became a steady market for the finished goods of European industries. As a result, colonial India had a full share in the Depression which started in October 1929, and her public finances, as well as the private finances of its people, were seriously affected by the slump. This is the subject of S.M. Mishra's essay which argues that a deep-seated agricultural depression coincided with the world financial collapse, and the primary producing economies like India experienced a huge slump in prices. Within the span of five years after 1928/9, Mishra observes that the value of India's foreign trade was nearly halved and recovery thereafter was sluggish and restricted. It is a truism, he says, that cumulative causation played an important role in economic affairs, but the perception of such processes is very often difficult for those who are involved in them as information is at a discount and preconceived notions prevail. The agriculturist's was a perennially distressed lot as his daily avocation was akin to hard labour and necessitated constant vigil against both natural elements and living beings. The income of the peasants fell while rent and revenue demands and debt service remained as before. The agriculturist was oppressed by his heavy debt. That was the real cause of his trouble. He did not really earn for himself, as a very large proportion of the money that he got for his crops went to pay the interest on the money he had borrowed. It was the usual charge against landlords and moneylenders that they deprived the agriculturist of his lands and contributed largely to his misery. An examination of the position of the government as a landlord undoubtedly showed that the policy which the government followed, with regard to assessment and realization of revenue, also contributed greatly to the indebtedness of the agriculturists. While the government as a landlord was a creditor of the agriculturist and also a guarantor of agricultural or taccavi loans, it held the prominent position of a creditor by the side of the moneylenders. It can, therefore, be easily imagined how the government as creditor contributed to and even aggravated the plight of the agriculturists. The government's deflationary policy to support the exchange rate intensified the impact of the Depression. The price movement in the 1930s, characterized by a sharp downward trend, had disastrous consequences for the peasantry. Because the prosperity of industry and other trades depended upon agriculture, unless the condition of the agriculturists improved, there was no hope of any betterment of the condition of industrialists as well as of other trades. Machine technology over time caused large-scale unemployment and distress, for example, the rice mill caused the dislocation of the extensive network of huskers, peddlers, carters and boatmen. Steam engines affected the oil and sugar industries while the old ox-driven oil mills were disappearing faster than the old sugar factories. As ever, the burden of the distress of the economic Depression was passed on to the rural poor. Mishra's study is based on a study of the debates by the then legislators in the Central Legislative Assembly of British India, and this may even have contemporary echoes and relevance.

The essay by Madhumita Saha goes beyond the colonial period and shows the phase of transition. Independent India had inherited the impact of the Great Famine of 1942 and the violent displacements caused by the Partition. The Grow More Food campaign, carried over from the last years of the British Raj, envisioned food self-sufficiency by the year 1951; yet, undernourishment continued to be a lingering problem to the exasperation of the members of the political and the research establishments. The worldwide discourse on hunger, Saha says, increasingly saw the deep involvement of the state machinery in the eradication of hunger in India. The effort of the political establishment to modernize the country underlined the necessity of having a well-fed citizenry; thus, the story of modernity in India as elsewhere became partially organized around the conquest of hunger. The national leadership and the planners were confident that hunger had to be banished if India has to come out of the 'waiting room' of development. The author shows how Government of India's strategy to eradicate hunger among its rural population gradually changed over the period since independence through the introduction of the Green Revolution technologies. Conceptualized largely as a major structural problem, indicating chronic social and economic inequity, the new government began to treat it as a technical problem of failing production, requiring technological intervention. Right after independence, the understanding of the Central Government as well as the Planning Commission was that as poverty was indubitably associated with rampant hunger, the task of eliminating it and improving agricultural production had to be integral to the larger goal of rural development. As the experts persuasively argued in favour of a technical solution to the food problem, rather than exploring its social and economic dimensions, the juggernaut of the Green Revolution technological package could roll out, even as it ignored the arguments of its critics, needs of small farmers, and any possible advantages of a research model that considered the social and material aspects of technology to be important. In reducing hunger as primarily a problem of low yield, the advocates of the capital-intensive agricultural model could set the direction of agricultural modernization in India along technocratic lines, away from its broader social and political underpinnings.

Saha's criticism is valid and shared by many. But was there any alternative? Population was multiplying at the Malthusian rate, economy was in shambles; what else could newly independent India have done? During the 1920s, some students of Professor Radhakamal Mukerjee surveyed the districts of Gorakhpur, Jaunpur, Banaras and Kanpur to analyse the effects of population pressure on agriculture and the social economy. The results were disconcerting.[25] Yet there were well-meaning people like Albert Howard who favoured and strongly pitched for organic farming and the use of plant science for improvement.[26] He was critical of Leibig's classic on agricultural chemistry that had appeared in 1840 and asked, 'Is there any method other than that of the single subject by which science can deal with the problems of the cultivator?'[27] Other scientists like Harold Mann and D. Clouston would put the onus on the farmers at large, dubbing them 'hopelessly conservative and prejudiced'.[28] Another scientist Martin Leake felt, 'we cannot, and never will be able to, attain that much-desired condition because a new aspect is arising every day.'[29] All these experts held high positions as head or director of agriculture departments in different provinces. And despite their advice and work on the field, foodgrain yield per acre declined 0.18 per cent annually for the period 1891–1947, while the non-foodgrain yield grew at almost 1 per cent annually.[30] The cash crops thus took a heavy toll on the production of foodgrains. Yet some scholars feel that even if the British Government had given a big funding and push to agricultural farms and research, there would have been no Green Revolution in foodgrains before 1947.[31] The problem was much more than that of funding or culture. In the midst of World War II, a British physiologist, A.V. Hill visited India and diagnosed India's problems as primarily biological in nature. He talked of a quadrilateral dilemma, i.e. population, health, food and natural resources. To him, the fundamental problems of India were 'not really physical, chemical or technological, but a complex of biological one referring to population, health, nutrition and agriculture all acting and reacting with another'.[32] He was right. No doubt, a greater push to all kinds of biological and chemical sciences was required. But it had its own consequences which one faces in the twenty-first century. As Howard had warned, 'Artificial manures lead inevitably to artificial nutrition, artificial food, artificial animals, and finally to artificial men and women.'[33] The dilemma continues and the question of hunger still hangs over this vast country. In 2015, India tops the World Hunger List with 194 million hungry souls.[34]

Notes and References

1. For details, see the chapter on agricultural production in Irfan Habib, *The Agrarian System of Mughal India: 1556–1707*, 3rd edn, Delhi, 2014, pp. 1–67.
2. In March 1874, the Chief Magistrate of Tirhut reported:

    One sees starvation in all degrees, but the saddest sights of all are the little children and young men, among whom one sees a degree of emaciation which you could not believe possible—literally skin and bone. . . . Tragedies like these would make a great noise at home; but so many horrible things happen here that they create little sensation.

    Cited in Anonymous, *Undeveloped Wealth in India: The Ways to Prevent Famines and Advance the Material Progress of India*, London, 1875, p. 42.
3. Agricultural and Horticultural Society of India, *Transactions of the Agricultural Society of India*, vol. II, 1835, p. 211.
4. A. Sarada Raju, *Economic Conditions in Madras*, Madras, 1941, p. 56.

5. 'In the colonies the pattern of production was progressively lopsided, geared to the requirements of metropolitan economy (i.e. exports), and also providing a market for the products of metropolitan industry (i.e. imports). . . . It was a disarticulated generalized commodity production, characteristic of a colonial economy and not an integrated generalized commodity production characteristic of the metropolitan economy.' Hamza Alavi, 'India and the Colonial Mode of Production', *Economic and Political Weekly*, vol. X, August 1975, pp. 1235–62.
6. A planter-poet George Williamson, cited in B.B. Chaudhuri, 'Growth of Commercial Agriculture in Bengal, 1859–1885', *Indian Economic and Social History Review*, vol. 7, no. 2, 1970, pp. 211–51.
7. G.P. Majumdar, *Vanaspati: Plants and Plant Life as in Indian Treatises and Traditions*, Calcutta, 1927, pp. 220–4; idem, *Upavana Vinoda: a Sanskrit Treatise on Arbori and Horticulture*, Calcutta, 1935.
8. Wladimir Ivanow, *Concise Descriptive Catalogue of the Persian Manuscripts in the Curzon Collection of the Asiatic Society of Bengal*, Calcutta, 1926.
9. Home, Public, no. 10, 5 December 1799, National Archives of India, New Delhi (hereafter, NAI).
10. J.A. Voelcker, *Report on the Improvement of Indian Agriculture*, 2nd edn, Calcutta, 1897.
11. David Ludden, ed., 'Introduction', *Agricultural Production and Indian History*, Delhi, 1994, pp. 4–5.
12. Letter from W. Carey to Dr Ryland, 23 October 1820, Carey Library, Serampore.
13. K.L. Tuteja, 'Agricultural Technology in Gujarat: A Study of Exotic Seeds and Saw Gins, 1800–1850', *Indian Historical Review*, July 1990 and January 1991, pp. 136–51.
14. Col. Tyrrell wrote a note under the title 'Picking India's Pocket' in 1874 and commented, 'We have given no thought to the natives in conjunction with their country! It suited England; it suited English engineers; it suited English merchants; it suited the English people to pour her sons into India on high pay, to saddle her with £93,000,000 for railways, while her people die, calling for water.' Cited in Anonymous, *Undeveloped Wealth in India*, p. 42.
15. Note by Richard Temple, 1 February 1872, *Temple Papers*, IOR, MSS Eur. F 86/ no. 117, British Library, London. Also, see Eugene C. Schrottky, *The Principles of Rational Agriculture Applied to India and its Staple Products*, Bombay, 1876.
16. *Proceedings of the Agriculture Conference*, 5th meeting on 10 October 1890; Revenue Department, Agriculture Branch, nos. 7–11, May 1891, Pt. A, File no. 26, NAI.
17. *The Indian Agriculturist*, vol. XIII, no. 10, 10 March 1888.
18. *The Belgaum Samachar* (a Marathi weekly), 16 September 1907, Native Newspapers Report (hereafter, NNR), Bombay, 1907, p. 1315; *Sasilekha* (a Telugu daily from Madras), 20 January 1911, NNR, Madras, 1911, p. 159.
19. Note by J.W. Mollison, Inspector General of Agriculture, 14 November 1907, Revenue Department, Agriculture Branch, General, nos. 2–9, December 1907, Pt. A, File no. 240, NAI.
20. *Amrita Bazar Patrika*, 24 November 1911, NNR, Bengal, no. 48, 1911, p. 530.

21. Revenue Department, Agriculture Branch, no. 8, September 1902, Pt. B, File no. 103, NAI.
22. *Desabhimani*, Cuddalore, 15 July 1911, NNR, Madras, no. 30, 1911, pp. 1084–5.
23. For example,

    year after year the hard-earned money of the poor subjects has evaporated like vapour over agricultural research but the people in return have got nothing . . . Bengalis do not want the lamp of Aladin. They will be satisfied with bare rice and vegetables. What is the use of adding insult to injury by speaking to them of research and such other things.

    The Ananda Bazar Patrika, 19 July 1922, NNR, Bengal, no. 30, p. 596.
24. Sam Higginbottom, *The Gospel and the Plow*, London, 1921.
25. 'Where man breeds like field rats and rabbits without provision even wholesale emigration or industrialization become mere palliatives.' Radhakamal Mukerjee, 'Introduction', in Jai Krishna Mathur, *The Pressure of Population: Its Effects on the Rural Economy in Gorakhpur District*, Allahabad, 1931, p. vii. Also, see Babu Ram Misra, *Economic Survey of a Village in Cawnpore District*, Allahabad, 1932.
26. Albert Howard and G.L.C. Howard, *The Application of Science to Crop Production*, London, 1929.
27. Albert Howard, 'Agriculture and Science', *Agricultural Journal of India*, vol. XXI, 1936, pp. 171–82.
28. H.H. Mann, 'The Introduction of Improvements into Indian Agriculture', *Agricultural Journal of India*, vol. II, 1909, p. 7; D. Clouston, 'The Development of Agriculture in India', *Agricultural Journal of India*, vol. XIX, 1929, pp. 164–5.
29. H. Martin Leake, *The Foundations of Indian Agriculture*, Cambridge, 1923, pp. 35–6.
30. George Blyn, *Agricultural Trends in India, 1891–1947: Output, Productivity and Availability*, Philadelphia, 1966, p. 151.
31. Carl E. Pray, 'The Impact of Agricultural Research in British India', *The Journal of Economic History*, vol. 44, June 1984, pp. 429–40.
32. A.V. Hill, *The Ethical Dilemma of Science and Other Essays*, New York, 1960, p. 375. Also, see A.V. Hill to John Mathai, 5 February 1954, *Hill Papers*, AVHL II, 4/79, Churchill College, Cambridge.
33. Albert Howard, *An Agricultural Testament*, London, 1940; repr. Goa, 1956.
34. Food and Agriculture Organisation, 'The State of Food Insecurity in the World 2015', *The Hindu*, 29 May 2015, p. 8.

2

Science in Agriculture: A Study in Victorian India

Deepak Kumar

> It is impossible not to deplore the same defective state in the agricultural, as in every other science in this country. Look where you will, and you find the same results—poverty, inferiority, degradation in every shape.
>
> —H. PIDDINGTON, *On the Scientific Principles of Agriculture*

> At his best the Indian raiyat is quite as good as, and in some respects, the superior of the average British farmer, whilst at his worst it can only be said that this state is brought about largely by an absence of facilities for improvement which is perhaps unequalled in any other country.
>
> —J.A. VOELCKER, *Report on the Improvement of Indian Agriculture*

The British had arrived in India as traders but soon evolved into the largest zamindars on the planet. Thanks to the grant of *diwani* and several other settlements which they had gained after waging so many wars, they now had huge tracts of land and a large population dependent on it. The new Raj could not afford a sharp break from the past but had to gradually build upon the then existing culture and practices. Indians always evoked a mixed, though curious, response from their colonizers. Medlicott, a noted geologist, wanted to wait until 'the scientific chord among the natives' was touched, and added, almost contemptuously, 'if indeed it exists as yet in this variety of the human race'.[1] But there were also others like E. Buck, a Civil Service Officer, who frankly accepted, before the Famine Commission of 1880, that 'for one thing in which we can beat the native, he can beat us in a hundred things'. D.B. Allen wrote:

---

* This is a slightly revised version of an article originally published in A. Rahman, ed., *Science and Technology in Indian Culture*, New Delhi, 1984, pp. 189–216. Extensive archival research for this article was done in the late 1970s, but the theme and its treatment seem to remain of contemporary relevance.

It is easy to call the Indian ryot slow and prejudiced, but when once a process is proved to be advantageous, he will be quick enough to adopt it. It is not strange that he is slow to hear the agricultural missionaries who visit the fair in the North-West Provinces, armed with improved ploughs and new-fangled water-wheels; for they can only talk, or if they do carry out successfully some simple experiment the result is sure to be ascribed by the wondering audience either to Jadu or Masala. But when they see their own Maharaja trying these experiments on a large-scale, there will be neither hesitation nor prejudice; for he will speak to them in a language that all understood, the language of the pocket.[2]

The Cash Crop Boom

Whether invective or adjective, in both cases it meant due recognition of India's potentialities as an agricultural country. And to the new rulers, the potentialities basically meant cash crops and the profits that accrued. After all, colonialism was no philanthropy and the British did not come here to fight famines. They wanted Indians to grow cotton, jute, opium, tobacco, indigo and tea. This was the reason the colonies were called plantations. The Indian cultivator had limited use of these articles of commerce; he had practical knowledge and experience but no capital. So the colonizers asked the British merchants and the trading companies to invest in cash crop-oriented agriculture. In 1839, a pamphleteer calculated how the British mind worked:

Let us now take a single article of native consumption, tobacco, which is raised by advances from native capitalists, and see what it may fairly say that they all expend money in tobacco, and the poorest not less than two *annas* per month upon it. This is sixty millions of rupees per annum for the cost of tobacco; and if we call only one-third of this advance to the cultivator, here are twenty millions of rupees advanced for a single product, in which from the first ploughing to the sale of the tobacco, at least fifteen months elapse before a return is obtained.[3]

Such demands for sinking more capital in agriculture were made throughout the nineteenth century. An influential journal thus exhorted the British capitalists to 'embark some portion of your redundant riches in speculations of Indian agriculture: grow there sugar, rice, cotton, tea, indigo, even wheat, enourage the silkworm, and culture of tobacco, opium, etc.'[4] Cash crops were the favoured ones. In 1803–4 at Malda, the rates of profit attending the cultivation of cash crops per acre were Rs.3 to Rs.7. 8*an.* for hemp, Rs.6 to Rs.9 for cotton, Rs.15 to Rs.18 for mulberry, Rs.9 to Rs.15 for sugar cane, and 8*an.* to Rs.6 for indigo. In contrast, rice fetched Rs.3 to Rs.4. 8*an.*[5] In 1793, net profit from a bigha of wheat yielded only 7*an.*[6]

Hemp, thus, was quite advantageous compared to wheat. Its economic as well as military significance lay in its use for the rigging of ships. Russia had to stop its supply during the Napoleonic wars, and the British shipping greatly suffered as the price of hemp which in 1792 was only $25 per ton, rose to $118 in 1808.[7] India was, therefore, looked towards for a supply of hemp.

Apart from hemp and fibre producing plants, others which received constant attention were sericulture, cotton and tea. Early researches in sericulture were made by W.M.H. Smith in 1814.[8] But the first clear statement of government's policy came in June 1887 when it promised to aid sericulture 'in legitimate directions such as instituting enquiries into diseases, difficulties connected with the propagation of the mulberry and general administrative facts which are impeding or may help to promote the silk industry of India.'[9] Next year, N.G. Mukherjee, a Cirencester scholar, was commissioned by the government for the investigation of silkworm disease in Bengal.[10] That year, he submitted a note on the decline of silk trade in Bengal and pointed out that while European, Japanese or Chinese silk sells at 45 francs per kg., Bengal silk sells at 32 francs. 'If Pasteur's system is introduced in the country and healthy seed brought within reach of the peasantry, there will be no doubt an increase in the production of cocoons and silks. I believe the Government has been well advised in deciding to establish a sericultural laboratory.'[11]

But the Director of Agriculture, Bengal, would not pay heed to expert advice. He wanted Mukherjee to concentrate exclusively on the eradication or mitigation of the disease of pebrine. And only after that he would consider 'whether laboratory for the investigation of other sericultural questions should be established'.[12] Similarly, when in 1895, Mukherjee asked for the introduction of agriculture teaching in the schools of silk districts, the D.P.I. of Bengal, A. Croft, scorned at and wanted sanitation to be taught, not sericulture.[13] The Government of India obviously lacked coherence and the will to act. Having permitted Mukherjee in 1888 to build a sericulture laboratory at Berhampur, the Imperial Government transferred its control to Bengal Government which in turn handed it over to three silk tycoons. Mukherjee was made accountable to a committee of merchants and not to the government. Such an abdication of the government's responsibility irked even the British Silk Association whose President T. Wardle wrote, 'If the Government of France and Italy have for so many years seen the necessity of preserving their respective silk industries by State watchfulness and nature, I feel certain that they are still more required in India. Were I in Parliament, I would move for a Commission on this subject.'[14]

But the government took them coolly in its stride.[15] Henceforth, sericulture was to remain a purely private concern. At the turn of the century,

M/s Tata & Sons successfully started a silk farm at Bangalore for the introduction of Japanese methods of sericulture. After J.N. Tata's death in 1905, F.G. Sly, the Inspector General of Agriculture, apprehended that the Bangalore farm would either be abolished or turned into a commercial undertaking and would not be available for experiments or as a training school.[16] Hence he sanctioned a similar farm with one Japanese expert at Pusa, where an agricultural research institute had already come up.

Experiments both in cotton and tea were carried on simultaneously. The government established experimental farms for the cultivation of cotton, as well as gardens for the cultivation of tea plants. Those for cotton, however, at the outset, received the most careful attention. Pecuniary advances were made to individuals, seeds were procured from Egypt, Brazil and North America, and saw-gins were sent to India. Several American planters were employed for instructing and superintending the cultivation, and large prizes were offered for the best samples of cotton. Lord Auckland looked upon the growing interest in tea and cotton as 'one fortunate consequence of the state of our Chinese relations' and wanted cotton to replace opium.[17]

American planters were brought to nurse experimental cotton farms. But this was doomed from the very beginning. The Superintendent of American Cotton Project, J.H. Pearly, frankly confessed that their object in coming to India was to make money which they found they could more easily accomplish in their own country.[18] C.G. Jackson, Collector of Agra, while lamenting the failure of America, furnished detailed accounts of the native modes of cotton cultivation.[19] A sample of cotton grown at Hazareebagh in 1836, from the superiority of its quality, led the Committee of Commerce and Agriculture of the Horticulture Society of Calcutta to suppose it to be the produce of Sea Island seed, whereas it was produced from Egyptian seed. A sample grown near Bombay was pronounced quite equal to New Orleans cotton.[20] Only problem was that the natives did not have the purchasing capacity to go for new machinery. Instead of enabling the Indians to go for new technology through financial assistance, the government tried to bring down the level of technology itself to suit the so-called Indian tastes and capabilities. In 1840, for instance, the Court of Directors placed $100 at the disposal of the Liverpool East India Association 'to be applied in rewards for the invention of the best instrument for cleaning cotton adapted to the use of the natives of India'.[21]

In a petition to Parliament in 1861, the Cotton Supply Association of Manchester demanded three specific improvements: the rapid promotion of public works to facilitate cotton transportation by rail and water; the sale of land at such low cost as to encourage English settlers to develop agriculture in India; and the improvement of Indian laws and law courts to allow binding commercial contracts between the British investor and speculator and the Indian native. When the Indian Government protested that it was not its function to intervene actively to promote cotton cultivation, it met a blistering attack from John Cheetham, the Chairman of the Cotton Supply Association. If the government aided the producers of opium and tea, why should they baulk at following the same policy towards cotton?[22] The obviously insufficient transportation facilities were bitterly denounced by the irate manufacturers. Merchants gathered statistics which showed that construction had been undertaken for military reasons and not for commercial benefit. Empire, in the eyes of Manchester, was not only a matter of convenience, but essentially a business venture.[23]

Mounting pressure from the British cotton tycoons forced the Government of India to initiate a vigorous cotton improvement programme. Cotton enthusiasts like Forbes, Ashburner and Rivett-Carnac were given charge of Dharwar, Khandesh and Central Provinces respectively. But like the earlier projects of 1840s, the efforts of 1860s also failed mainly because they were made without sufficient botanical knowledge or the necessary market research. Later on, in 1890, Dr Voelcker specifically called for the association of expert botanists in cotton experiments.[24] The cultivation and marketing of existing varieties produced a relatively stable and acceptable rate of return to ryots, moneylenders and dealers. New and untested varieties involved different methods of cultivation and greater labour input, without a higher level of output or profit, and with the risk of severe loss to each of these classes.[25]

Last quarter of the nineteenth century saw the closure of several experimental farms. But private farms patronized by cotton mills proved remunerative. For example, Mungeli farm at Bilaspur of the government had to be abandoned,[26] whereas the nearby Khyragarh and Nandgaon cotton farms owned by Bengal-Nagpur Cotton Mills Company produced 6,00,000lb.[27] Why this difference? Perhaps because the mills could procure cotton by advancing seeds to cultivators who cultivated them through traditional techniques, while the government exercise in Mungeli failed because the government was enamoured of imported technology and ideas. The ever-growing Indian mills did not wish for a superior or long staple but only a pure one, and this indirectly discouraged improvement.[28] Voelcker blamed the traders, not the cultivators. He asked, 'If the trade really complains of the inferiority of the short-stapled cotton why does it continue to use up readily every bit of cotton the ryot can produce? If the cry for long-stapled cotton is so urgent, why will the trade, not give more for it and refuse to take the short stapled?'[29] Cotton improvement programmes were thus designed to provide a resource desired by Lancashire manufacturers, who themselves were unwilling to make technological or market adjustments in their industry, which would have made Indian cotton more acceptable to them.[30] But the policymakers hardly learnt from past experiences. In 1904, the Secretary of State advocated a specialized section for cotton in the Department of Agriculture, more seed farms, and even importation of experts from America.[31] Such a mercantilist approach not only affected cotton cultivation but also hampered the growth of textile technology itself around Indian mills.[32]

Like cotton, tea generated a lot of enthusiasm. Lt. Col. Kyd had grown China tea in his botanical garden in 1780s. But the authorities probably discouraged further experiments as it presented the possibility of a rival to the China tea trade, which was a source of much wealth to the East India Company.[33] When tea was first located in Assam in 1825, Dr Wallich, the official botanist, hesitated to accept it as one of the real tea species.[34] Bentinck showed keen interest in the possibilities of extensive tea cultivation as he had done in steam transport and coal explorations. A Tea Committee was formed in 1835 and one Mr Gordon was sent to China to procure seed, plants, and men experienced in all the operations of tea planting and tea making, and for this purpose a credit of $25,000 was placed at his disposal.[35] The government asked for a rough map of the tea-bearing areas in Assam.[36] In 1835, the government made the first attempt by establishing an experimental garden at Lalitpur in Assam; it failed.[37] Dalhousie paid greater attention.[38] In 1855, indigenous tea was found in Cachar; in 1856, in Sylhet. Kumaon started tea planting about 1850, Darjeeling in 1860, Neelgiris in 1862, Chittagong in 1864 and Chotanagpur and Ceylon in 1872.[39] In 1869, Dr Thudicum experimented on the production of new beverages by treating a decoction of tea with yeast, sugar and alcohol.[40] A.W. Blyth experimented with tea leaves and established a process through which anyone could know whether the merest fragment of a plant belonged to the 'theine' class or not.[41] But nothing was done to control the scourge of blight. The Agricultural and Horticultural Society made some slight attempt in this direction, but broke down for want of funds, being unable, therefore, to procure from England, as they had desired, a skilled entomologist. The government promised aid, but it never went beyond this stage.[42]

Indigo remained unrivalled and, therefore, its scientific culture uncared for till the end of the nineteenth century, when Germany perfected its synthetic counterpart. The German feat sent shivers among the British planters and brokers in India who hastily responded by forming an Indigo Improvement Syndicate. An experimental farm at Dalsinghsarai was started in July 1899 for conducting agricultural experiments in manuring, cultivation, etc., of indigo. The Syndicate frankly regretted that 'in more prosperous times no efforts were made in the direction of scientific improvements, and the urgency of prompt action, now that the synthetic indigo is making such rapid strides, is all the greater'.[43] George Watt, the economic reporter to the government, called it 'a disgrace to the industry that so little should be known of the botany and agriculture of the plant upon which so much capital has been invested'.[44] The Syndicate annually spent Rs.28,000 on research and asked the government to contribute Rs.40,000 per annum for at least three years. The Lieutenant Governor of Bengal granted Rs.50,000 a year on the condition that the Syndicate would also contribute Rs.75,000, the whole, Rs.1,25,000, being devoted to the furtherance of researches.[45] Bloxam and Leake were sent to Dalsinghsarai in 1902 but they returned to England within two years, as they found the Clothworker's Research Laboratory of the University of Leeds more appropriate for such researches.[46] Synthetic indigo had depressed the British planters so much that Curzon thought of even replacing indigo by sugar cultivation in Bihar.[47]

Like indigo, tobacco got a rather belated attention. The area under this staple in 1902–3 was 9,17,792 acres. 1,01,500 cwt. of unmanufactured tobacco valued at Rs.12,86,000 and 900 cwt. of the manufactured product valued at Rs.8,10,500 was exported from India in that year.[48] The trade was thus lucrative, and like silk, it could have been more if pests and diseases were controlled. Curzon, therefore, considered 'the improvement of the curing process rather than the investigation of methods of cultivation as the most urgent requirement' and called for a scientific expert with a thorough knowledge of the latest methods of tobacco curing.[49]

Similarly, the possibilities of improvements in the milling and baking qualities of wheat, and not its cultivation methods, attracted Curzon's attention. For this also, he asked for a competent expert. After all, India stood fourth on the list of wheat producing countries. The total area under this crop in 1903–4 was 28,413,743 acres, while the quantity exported was 26,722,00 cwt. of a total value of nearly Rs.7,740,000.[50]

Response of the Raiyats

Maximum profit from agriculture was thus the hallmark of British policy. But how to achieve it? The half-starved raiyats, with their half-starved bullocks, working on a half-starved soil were not the men to compete with the millionaire farmers of Europe and America.[51] In Europe, agricultural instruction was given to the farmholders and the labourers alike. But in India, the British publicists wanted such education to be imparted not to the raiyats but to the rising generation of the zamindars.[52] With the possible exception of the Oudh talukdars, who held a few agricultural meetings during 1880s,[53] the zamindars, by and large, had no direct interest in increasing the productiveness of the land. On the contrary, under section 30(b) of the Bengal Tenancy Act, the landlord could claim an increased rent on grounds of the rise in prices of foodgrains, and as the prices could be made to rise more easily by making the supply to fail, than by causing the demand to rise, the landlord could very easily see that his interest lay in seeing the supply of foodgrains to fail by making the land produce less.[54] The zamindars were certainly not as ignorant as their raiyats. The economics of their time taught them to behave more as tax gatherers than as landlords.[55] The average Indian cultivators, on the other hand, had their own reservations, and perhaps valid ones. Piddington gives an interesting instance. Once he offered the seeds of Bourbon cotton to some raiyats who got quite convinced of its advantages but said that they dared not cultivate any new article. The reason assigned was that 'Our zamindar would make demands for beyond the profits'.[56] And how true they were!

Several British publicists and officials had almost rhetorically dubbed Indian peasants as conservative and resistant to change.[57] But the facts do not conform to it. The very first Agricultural Exhibition held at Alipore in January 1864 evoked an excellent response, and the authorities reported: 'The purchases of machinery and of agricultural implements made at the Exhibition by the native gentlemen from the moffusil afford an undeniable proof that there is no want of disposition to give those modern appliances a fair practical trial.'[58] Once a process was proved to be advantageous, the raiyat was quick enough to adopt it.[59] The success of Ms Thompson and Mylne's sugar pressing mill is a case in point. It was sold by thousands, showing that Indians had no objection to machinery when it paid them to use it.[60] But large steam-power machines or implements were not wanted.[61] Only small hand- or cattle-based implements would have found favour.[62] After all, from where could they have money to go for sophistication?[63]

Government's Responsibility

Obviously the onus lay on the British Government in India which itself was the largest estate holder in the world and its sole beneficiary.[64] Till the famines struck grievously in mid-1860s, calling urgently for the formulation of an agricultural policy, the government had tried to act by proxy, i.e. through the agency of the Agricultural and Horticultural Society. Government officials and influential zamindars were members of this Society and its funds came through private subscription, sale of plants and government aid. In 1830, the government gave Rs.20,000 to this Society in Calcutta, which was to be distributed to the most successful cultivators of cotton, tobacco, sugar, silk and other articles of 'raw produce'.[65] The Society picked up very fast and made ambitious plans. It wanted to be like similar societies in England. 'Of capital there is no want, of labourers there is no want, of flocks and studs there is no want. But there is want of improvement; there is a want of a race of zealous, enterprising professionals, capable of seizing in all these natural advantages turning them to the best accounts.'[66] Such was the spirit.

Basically engrossed in the acclimatization of several foreign varieties of fruits and vegetables,[67] the Society often assumed the role of an adviser to the planters as to where and how to invest capital in new agricultural ventures.[68] In 1837, it called for the establishment of an experimental cattle breeding farm[69] which could not materialize. But it persisted in its efforts and finally persuaded the government to establish a zoological garden for breeding purposes.[70] The Society experimented with flax culture[71] and tinkered with tea.[72] The Society initiated scientific debates and offered prizes on the two burning problems of the day, viz., cotton cultivation of exotic varieties[73] and the introduction of quinine-yielding cinchona.[74] During 1838–85, it distributed 41,586 plants of known economic value.[75] The Agricultural and Horticultural Society at Madras was equally vigorous and in that period sent out 1,17,640 rooted plants of *Fourcroya gigantea* alone for 'fibre' growing experiments,[76] and in addition imparted botanical instruction to the medicos there.[77] Similar societies at Lahore and Nagpur gradually became less like a society and more like agencies for supplying seeds.[78] So, in 1883, they were abolished and their gardens were taken over by the government.[79] A few private gardens, like the one at Bhagalpur devoted to arrow-root and coffee, served as model farms in district towns.[80]

These agencies, however, worked under severe limitations, and the country's agriculture would not have run with the help of such crutches. The crash came in the shape of famines, and it is no mere coincidence that the government would open its eyes only when goaded by the Famine Commissions of 1866, 1880 and 1901. Commercial interests were also at work, and their influence was perhaps decisive. For example, the Cotton Supply Association of Manchester wanted the creation of a Department of Agriculture so that concentrated attention could be given to cotton cultivation.[81] Lord Mayo accepted the suggestion, but his own council members, notably Richard Temple and Henry Durand, opposed it. For Durand, the creation of separate Secretariats for the Railways and Legislative Departments were of much greater and urgent necessity than that for agriculture.[82] Still the new Department was opened in 1871,[83] only to be closed in 1879 in the name of financial stringency. A second start was made at the behest of the Famine Commission of 1880,[84] followed by the establishment of such departments in each province. But the provincial departments were unable to make much progress mainly because they were not given a staff of specialists skilled in agriculture and its allied sciences.[85] They could not go beyond the collection of revenue data and famine relief operations.[86]

The government, thus, preferred to lean upon traditional agencies like the Agricultural and Horticultural Society than to innovate. The Finance Department had always been unhappy with the performance of these societies and had even asked for stoppage of grants to them in 1873.[87] The government did not oblige, but asked the societies to work 'rather for the advancement of naturalization of species and staples of general utility, than (as in some instance appear to have been too much the case), for the establishment of the station where the society garden is and the supply of fruits, vegetables, flower and seeds to the European subscribers.'[88] In 1885, the government emphasized 'the mutually supporting and harmonious relationship' between the agricultural departments and the agricultural societies.[89] The Society in Calcutta promptly agreed to conduct all experiments in economic products, and the government in return raised its 'grant' from Rs.2,400 to Rs.6,000 per annum.[90] The Society accordingly experimented with plantain, musk mallow, rhea, ochro, bowstring hemp, etc.[91] The objectives were:

to obtain precise and trustworthy details as to the cost of cultivation and produce per acre of fibre-bearing plants of a promising character, so that the Agriculture Department, may be able to form a decisive conclusion as to the prospects of a profitable exploitation of the plants in question; to secure a competitive trial of machines and processes for the extraction of the fibres, so as to raise a spirit of emulation which might, if not immediately, at all events at no distant date, induce competent and interested persons in either investing or bringing out to this country a machine or process which would not only be efficient in its working, but cheap, portable and simple of construction.[92]

The government was, thus, asking the society to look for financial gains, but on the other hand, in April 1884, it enthusiastically resolved, 'that in the management of the experimental farms, undue stress should not be laid on the financial results, no portion of the area being necessarily managed with the sole object of obtaining a net profit on the outlay'.[93] The dichotomy, however, was short-lived, for, the very next year, the government declared that it 'cannot sympathize with any measures which involve the conduct of "doubtful" experiments entailing either lavish expenditure or a loose and vague record of results'.[94]

Perhaps the full acceptance of the doctrine, that effective improvement in Indian agriculture depended mainly on the application of European science to Indian facts, dates from the time of Dr Voelcker's visit to India in

1890.[95] His object was to enquire into and advise upon the improvement of Indian agriculture by scientific means.[96] He found that technical knowledge of agriculture was 'the missing element' in the then existing agency called the Department of Land Records and Agriculture. He appreciated the view that Agriculture should form a department quite separate from that of Land Records and that its Director and Assistants should be experts, not civilians, on the pattern of the geological, botanical and meteorological departments. But he doubted its feasibility and, therefore, confined his job only 'to suggest what can be done rather than what ought to be done; to graft improvements upon the existing systems, rather than to suggest the subversion of the latter'.[97]

Being aware that India, unlike Britain, did not have a class of landowners with sufficient wealth and inclination to carry out scientific improvements, Dr Voelcker wanted the government to assume that responsibility.[98] He emphasized upon the utility of experimental farms, the need of agricultural chemists and the spread of agricultural education. But he was against technology importation[99] and had praise for the Indian cultivators and their implements. Voelcker wrote, 'He will indeed be a clever man who introduces something really practical.'[100]

Almost every important official commented upon the 'native'[101] phenomena. For example, C.P. Carmichael, a senior member of the Revenue Board of NWP, claimed that the native knew more about well construction than officers sent by the government and, on this basis, argued that the agriculturists did not want assistance from the Department regarding the silting and sinking of well.[102] Earlier in 1882, Capt. Clibborn had found that the zamindars preferred to dig their own wells and did not want government assistance, and even disliked taking taccavi loans for these purposes.[103] Clibborn and Carmichael aired pessimism, and were subsequently proved incorrect, for when the NWP Government offered cheap and effective well boring equipment, the response was tremendous.[104] In contrast, Dr Voelcker and Buck sound unusual and refreshing because they had an optimistic and positive outlook. Buck, for instance, frankly confessed, 'Hitherto our efforts to carry into effect many of the suggestions of the Famine Commissioners have been treated with apathy, if not with opposition, and Dr Voelcker's report will perhaps awaken the India office to a sense of their importance.'[105]

Agricultural Education

In no small part, the solution of agricultural problems lay in the provision and quality of agricultural education. Wellesley was perhaps the first Governor General to have talked of model farms as forming a branch of agricultural instruction.[106] Bentinck revived the idea but could not do more. In 1839, a contemporary bemoaned, 'In Germany, in servile Russia, in bigot Spain, distracted Italy, we find schools and professorships of agriculture, but in British India, "depending wholly on the soil for its revenue", there is nothing of kind'.[107] In 1860s, the Landholder's Association and the British Indian Association repeatedly asked for such education.[108] But the officials would only pay lip service and were profuse in regrets. Break was given by the Government of Madras which, in 1865, established an experimental farm at Saidapet 'to afford facilities for testing the merits of little known machines, implements, manures, crops, systems of culture, livestock, etc., believed to be suited more or less to the requirements of southern India'.[109] When its Superintendent, Robertson, pointed out how literary education had robbed agriculture of her best men and called for the establishment of an agricultural college, the Madras Revenue Board would not agree on the ground that even the so-called 'degraded' agriculture of India used to feed 30 million people, yield a considerable surplus of rice, cotton, sugar, oilseeds and indigo for export, and pay £4.50 million of land revenue.[110] But Robertson persisted and got an agricultural college opened at Saidapet in 1876.

That very year another agricultural college was proposed as an adjunct of the Patna College in Bihar. The then Bengal D.P.I. admitted that as he had no precedents to guide him, he had to run after European models. Cirencester was thought to be too theoretical while Massachusetts heavily experimental. So he preferred the Belgian example which could be partially imitated, partially because he had no hope 'of seeing any sufficient imitation of the beautiful laboratory of the spacious economic museum, and of the scientific equipment which the little kingdom of Belgium thinks it right, "regardless of cost", to give to its colleges'.[111] But not even 'partial' imitation took place; instead, an industrial school was opened at Patna and the scene shifted to Poosah estate in north Bihar where the government was experimenting with tobacco culture.[112] The Lt. Governor, R. Temple, wanted that farm to be kept permanently and made use for scientific experiments connected with physiological botany and agricultural chemistry. But the failure of tobacco experiments there made the officials despondent and so on the convenient pretext of the utility of spending money in attempts to teach the Indian peasants the agricultural sciences of Europe,[113] the Poosah estate was leased to Ms Begg, Dunlop & Co. for tobacco cultivation.

Richard Temple carried his enthusiasm for agricultural science to the Bombay presidency. In 1879, an agricultural class was opened at the Poona Civil Engineering College, and such classes were opened also at six of the principal *zilla* schools.[114] Temple thought of the native revenue officers, like *Mamlatdars* and *Karkuns*, as the most effective agents for diffusion of agricultural knowledge and wanted the Bombay University to conduct a degree course in agriculture.[115] But the University refused to concede more than a diploma. Later Dr Voelcker found the diploma course 'a sort of halfway house'; better than nothing, but not the equal of a degree.[116] These were meant for the natives to be placed in lower grades. Need was felt to secure in higher grades a proportion of officers with a scientific and practical training in agriculture. The Famine Commission of 1881 proposed first a theoretical training of I.C.S. probationers, and then the practical training of few of these at an agriculture college.[117] The Government of India rejected them on the grounds of 'impracticability' and the Secretary of State claimed that a superficial theoretical training would not only be useless but would be injurious to a civil servant, for it would divert his time and attention from the study of those subjects which are necessary for his efficiency as a judicial or administrative officer in India.[118]

Meanwhile, the Government of Bengal had created two special scholarships of $200 a year to be held for two and a half years by science graduates of the Calcutta University at the Royal Agricultural College, Cirencester. The aim was to have a team of experts as prelude to the establishment of colleges and schools 'for grafting on eastern practice, as far as may be found possible or desirable, the ascertained results of western research'.[119] It was a significant step; that too, taken by a local government and not the Government of India. When the Agriculture College of Salisbury requested the Viceroy to establish scholarships there for his Indian subjects, he simply declined.[120] Obviously, the Bengal example had failed to inspire the Imperial Government.[121] But even this experiment did not prove a happy one. A few Cirencester scholars diverted their attention to legal and other studies. The Bengal Government issued warnings. The scholars themselves felt unhappy when on returning home they found they could get nothing more than deputy collectorships.[122] So, in 1887, the government decided to discontinue the scholarship.[123]

The enthusiasm of mid-1870s was thus on wane in the mid-1880s. Except the Poona unit, which made important experiments in cotton and plough technology,[124] all other centres were going down. Saidapet, for example, notwithstanding the value of the theoretical and scientific instruction which it gave, could not turn out 'practical' farmers, and was, therefore, pronounced a failure by W. Wilson, Director of Agriculture, Madras.[125] This, however, does not mean that agricultural education and experiments had come to a dead end. Thanks to E.G. Buck, his hammerings had kept the issue very much alive.[126] In June 1888, the Government of India resolved to encourage the introduction of studies inclining to the application of natural science. Madras Government promptly took up the clue and in December 1888 appointed a committee to report on the working of Saidapet Agricultural College. The committee declared Saidapet to be absolutely essential and recommended provisions for instruction in veterinary and forestry. But the Madras Government could not stomach it, and instead sanctioned five agriculture schools.[127]

The waning spirit, however, got a boost from Dr Voelcker. His criticisms brought the whole issue under a sharp focus. He found that no encouragement was given to the pursuit of scientific investigation in India and that men who might have been original workers in science had to abandon it for the duties of school inspectors.[128] For the 'unsettled state' of Saidapet College, he laid responsibility on the constant change of policy pursued by the Madras Government.[129] The syllabus was found to be modelled on English and not on Indian experiences. Thus, practices such as 'paring and burning' and 'warping of land' were mentioned; manures such as sulphate of ammonia, dried blood, soot and artificial manures, none of which had any place in Indian agriculture, were introduced; the requirements of 'fattening animals' were supposed to be learnt and this in a country where no fattening of animals whatever was carried on. On the other hand, many subjects of special interest to Indian agriculture were omitted, such as canal and well irrigation, oilcake refuse, ghee, etc.[130]

To take advantage of Dr Voelcker's presence and advice, the Government of India called for a Conference of Directors of Agriculture in October 1890.[131] The Conference discussed inter alia whether special teaching in agriculture was desirable or if it was to proceed from above downwards or from below upwards.[132] Buck favoured the latter, i.e. not to go for agricultural colleges and schools but only teach, in an elementary fashion, how a crop grows. This was what he called 'the first process in the cultivation of the agricultural intellect, the first manuring of the mental field in which agricultural progress is to be developed'. He came to this conclusion only because of financial constraints. Otherwise, he very much desired early establishment of high-class agriculture colleges and the education of natives at Saidapet and Dehra instead of Cirencester.[133] Another view was put across by the Agriculture Commissioner of Madras, H.F. Clogstoun, who opposed Buck and advanced downward filtration. He said, 'the source from which the teaching power springs must be put on a higher level and we can never look to the stream to flow healthily throughout the country unless the fountain head be pure'.[134] The Conference finally recommended that higher agricultural education should not be provided by special institutions but be grafted to the existing ones, and that the claims of men with training in scientific agriculture be freely recognized in the revenue and cognate departments.[135]

These proposals were considered by another conference in 1893, where again Buck's views prevailed and the very next year the Government of India resolved that 'the question is one which cannot be forced, but should be dealt with gradually, and that greater success is to be expected from making instruction in the rudiments of agriculture, part and parcel of the primary system of instruction than from teaching it as a subject apart from the general educational programme'.[136] It was a carefully chosen dilatory move on the part of the government. Instead of being a specialized study, where enquiry and experimentation was followed by application, agriculture resulted in becoming a part of general school curriculum, and that too, as an optional subject.[137]

Until 1897, the Government of India held the view that high-class educational institutions professing to teach agriculture should be national rather than provincial, and that one or two national colleges would meet the needs, then existing, of all India. But the provincial governments were clamouring for college-level teaching in their respective areas. They argued that education in an agricultural college was as good training for state officials in the land revenue and cognate services as training in an arts college. The Imperial Government got attracted to this logic and proposed (vide Resolution, 20 March 1897) the establishment of four high-class colleges at Madras, Bombay, Calcutta and some place in northern India.[138]

One of the results of this resolution was the starting of agriculture classes at Shibpur Engineering College in 1899.[139] In some *zilla* schools also, agriculture classes were opened but they soon collapsed as agriculture was not included in the syllabus for the university entrance examination.[140] The new regulations of Calcutta University did not provide for degrees in agriculture. It was suspected that affiliation of agricultural colleges to the universities would tend to accentuate the tendency to regard these institutions as avenues leading to government service and thus it was required to place them still further outside the practical scope of persons who would turn the training received to profit in private farming or estate management.[141] This was the feeling of the Simla Conference on Agricultural Education held in 1901. It even recommended vernacular to be substituted for English as the medium of instruction and, thereby, the government hoped to 'popularize the study of agriculture amongst those who live by its pursuit and are now debarred from it by their ignorance of English'.[142]

At Kanpur grew an agriculture school out of a training school for Qanungos.[143] Taking advantage of the resolution of 1897, the NWP Government asked the Allahabad University in 1901 to raise the status of that school to that of a college and the University Senate readily agreed.[144] Basic objects were threefold: to train teachers, to give instruction to landowners, and to maintain a supply of revenue officials. Research work was to be encouraged and the award of postgraduate studentships, even doctoral degrees, was contemplated.[145] Earlier, in 1897, the Government of India had acknowledged the desirability of an agricultural college in NWP, but when the proposal came officially, it lost enthusiasm. J.B. Fuller agreed that such a college would have a good chance of attracting a fair number of students, but at the same time condemned it as 'premature'.[146] He held that its course of study would be of no value in training men to manage landed estates, and that the revenue officers did not need acquaintance with science so much as with practical agriculture. So, in place of the college, Fuller advocated the establishment of what he called a 'Zamindari School' and the Viceroy concurred.[147] Behind this refusal perhaps lay the spectre of cost, which, as Fuller calculated, would have amounted to Rs.50,000 a year. Similarly, in 1903, when the Madras Government asked for Rs.11,700 per annum to create rural teaching posts for the students of Saidapet, as an incentive to agriculture education, the Government of India refused.[148]

Lack of an all-India policy led to certain management anomalies. The Saidapet and Poona Colleges, for instance, were administered by the Education Department, while the Kanpur and Nagpur Schools were under the Agriculture Department. But none of them provided a complete agricultural education.[149] Though the Resolution of 1897 had talked of four high-class colleges, the Government of India subsequently narrowed its choice to only one institute, which would serve as the nucleus for agricultural education and research. But who was to foot the bill? The Imperial Government was reluctant and the provincial ones used to spend on agriculture only in a piecemeal fashion.[150]

The windfall came when, in January 1903, an American philanthropist Henry Phipps gifted $100,000 to Curzon for the establishment of 'a laboratory to determine the economic value, and the medicinal qualities of the plants of India—or to be used in any other way that promises enduring good to India'.[151] Ibbetson favoured a laboratory for indigenous drugs while the Home Secretary, H.H. Risley, wanted the 'Pasteur Institute' to be assisted with Phipps's money.[152] It was finally decided to establish a laboratory for agricultural research, designed to form what Curzon called a 'Centre of Economic Science'[153] and also a Pasteur Institute in south India. The former was Curzon's favourite,[154] while the latter was Risley's. Mr Phipps made a further offer of $50,000 and the proposal for agricultural laboratory was then expanded to include an agriculture college and experimental farm also.[155]

Apart from Phipps's donation, the interest of the indigo lobby was also working behind the creation of a research institute. A village in north Bihar was selected as the site because of its proximity to the planters and it was called Pusa (Phipps of USA). This site was probably known as Poosah earlier. It was a strange but useful coincidence. As the then I.G. of Agriculture noted, 'a research and experimental station in Bihar is likely to be more successful than in any other part of India, because useful results will be assimilated by the European farmers there and filter to the natives around them and then I hope to other parts of India'.[156] Not only this, B. Coventery, who owned the

Dalsinghsarai indigo experimental farm, was made the first director of the Institute and also the principal of the college at Pusa. Interestingly enough, he possessed no university degree in agriculture and his qualification was officially described as, 'skilled practical acquaintance with agriculture, energy and administrative capacity, and sufficient general knowledge to be able to comprehend and utilize the results achieved by scientific experts'.[157] What a smokescreen to hide his interests as a seasoned indigo planter!

The government was quite aware of the possibility of being accused 'of doing a job for the sole benefit of European Industries' (i.e. indigo).[158] Laying the foundation stone of the agriculture college at Pusa on 1 April 1905, Curzon himself talked of the German synthetic indigo as 'blue terror' and hoped that, with the help of researches, the Bihar indigo would beat 'the finest product of Teutonic synthesis'; but he added, 'price, however, is the determining point. Science has got to help you to bring it down'.[159] Finance for such researches posed no problem. Earlier the Bengal Government had already granted Rs.50,000 a year to the Indigo Syndicate, another Rs.15,000 were saved by abolishing the agricultural unit in Shibpur, and now the whole money was pumped into Pusa.[160]

The Pusa spirit led the Curzonian team to make more ambitious plans. In October 1905, the government proposed 'to establish in each important province an agricultural college and research station, adequately equipped with laboratories, and the staff consisting of an expert agriculturist, economic botanist, agricultural chemist, entomologist and mycologist, etc.'[161] But this idea was scuttled at the India Office. Morley vetoed it by advocating what he called, 'a policy of cautious advance'.[162] Minto, the new Viceroy, however, persisted; he argued 'that we are greatly behind other countries in the application of science to agriculture, and now that funds are available we consider it one of our first duties to remove this "reproach". The total expenditure proposed by us is insignificant compared with the interests involved and work to be done. We strongly urge for sanction to full 30 lakhs.'[163] But Morley refused to budge and sanctioned only 4 lakhs.[164] Minto again approached Morley, citing the examples of colossal losses from wheat rust, sugar cane disease and cotton pests, for the appointment of mycologists and entomologists in the provinces.[165] But Morley refused, calling the proposal 'premature'.[166]

Reflections

Official attitudes to agricultural problems unfolded themselves in three distinct phases. The first one, covering the period 1840–70, was basically a phase of 'masterly inactivity' on the part of the government and of 'superactivity' on the part of private societies and the evergrowing band of planters with their exotic varieties. The luxuriance of the tropical virgin forests appealed to the colonizing enthusiasts, chiefly indicating possibilities of its succession by equal luxuriance of plantations controlled by planters. They imagined the 'jungle', 'bush' and 'shrub' being replaced by fields of sugar, cotton or tea.[167] Heyday for the cash crops had arrived.

The second phase, the last two decades of the nineteenth century, was full of hectic but confused activities. This period saw several well-intentioned scientific men like Benson, Bourbell, etc., trying to impress upon the government without success. Benson experimented with agricultural implements and, at his behest, several makers of British implements showed interest in the Indian market.[168] Bourbell surveyed mechanical improvements in the United States and wanted turbine water wheels and machines, for cleaning flax, cotton, etc., to be introduced in India.[169] Buckingham pleaded for entomological researches.[170] Private bodies like the Indian Tea Association and the Bengal Silk Committee showed more interest in such studies than the Government of India. When in 1892, the Indian Museum submitted a scheme for similar investigation, the government refused on the usual grounds of meagre finances.[171] The government would, thus, always recognize the importance of science in agriculture, but would never approve of 'any large outlay upon them which must, however, useful in its remote results, be immediately unremunerative'.[172] The Resolution of 1881 asked the agriculture departments to be conscious only of revenue records and look upon 'agricultural improvements' as 'subsidiary duty'.

In sharp contrast to the pessimism of senior officials like Durand, Carmichael, Morley and the scepticism of Finance Department, stands the enthusiasm of Richard Temple, E.C. Buck, Voelcker and Curzon. But streaks of negativism can be spotted even in Buck. There were political reasons behind his support for agricultural education. He himself admitted,

I believe that the position into which the educational system is drifting is, politically, most dangerous. We are overcrowding the learned professions and government offices with university scholars, and the residue, increasing enormously every year, go in the ranks of discontents. At the agricultural conferences I spoke strongly on the subject. I begged Dr Voelcker to take it up, and much of what he wrote was at my dictation.[173]

This perhaps might be the reason why Buck advocated the spread of agricultural education from below upwards. He preferred primary education, because higher learning, even in agriculture, would have created more awareness among the natives and thereby fuelled discontent. The ghost of Charles Wood who in 1864 had advocated emphasis on primary education alone had not died.[174]

Curzon struck a different note. Perhaps he was not that afraid of political discontentment. He had economic reasons and looked upon agricultural education as investment. At the Director of Public Instruction's Conference in September 1905, he said, 'agriculture is the first and capital interest of India, and agriculture, like every other money-earning interest, must rest upon education'.[175] Earlier, in 1901, the Simla Conference on Agricultural Education had condemned Saidapet and Poona for being too 'theoretical', rather than 'utilitarian'.[176]

Voelcker had a sympathetic understanding of the Indian peasants' needs and their capabilities. He was against the import of foodgrain technology or expertise, and wished for native experts, trained in India, and not in England.[177] But, he wanted this to be done within the existing framework.[178] He found that though the Indian students were quite capable of passing any examination with the help of marvellous retentive memories, 'they could not display a practical understanding of the subject'. And that is why he did not expect great results to follow at once with the introduction of teaching of agricultural chemistry.[179] Still he was not a pessimist and called for perseverance and patient efforts. Even then, his report was often misread. His call for 'improvement from within' was interpreted by the editor of the *Madras Mail* to mean that 'there must be, "amongst the cultivators themselves", cooperation and organization, mutual instruction and mutual experimentation',[180] as if the government, even though alien, did not come under the term 'within'.

The post-Voelcker era saw a better appreciation of the native problems. Earlier, most of the officials and planters started with the assumption that the agricultural methods of India were bad, and that they must make room for European methods. Their vital error lay in ignoring the then existing science of agriculture within the Indian customs and practices.[181] They always ran after European models. The Superintendent of Saidapet farm, C. Benson, justified major emphasis on theoretical instruction there on the grounds that Professor Jorgensen had done the same thing for the Royal Agriculture College at Copenhagen. Professor Jorgensen had founded the Danish College 'to give only a theoretical education, not making any attempt at all to offer any kind of instruction in practical farming'.[182] What was suitable for Denmark was, thus, held good for Madras also. And in this lay the strength as well as the basic weakness of those agricultural enthusiasts.

In the white colonies, the farmers had some capital, but little experience; in India, people had no capital, but 'a wealth of experience that sometimes turns the wisdom of the West to foolishness'.[183] Thiselton-Dyer regretted that India had no such men as the Earl of Leicester who, without any government assistance, had carried on experiments on a gigantic scale.[184] But was the extortionist zamindari system, which the British had introduced in India, geared to produce such men?

Within their own limitations, several landholders did respond quite favourably—whether it be the acceptance of Behea sugar mill, or the well boring scheme in NWP, or education in agricultural schools. At Poona College in 1905, 19 students out of 29 belonged to the landowning classes.[185] This response came in spite of the paucity of textbooks and the problems in mediums of instruction. Voelcker favoured teaching in the vernacular,[186] while Moreland rejected it as totally unsuitable for scientific expressions.[187] This vacillation prevented the percolation of agricultural education to lower classes. The British had all along been very particular about not disturbing the existing class relations. Upper classes were their natural allies. R.J. Henry argues that it was through this class understanding that the Raj (which feared another 1857) prevented a cultural clash in the accepted sense of a tradition opposing modernity.[188]

The government looked to the middle class for initiative. But why should the *bhadralok*, so fond of quoting Milton and Shakespeare, condescend to take an agricultural degree? The education system led to the acquisition of literary, rather than of scientific tastes, 'tastes which are best satisfied by the profession of the lawyer, teacher or the government official'.[189] Even those who took to higher education in agriculture had to work under unhappy circumstances. N.G. Mukherjee, for example, was placed under a committee of merchants who did not allow him a freehand in sericultural researches.[190] When the agricultural classes at Shibpur were abolished, its teacher, D. Dutta, was not absorbed by the Pusa Institute which, for obvious reasons, preferred the staff of a private indigo farm.

Dissemination of scientific agriculture through an agricultural journal was also thought of by the educated natives. In 1885, G.C. Bose, J.N. Dey and H. Patra brought out *The Indian Agricultural Gazzette*.[191] It was an entirely Indian enterprise and bereft of any government support, it was closed in 1889.[192] Even the influential editor of *The Statesman*, Robert Knight, who also published an agricultural journal, *The Indian Agriculturist*, complained of 'weak support of government' and his request to the government for a raise in subscription met with refusal.[193]

The rising Indian bourgeoisie was not at all oblivious to the importance of science in agriculture. The Bengal Chamber of Commerce showed interest in the reorganization of agriculture departments.[194] At Belgaum, Kirlosker was getting famous for his chaff-cutters and other agricultural implements.[195] Tata initiated sericultural experiments at Bangalore.[196] In 1903, the native luminaries of Bombay submitted a scheme for agricultural improvements in their presidency. They alluded to the US Department of Agriculture which comprised of six divisions, with hundreds of scientists working in each division and lamented that the Bombay Government did not employ even one purely scientific worker.[197] They called for the establishment of an agricultural institute with affiliated farms.

But it would have initially cost Rs.5 lakhs, and this was sufficient to frighten the government, even though all officials ostensibly lauded the objects and plans of the memorialists. Making a polite refusal, Bombay's Revenue Secretary, R.A. Lamb, clarified, 'the Imperial Government has made substantial contributions to expenditure under medical, education, civil works and revenue settlements. The Governor in Council, therefore, feels bound to restrict proposals for expenditure on agricultural improvement to the amount required to meet only immediate needs and to gain objects of which the advantage is clear.'[198]

No better confession could have been made. The problem was, thus, not cultural stagnation or social conservatism of the Indians, rather it was finding economically viable appropriate technological solutions. The government first ignored it and later tried to use science as a crutch, which could, at best, only provide a halting gait.

Notes and References

1. Revenue Department, Agriculture Branch (hereafter, Revenue Agriculture), Survey, no. 25, September 1880, National Archives of India, New Delhi (hereafter, NAI). All archival references unless otherwise indicated are from the NAI.
2. D.B. Allen, 'Demonstration Farms', *Indian Agricultural Gazette*, 31 July 1885, pp. 5–6.
3. H. Piddington, *On the Scientific Principles of Agriculture*, Calcutta, 1839, p. 5. Similar minds ran the opium business in China, obviously at the cost of more pressing needs of the Indian agriculturists.
4. *Alexander's East India and Colonial Magazine*, vol. VIII, no. 48, November 1834, p. 430.
5. R.Wissett, *On the Cultivation and Preparation of Hemp*, London, 1804, p. 113.
6. Ibid., p. 118.
7. Home, Revenue, nos. 1–10, 13 April 1840, NAI.
8. Home, Public, no. 33, 5 July 1814, NAI.
9. Revenue Agriculture, nos. 11–16, May 1888, West Bengal Archives (hereafter, WBA).
10. Ibid., nos. 48–51, May 1888, Pt. B, WBA.
11. Revenue Agriculture, nos. 12–24, March 1890, Bihar State Archives (hereafter, BSA).
12. Ibid.
13. Revenue Agriculture, nos. 7–9, March 1896, WBA.
14. Revenue Agriculture, nos. 12–24, March 1890, BSA.

15. Ibid.
16. Revenue Agriculture, nos. 2–5, April 1905, File no. 38.
17. Home, Revenue, nos. 1–4, August 1839.
18. Home, Revenue, nos. 2–5, 28 February 1842.
19. Home, Revenue, no. 13, 28 June 1845. Appendix B gives a memorandum by Bisheshwar Dayal on his experimental cotton plantation in 1844.
20. W.N. Lees, *Tea Cultivation and other Agricultural Experiments in India*, Calcutta, 1863, p. 100.
21. Revenue Agriculture, Fibres and Silk, nos. 18–27, 22 June 1840.
22. T.J. O'Keefe, 'British Attitudes towards India and the Dependent Empire, 1857–1874', unpublished Ph.D. thesis, University of Notre Dame, Indiana, 1968, pp. 134–6.
23. Ibid.
24. Note on the improvement of the cotton industry in Berar by Voelcker, 2 October 1890, Revenue Agriculture, Fibres and Silk, nos. 11–17, February 1891, Pt. B.
25. Peter Harnetty, *Imperialism and Free Trade*, Manchester, 1972, pp. 99–100.
26. *Revenue Administration Report*, C.P. (Chhattisgarh Division), 1903–4, para 6.
27. F.G. Sly, Agricultural Commissioner, C.P., *Report on the Development of Land Records and Agriculture, 1900–1901*, para 61.
28. George Watt, *Memorandum on the Resources of British India*, 1894, p. 10.
29. Revenue Agriculture, Fibres and Silk, nos. 11–17, February 1891, Pt. B.
30. Harnetty, *Imperialism and Free Trade*, p. 100.
31. J. Brodrick to Curzon, 3 March 1904, *Curzon Papers*, NAI Microfilm no. 1633.
32. Between the two world wars, research funds for British textile industry were funnelled through government agencies, corporate research association, private firms, and the education system. In contrast, research in India was conducted mainly through a trade association laboratory at Matunga, and to a small degree by individual manufactures. Heavier reliance was put on imported efforts, rather than comprehensive original study.
33. *The Tea Cyclopaedia*, Calcutta, 1881, pp. 9–10.
34. Lees, *Tea Cultivation and other Agricultural Experiments in India*, p. 36.
35. Ibid., p. 11.
36. Home, Revenue, Governor General, nos. 18–20, October 1839.
37. *The Tea Cyclopaedia*, p. 10.
38. Dalhousie's Minute, 28 February 1856, *Selections from the Records of Government of India*, no. XIV, Calcutta, 1856, pp. 34–5.
39. *The Tea Cyclopaedia*, p. 11.
40. Ibid., p. 7.
41. Ibid., p. 34.
42. Ibid.
43. Revenue Agriculture, nos. 1–8, May 1901, BSA.
44. Ibid.
45. Ibid.
46. *Nature*, vol. LXXVIII, 30 June 1908, p. 295.

47. Letter from Curzon to Hamilton, 28 February 1901, *Curzon Papers*, Microfilm Roll 2, nos. 1630–43.
48. Revenue Agriculture, no. 44, November 1905, File no. 81.
49. Ibid.
50. Ibid.
51. Revenue Agriculture, no. 7, August 1904, Pt. B, File no. 70.
52. *The Indian Agriculturist*, vol. XIII, no. 32, 11 August 1888, p. 436; Piddington, *On the Scientific Principles of Agriculture*, pp. 11–12.
53. Revenue Agriculture, *Agriculture & Horticulture*, no. 47, April 1880; Revenue Agriculture, *Agriculture & Horticulture*, nos. 1–3, June 1881, Pt. B.
54. Dr Datta, In-charge of Shibpur experimental farm, quoted in Revenue Agriculture, no. 7, August 1904, Pt. B, File no. 70.
55. This explains why out of 84 papers published in the volume of the Transactions of the Agricultural and Horticultural Society of India (1838–9), only 2 were contributed by the natives; Piddington, *On the Scientific Principles of Agriculture*, p. 10.
56. Ibid., p. 18.
57. *Calcutta Review*, vol. 80, no. 160, 1885, p. 474.
58. Home, Public, nos. 32–5, 29 July 1864.
59. *Calcutta Review*, vol. 81, no. 162, 1885, p. 428.
60. D.O. Letter from E.C. Buck to the Editor of *Implement and Machinery Review*, 14 June 1884. Revenue Agriculture, nos. 4–21, October 1884, K.W., File no. 84.
61. In 1880, the Agriculture Department of NWP decided to purchase a steam tractor for the eradication of *Kans* weed in Banda district. By the time the tractor arrived, the dry season had already eradicated the weed. The cost of operating that tractor for any other purpose was unacceptably high. As a result, the Department was forced to break-up and sell the tractor on the spot for whatever price its parts could fetch. This misconceived attempt to solve an agricultural problem with modern technology cost the Department half its total annual budget. See R.J. Henry, 'Technology Transfer and its Constraints: Early Warnings from Agricultural Development in Colonial India', in *Technology and the Raj*, ed. Roy MacLeod and Deepak Kumar, New Delhi, 1995, pp. 25–50.
62. Note on Agricultural Implements, E.C. Buck, 7 February 1883; Revenue Agriculture, nos. 14–32, March 1883.
63. Ibid. The cost of Behea sugar mill was Rs.80, of a winnower Rs.85, of Swedish plough for black cotton soil, it was Rs.35 to Rs.50, and of double stilted Swedish plough for deep tillage Rs.35. Kaiser plough for light soil was for Rs.6, wheat thrasher for Rs.42 and *jowar* thrasher for Rs.85. But for their price, these implements would have got a better response.
64. *The Indian Agricultural Gazette*, vol. I, no. 5, 1885, p. 104.
65. *Calcutta Gazette*, 20 May 1830.
66. Records of the Agricultural and Horticulture Society of India MS, May 1837, pp. 29–30, Agricultural Society, Calcutta.
67. Ibid., February 1827, p. 115.

68. Ibid., 13 May 1835, unpaginated.
69. Ibid., May 1837, p. 25.
70. Ibid., 28 January 1885, p. 156.
71. Dwarkanath Tagore, an effective member of the society, offered his extensive premises at Manicktoila, rent-free, for a school where a limited number of native workmen could be trained in flax culture. Home, Revenue, nos. 14–17, 22 November 1841.
72. *The Tea Cyclopaedia*, p. 34; Records of the Agricultural and Horticulture Society of India, 26 February 1890, p. 275.
73. Revenue, Land Revenue, nos. 54–5, May 1861, WBA.
74. Home, Public, no. 60, 7 November 1856.
75. Revenue Agriculture, nos. 3–9, October 1887.
76. Ibid.
77. Revenue Agriculture, *Agriculture & Horticulture*, nos. 36–42, September 1879.
78. Revenue Agriculture, nos. 103–5, February 1886, Pt. B, File no. 44. This was true of even the society at Calcutta, and some private nurseries and agencies once petitioned the government not to patronize that society which had become a purely commercial organization. See Records of the Agricultural and Horticultural Society of India, 28 June 1899.
79. Revenue Agriculture, nos. 1–10, September 1882, File no. 6.
80. J.D. Hooker, *Himalayan Journals*, vol. I, London, 1854, p. 91.
81. Home, Public, no. 92, 2 April 1870; July 1871, no. 83.
82. Minute by Durand, 31 March 1870, Home, Public, no. 100, 9 April 1870.
83. Home, Public, no. 134, 6 June 1871.
84. Home, Public, no. 97, July 1881.
85. F.G. Sly, 'The Departments of Agriculture in India', *The Agricultural Journal of India*, vol. I, Calcutta, 1906, p. 3.
86. Some have justified such a myopic policy. P.K. Jain, for instance, writes:

    if the Department had followed the policy, not unnaturally dictated by certain sections of the public and the press, in the direction of anything like lavish expenditure on attempts at agricultural improvement, which could only have been of a crude and unscientific character, instead of directing their chief efforts to administrative duties of an immediately useful kind, they could not have overridden the storms which were raised by the financial difficulties occurring at intervals since 1881.

    P.K. Jain, 'The Indian Agricultural Service and the Development of Agriculture in India from 1898 to 1974', unpublished Ph.D. thesis, Agra University, 1975, p. 13.
87. Revenue Agriculture, *Agriculture & Horticulture*, nos. 21–3, May 1873.
88. Ibid., nos. 9–23, December 1873.
89. Records of the Agricultural and Horticultural Society of India, 28 January 1885, p. 157.
90. Revenue Agriculture, no. 12, March 1886, File no. 67. The society objected to the use of the term 'donation' or 'grant'. At a special meeting it resolved that 'a more accurate description would be payment made as remuneration for services

44 *Deepak Kumar*

       to be rendered by the society to Government.' See Records of the Agricultural and Horticultural Society of India, 28 June 1899, p. 421.
91. Revenue Agriculture, Fibres and Silk, nos. 1–2, March 1886, File no. 10.
92. Ibid., nos. 23–5, May 1887, File no. 17.
93. Revenue Agriculture, no. 56, April 1884, File no. 97.
94. Ibid., nos. 12–14, February 1886.
95. Revenue Agriculture, no. 44, November 1905, File no. 81.
96. Ibid., nos. 1–8, February 1890, File no. 21.
97. J.A. Voelcker, *Report on the Improvement of Indian Agriculture*, 2nd edn, Calcutta, 1897, pp. 304–5.
98. Ibid., p. 313.
99. Ibid., p. 227. Voelcker narrates an interesting instance:

> Near Ferozpur, Mr E.B. Francis showed me some light sandy land on which when a shower of rain falls soon after sowing a crust is very apt to form, so that the young shoots cannot force their way through it. When it forms, the people habitually re-sow the crop, for they have no implement corresponding to a harrow. I have instanced how careful the Behar indigo planter is to break up this crust the instant it forms using a bullock-rake or harrow having spikes some 8 inches long and penetrating about 2 inches into the soil. An implement of this kind if introduced at Ferozpur would entirely dispense with the necessity of re-sowing. The improvement here would consist in a transference of native methods, not an importation of foreign ones.

100. Ibid., p. 223.
101. The use of the term 'native' in government documents seems to be an offshoot of racial consciousness. It was chiefly used for 'Non-Caucasian people of inferior civilization, but often semi-humorously for the inhabitants of any region spoken of as if strange'. See *Webster's New International Dictionary of English Language*, 2nd edn, Massachusetts, 1958, p. 1630. In this paper, this term has throughout been used with a view to keep its 'original flavour'.
102. NWP & O, Revenue Proceedings, no. 83, January 1884.
103. Ibid., pp. 8–9. The zamindars were very reluctant to deal financially with the government, because of the government's rigid enforcement of revenue demands, which forced them to alienate their land.
104. NWP & O, Revenue Proceedings, nos. 12–18, July 1885.
105. E.C. Buck to James Cairn, 25 January 1891; Revenue Agriculture, nos. 7–11, May 1891, K.W. File no. 26.
106. J. Long, 'Introduction', *Adam's Report*, Calcutta, 1868, p. 35.
107. Piddington, *On the Scientific Principles of Agriculture*, p. 8 (emphasis as in original). A century later, Dr Nilratan Sarkar voiced similar feelings on agricultural education, 'It is necessary for Edinburgh. It is necessary for almost every university in America. But it is not necessary for Calcutta. The Bhadraloks of Bengal should always remain Bhadraloks. They should not have anything to do with ploughshares.'
108. Long, 'Introduction', *Adam's Report*, p. 36.
109. Revenue Agriculture, nos. 13–14, January 1884, Pt. B.

110. *Madras Revenue Board Proceedings*, no. 2410, 25 November 1873. Quoted in R. Ratnam, *Agricultural Development in Madras State prior to 1900*, Madras, 1966, pp. 390–2.
111. H. Woodrow, DPI, Bengal, to Secretary to Government of Bengal, 8 March 1876; Revenue Agriculture, nos. 17–19, April 1876, BSA (emphasis added).
112. *Report on the Administration of Bengal, 1875–6*, Calcutta, 1877, p. 166.
113. *Report on the Administration of Bengal, 1876–7*, Calcutta, 1878, pp. 138–9.
114. Home, Education, nos. 27–30, March 1879.
115. Ibid. The Famine Commission of 1881 held similar views. Home, Public, no. 89, July 1881.
116. Voelcker, *Report on the Improvement of Indian Agriculture*, p. 384.
117. The commissioners thought that Indian languages, Indian law and political economy were subjects in which the requirements might be abridged, and of which some might be made optional, their place being taken by agricultural and organic chemistry and botany. Home, Public, no. 89, July 1881.
118. Home, Public, no. 182, June 1884.
119. *Report on the Administration of Bengal, 1879–80*, Calcutta, 1880, p. 495. The catch, however, lay in the parentheses which made the whole thing dependent upon 'possibility' or 'desirability'.
120. Revenue Agriculture, *Agriculture & Horticulture*, nos. 34–5, January 1881, Pt. B.
121. Ibid.
122. Revenue Agriculture, nos. 76–80, March 1887, BSA.
123. Ibid., nos. 112–13, April 1887. This was an extreme step, like chopping the head off as a remedy for headache.
124. Revenue Agriculture, nos. 20, 127, March 1883.
125. Ibid., nos. 13–14, January 1884, Pt. B. Wilson distinguished between the art and science of agriculture and wanted the former to be given precedence which Saidapet was not doing. Naturally, he was indignant and condemned the Saidapet system as 'a process very similar to that of attempting to acquire a language from its grammar instead of learning the grammar from the language'.
126. E.C. Buck's note, 10 January 1886, Home, Education, nos. 14–88, October 1897, Pt. B.
127. Home, Education, no. 6, October 1895.
128. Voelcker, *Report on the Improvement of Indian Agriculture*, p. 331.
129. Ibid., p. 392.
130. Ibid., p. 391.
131. Revenue Agriculture, no. 6, June 1890, File no. 63.
132. Ibid., nos. 7–11, May 1891, File no. 26.
133. Ibid., Presidential address, 6 October 1890.
134. Ibid., *Proceedings of Agricultural Conference*, 5th meeting on 10 October 1890.
135. Voelcker, *Report on the Improvement of Indian Agriculture*, pp. 382, 397.
136. Education Resolution, 7 September 1894, Home, Education, no. 6, October 1895; Revenue Agriculture, no. 21, January 1896, WBA.

137. J. Mangamma, 'Technical and Agricultural Education in Madras Presidency, (1854–1921)', unpublished Ph.D. thesis, University of Delhi, 1971, p. 80.
138. Home, Education, nos. 57–9, July 1903.
139. Revenue Agriculture, nos. 21–6, June 1897, File no. 48.
140. Revenue Agriculture, no. 7, August 1904, Pt. B, File no. 70.
141. *Third Quinquennial Review of Education in Bengal*, Calcutta, 1907, p. 92.
142. Home, Education, no. 47, November 1901.
143. Revenue Agriculture, no. 8, September 1902, Pt. B, File no. 103.
144. Home, Education, nos. 57–9, July 1903.
145. Revenue Agriculture, no. 3, December 1901, File no. 89.
146. Note by J.B. Fuller, Secretary, Department of Revenue, Agriculture, 4 September 1901, Revenue Agriculture, no. 3, December 1901, File no. 89.
147. 'The idea of a college seems to me a mistake, and would probably only repeat the failure of Madras.' Note by Curzon, 24 October 1901, Revenue Agriculture, no. 3, December 1901, File no. 89.
148. Home, Education, no. 116, July 1903.
149. Ibid.
150. Bengal Government, for example, would profess to serve the cause of agriculture by giving aid either to the Bengal Silk Committee, or even to private estate-holders like the Maharaja of Dumraon. Revenue Agriculture, nos. 7–11, May 1890, Appendix D, File no. 26.
151. Jay Phipps to Curzon, 30 January 1903, *Curzon Papers*, no. 10, Microfilm Roll, nos. 1630–43. Later on, when this money was used for an agricultural institute in north Bihar, Thiselton-Dyer prophesied that it would someday become the Rothamstead of the East, but added: 'The characteristic irony, I might almost say cynicism, of the British race is that it should own its foundation in great part to the large-minded munificence of an American gentleman'. *Nature*, vol. LXXIII, 19 April 1906, p. 587.
152. Home, Medical, no. l, March 1903, Deposit Collection.
153. Ibid.
154. Curzon knew the economics of agriculture. Earlier while defending the Punjab Land Alienation Bill on 27 September 1899, Curzon said, 'There is no country in the world that is so dependent upon the property of the agriculture classes in India. There is no government in the world that is so personally interested in agriculture as the Indian Government. We are, in the strictest sense of the term, the largest Landlords in creation.' *Curzon, Speeches*, vol. I, Calcutta, 1900, p. 123.
155. Curzon to Phipps, 20 February 1903, *Curzon Papers*, no. 10, Microfilm Roll, nos. 1630–43.
156. Note by J. Mollison, l4 February 1903, Home, Education, no. 116, July 1903.
157. Revenue Agriculture, nos. 14–18, September 1905, File no. 193.
158. Denzil Ibbetson to the Lt. Governor of Bengal, 10 March 1903, Home, Education, no. 116, July 1903.
159. *Curzon, Speeches*, vol. IV, Calcutta, 1906, pp. 116–17.
160. Home, Education, nos. 36–7, December 1903, Pt. B.

Science in Agriculture 47

161. Revenue Agriculture, no. 44, November 1905, File no. 81.
162. Revenue Agriculture, no. 1, August 1906, File no. 66.
163. Telegram from Viceroy to Secretary of State, 19 February 1906, Revenue Agriculture, no. 2, August 1906, File no. 66.
164. Revenue Agriculture, no. 5, August 1906, File no. 66.
165. Revenue Agriculture, no. 7, August 1906, File no. 66.
166. Revenue Agriculture, no. 38, January 1907, File no. 14.
167. R.H. Wallace, 'Agricultural Education in Greater Britain', *Journal of the Society of Arts,* vol. XLVIII, no. 2468, 9 March 1900, p. 326.
168. Revenue Agriculture, nos. 116–17, January 1883, Pt. B, File no. 38. In the middle of the nineteenth century, with the growth of steel industry, a variety of agricultural implements were manufactured in England. Agricultural enthusiasts, without examining the conditions prevalent in India, imported agricultural implements from England for the experimental farms in India. It was not realized that novelty is not necessarily an improvement. The result was that large dumps of agricultural implements and machinery got accumulated at these farms.
169. Revenue Agriculture, no. 17, February 1883, Pt. B, File no. 32.
170. Ibid., nos. 18–34, August 1884, File no. 89.
171. Ibid.
172. Note by E.C. Buck, 10 January 1886, Home, Education, nos. 14–88, October 1897, Pt. B.
173. E.C. Buck to the P.S. to Viceroy, 12 September 1895, *Elgin Papers,* Microfilm no. 2032.
174. Charles Wood had perhaps similar motives while writing his famous 'Despatch on Education' in 1854.
175. C.S.R. Rao, ed., *Notable, Speeches of Lord Curzon,* Madras, 1905, pp. XVI, 404.
176. Home, Education, nos. 57–9, July 1903.
177. Voelcker, *Report on the Improvement of Indian Agriculture,* p. 306.
178. His recommendation that agricultural education be grafted on existing institutions implied a veto for the five special agricultural schools proposed by the Madras Government in 1889. Home, Education, no. 6, October 1895.
179. Voelcker, *Report on the Improvement of Indian Agriculture,* p. 319.
180. H.K. Beauchamp, 'Agricultural Associations in India', *Proceedings of the First Industrial Conference,* Benaras, 1905, p. 11.
181. *Indian Engineering,* vol. XXI, 15 May 1897, p. 381.
182. Revenue Agriculture, *Agriculture & Horticulture,* no. 6, October 1880, Pt. B, File no. 46.
183. Note by W.H. Moreland, 4 May 1905, Revenue Agriculture, nos. 12–44, November 1905, File no. 81.
184. *Journal of the Society of Arts,* vol. XXVIII, no. 2468, 9 March 1900, p. 341.
185. Revenue Agriculture, nos. 11–21, January 1905.
186. Voelcker, *Report on the Improvement of Indian Agriculture,* p. 388.
187. Moreland cited an example of such 'hybrid stuff' from a contemporary work on agricultural chemistry: *Kuch darkhtan nitrates, ammonia aur kuch amides*

kha sukh-te-hein, aur chunki, ammonia se nitrates juldi bante hain, galiban darkhtan apane kul nitrogen nitrates ke shakal main khate hain. Revenue Agriculture, no. 8, September 1902, Pt. B, File no. 103.
188. Henry, 'Technology Transfer and its Constraints', pp. 25–50. This might be true as for the landowning classes. But with regard to the urban bourgeoisie, the clash did occur. After all, what did Mahendra Lal Sircar and his Association for Cultivation of Science represent?
189. Note by E.C. Buck, 10 January 1886, Home, Education, nos. 14–88, October 1897, Pt. B. Buck noted: 'Even the sons of Baniyas (grain-dealers), if they attend our schools, become sometimes so demoralized as to despise the paternal trade, and consider that the education which they have received is too good to be thrown away on keeping an account book in bad Kaithi or Mahajani penmanship.'
190. At the Agricultural Conference of 1890, T.H. Middleton of Baroda College pointed out: 'Salaried men have to do what their employers ask them to, and when no good results follow their work, it is very often because of the multitudinous duties they are expected to perform.' Revenue Agriculture, nos. 7–11, May 1891, File no. 26.
191. *The Indian Agricultural Gazette*, vol. I, no. 1, April 1885.
192. *The Indian Agricultural Gazette,* vol. IV, no. 46, March 1889.
193. Revenue Agriculture, nos. 18–19, September 1885, Pt. B, File no. 124.
194. Revenue Agriculture, no. 9, August 1900, File no. 67.
195. *Report on the Agricultural & Botanical Stations in the Bombay Presidency for 1905–6*, Bombay, 1907, p. 50.
196. Revenue Agriculture, nos. 2–5, April 1905, File no. 38.
197. Revenue Agriculture, nos. 11–21, January 1905. The signatories were G.K. Gokhale, J.N. Tata, Pherozshah Mehta, W. Greaves, Marshall Reid, V.D. Thackersay, N.G. Ghorpade and Jamsetjee Jeejeebhoy. Had the Government of British India imported American ideas, then at least the institutional framework for scientific research would have been established. As often happens, a process of 'Indianization' would then have followed.
198. Revenue Agriculture, no. 14, January 1905.

3

Contextualizing Modern Science in Agriculture in Colonial Bengal 1876–1928: A Case of Productionist Discourse

Arnab Roy

The present study investigates the context in which the initial attempt to modernize agricultural production in colonial Bengal took place. The term 'modern science' is used to denote the development of science and technology in the modern world, especially in Europe in the post-Industrial Revolution period. The remarkable growth in the different fields of science became evident for the first time in the history of human civilization. Technological innovation and scientific invention helped in the growth of the modern capitalist world.[1] In a particular historically specific situation, the growth of capitalism and colonialism went hand in hand. The ferocious desire of the colonial merchants to capture markets in the poor world resulted in the spread of modern science and technology. In the same fashion, the emergence of modern science in different areas can be understood from the exhibit of colonialism in India. From surveying to engineering, knowledge and application of modern science became inevitable. The term 'agriculture' explains the cultivating practices of the Indian farmers in the colonial time. The expression here is more specifically used to connote the production of food and cash crops in contemporary colonial Bengal which included rice, jute, tobacco, wheat, sugar cane, etc. This essay investigates the early discourse on the beginning and entry of modern science into agricultural production of colonial Bengal. We concentrate on the period between 1876 and 1928. The essay, on the one hand, takes into account the role and attitude of the colonial masters and on the other hand, engagement of the Bengali literati in the introduction of modern scientific knowledge in agricultural production.

Historiography and Sources

Historiography on agriculture from the view point of science is limited. One of the primary inclinations of the historiography is to deal with the institutionalization of knowledge. The historiography of science in general and historiography of agriculture in particular, have tended to focus on the institutionalization of scientific knowledge.[2] Historians have not taken into account the engagement of different sections of the Indian population in internalizing the traditions of modern science into agriculture. This reading seems to be obvious as there has been a tendency in scholarly writings to focus either on the role of a handful of individuals or of the colonial government to account for the development of modern agricultural tradition. Such an understanding has evidently placed the collective role of the Indian people at the periphery in the politics of knowledge production. On the other hand, as exception to the primary trend of historiography, some scholars have studied the issues of technology transfer in agricultural production due to the use of new methods, tools and techniques, etc.[3]

There is no denying the fact that most of the scholarly works on the history of science and technology in colonial India has been based on English language sources. It could be argued that the use of the English source materials seems to be obvious as the subject of the study has been one which came from the country of the colonizers. But vernacular source materials appear to be important for our purpose as it would bring out the Indian engagement in internalizing modern scientific knowledge in agriculture. Moreover, politics of knowledge production and dissemination might not have been restricted only to the written sphere of English language. Thus, a study, consisting of both the English and vernacular Bengali source materials, seems to be significant to understand the nature of the debate between the colonizers and colonized Bengali literati over rural issue like agricultural modernization.

Location and Time Frame of the Study

The investigation is primarily confined to colonial Bengal. Bengal was one of the first provinces of colonial India where the East India Company instituted a land revenue policy in the last half of the eighteenth century in the form of Permanent Settlement that resulted in the absentee zamindari system, which continued until the middle of the twentieth century. Thus, any technological development that might have taken place within the limit of existing production relations was dominated by the rural landlords. It could be argued that any technological innovation and critical assimilation of scientific knowledge that might have taken place in agriculture, in a situation where there was a strong presence of a landlord class that dominated the rural scene. On the contrary, it was agriculture where majority of the population was engaged. Thus, the case of modern agricultural development in colonial Bengal calls for a unique study. The year 1876 marks the formal emergence of the indigenous scientific community, with the beginning of the Indian Association of Cultivation of Science in Calcutta. At the same time, the last quarter of the nineteenth century saw sporadic measures to promote scientific development in agricultural production. The Royal Commission of Agriculture came into being in 1928. In the present essay, we will take up only some of the themes related to the introduction of modern agriculture in the time period between 1876 and 1928, as investigating complete agricultural developments after the constitution of the Royal Commission requires a full-length study.

Colonial Intervention in Agriculture

In the newly instituted Permanent Settlement in 1793, the land revenue was set at a fixed rate. In turn, the Company gave the landlords the legal right of ownership, and thus, a system which was far easier to apprehend than the earlier ones. This system created a class of absentee landlords, which was qualitatively different from that of the zamindars in the pre-colonial times. British officers thought that the creation of the landholder class with Permanent Settlement would strengthen colonial rule and the same class would undertake the responsibility of improving agriculture. The model for the development of Indian agricultural production followed from the exemplary British landlord who was responsible for innovative development of British agriculture. But this exercise, to depend upon the landlord for the improvement of Indian agricultural production, did not bring any drastic alteration in agriculture in the form of large-scale agricultural production and improved techniques. Agricultural production continued to be village-based and neither the zamindars, nor the jotedars, the real landlord of the villages, cared about improving the farming with new inputs.

Apologists of colonial rule, in the later years, argued that these absentee zamindars were 'mistaken' for the English landlords.[4] Was the identity of the Bengal zamindars mistaken for the English landlords? This false reading was challenged by the left intellectuals and politicians in the later years, especially from 1940s. Rajani Palme Dutt saw it as a conscious move on the part of the colonial officials to create 'a new class of landlords after the English model as the social buttress of English rule'.[5] M.N. Roy also saw it as a calculative effort on the part of the Company to divert the flow of capital and energy of the trading class to the investment on land, thereby eliminating the possibility of the trading class, which 'under normal circumstances would have proved to be the forefather of the Indian bourgeoisie'.[6] It seems that the colonial government's rhetoric about mistaken identity was problematic but the enactment of Permanent Settlement gave rise to a landlord class for the political benefit of the colonial authority. The newly formed zamindari class subsequently did not see itself as a responsible agency for promoting modern agriculture. They invested capital in land and not in agriculture. As a consequence of the low investment in agriculture, productivity suffered a setback. The actual cultivating class was not left with any resource to improve the conditions of land, but they had to carry out the burden of this oppressive system.

Against the backdrop of lower level of productivity, the colonial government began to see itself as responsible for bringing about changes in agricultural production. Lord Mayo's remark is worth recall here as he noted that 'the Indian agriculture is in a primitive and backward condition' and 'the duties which in England are performed by a good landlord, fall in India, in a great measure, upon the Government'.[7] Schrottky identified Indian agriculture in a 'neglected' state of affairs which was 'carried on very much in the same way now as it was two or three thousand years ago'.[8] The responsibility of the 'good landlord' was even recognized by the Famine Commission as it commented that 'a necessity is laid upon the State to set before the people example of better practice, both in the art of agriculture and in the breeding and management of livestock'.[9] Even at the end of the nineteenth century, Lord Curzon, while defending the Punjab Land Alienation Bill on 27 September 1899, said, 'There is no country in the world that is so dependent upon the prosperity of the agricultural classes in India. There is no Government in the world that is so personally interested in agriculture as the Indian Government. We are, in the strictest sense of the term, the largest landlords in creation.'[10]

Thus, it seems that the colonial authority, despite having an interest in agriculture from the time of the Raj, started to assume the role of the 'good landlord' and saw itself responsible for bringing about modern scientific investigation into existing agricultural practice only from the last quarter of the nineteenth century. It is not entirely sure as to what was meant by 'primitive and backward condition' in Indian agriculture, as noted by the colonial administrators. It might be possible that the term was used to refer to the existing agricultural practice in contemporary India, which one could assume to be different from the practice that prevailed in the West for varied economic, political and social reasons. It could also be probable that Indian agriculture was termed as 'primitive' because of its low level of productivity.

Why did the colonial masters suddenly assume the role of a 'good landlord' to promote modern agriculture from the last quarter of the nineteenth century? The importance of agriculture as a source of surplus extraction in the form of revenue had always been felt, especially from the late eighteenth century onward.[11] The colonial government might have taken up the self-assigned role to promote modern agriculture, long before the last quarter of the nineteenth century. How can one interpret this inordinate delay? Did it take almost a century for the colonial government to identify the Bengali zamindars as inconsequential in promoting modern agriculture? There is no denying the fact that agriculture had always remained a source of revenue for the Company. A new development took place from the middle of the nineteenth century. From this time, at an ever increasing pace, agriculture became exposed to commercial influences, owing to development in the advanced capitalist countries (however, this is not to suggest that agriculture was never commercialized before). Colonial intervention in agriculture could well be understood from their urgency to indulge in cash crop cultivation.[12] Thus, for instance, one might perhaps look into the case of cultivation of tea. The very first attempt to cultivate tea took place in 1835. Within four decades, vast places of eastern India had come under tea cultivation. The demand for Indian tea in the international market was evident by 1880s:

Indian tea has always realised the highest prices in the London market. In 1876 the average price per lb. of Indian tea in bond was Is. IId (IId is two pence), whilst that of China was only Is 3d. . . . The principal Excellency of Indian tea has always been its known purity. . . . If our tea deteriorates, the demand cannot continue. Indian tea is like a child striving against a giant; and it behoves all planters to do their utmost to defeat the monster (China) by hook or by crook, noticing all the defects and weak points in him and striving to prevent a similar appearance in Indian tea.[13]

The cultivation of other cash crops like jute, indigo, cotton, etc., received similar attention from the colonial authorities. The demand for commercial crop was also noticeable in the reports of the Indian newspapers. Thus, on one occasion, the Marathi periodical *Kesari* commented, 'the Egyptian cotton produced in Sind was not offered to public for sale either at Karachi or Bombay, but was brought outright by a single European firm from the Agricultural Department and exported to England'.[14]

Continuous presence of famines throughout colonial times, looming large in the Indian scenario, was yet another reason for the colonial power to reflect on improving Indian agricultural conditions. In fact, famine had been a periodical phenomenon of the British rule. Even before the colonial rule, there were instances of famines in India. But the famines, which British India witnessed, superseded all earlier famines in sheer magnitude of holocaust and devastation. In its early phase of rule, the colonial government had not engaged with scientific and technological development of agricultural production. Agricultural production remained static. But the demand for land revenue had increased manyfold. The cultivators were the worst sufferers of this policy. They could hardly retain any foodgrains for their own consumption during the time of natural calamities. As a consequence, famine had become endemic in the colonial period.[15] Six subsequent famines took place between 1876 and 1878, which claimed the lives of more than 60 million people. This periodical existence of famines led to the formation of the Famine Commission of 1880. The devastating nature of famines affected agriculture which was the primary source of surplus extraction for the colonial government. Thus, it was apparent that the colonial authority focused on the development of Indian agricultural production since the presence of famines started affecting agriculture.[16]

Was it only the commercial purpose that compelled English rulers to intervene and adopt an agricultural discourse against the backdrop of famine? It is undeniable that there were peasant rebellions throughout the colonial rule. Especially from the nineteenth century, these movements were widespread.[17] The origin of the bulk of the peasant movements in British India resulted from the colonial policy of instituting a land revenue system that took away even minimum subsistence from the actual tillers. Colonial authority was not oblivious to the fact that widespread peasant mutiny might challenge the colonial order if nothing is done in that direction. In a letter to W.R. Robertson, Principal, Saidapet Agriculture College, F. Nightingale commented,

If there is anything that England wants to know about and knows nothing about, it is agriculture in India. And when one thinks that we take 20 million land revenue out of India's agriculture and give nothing back, one almost wonder that there is not an universal agrarian mutiny. But the day cannot be far distant if we still continue doing nothing.[18]

In this condition, the colonial masters had to initiate an agricultural discourse, or the image of the self-imposed legitimacy and superiority of the colonial rule would have been damaged. It could also be argued that describing Indian agriculture as 'primitive' and thereby, devising the role of a 'good landlord' might have also served the inherent colonial logic of assuming a superior identity over the natives.

To sum up, it could be argued that against the background of famines, colonial authority could not leave the task of improving agriculture in the hands of absentee zamindars. They had to take up a discourse on agriculture either to safeguard profit from land revenue or to divert the peasant discontent or to create a cultural and social supremacy of colonial rule by terming Indian agriculture as 'primitive and backward'. Now we turn our attention to understand the nature of this agricultural discourse, initiated by the colonial authority.

Bringing Science into Agricultural Production

What was the colonial authority's idea of improving agricultural production? The knowledge of modern science was considered to be an important factor in posing challenge to the adverse situation that Indian agriculture was facing. It appears that the critical assimilation of modern scientific knowledge came to be recognized as taking a central position in the discourse of the colonizers to address the poor levels of production in Indian agriculture. One of the early observations of using modern science in agricultural production was voiced by Lord Mayo, the fourth Viceroy of India (1869–72). He commented, 'when the light of science has been properly brought to bear upon Indian Agriculture, the results will be as great as they have been in Europe'.[19] Eugene C. Schrottky, one of the colonial administrators, in his work titled, *The Principles of Rational Agriculture Applied to India and its Staple Products,* commented on the need to use science for the development of the state of affairs in agriculture that prevailed in India. He noted that 'considering the rapid advances agriculture has made in Europe during the last century—advances which have raised this pursuit to the rank of a science—it is astonishing that not one of the principles on which alone it can be carried on profitably, and without injury to future generations, has found its way into India'.[20] Famine Commission was also not oblivious of the fact of combing modern scientific knowledge with agricultural production:

On the importance of agriculture knowledge both to the governing class and the governed, it is unnecessary for us to dilate; how essential we think it that technical knowledge should be called in to enable the productive powers of the soil to be applied in the most effective manner, not merely to add the wealth of the country, but to secure a food-supply which shall keep pace with the increase of population.[21]

One of the important figures in the agricultural discourse in colonial India was Dr Voelcker who was sent to India as an agricultural chemist to investigate and comment on the 'improvement of the Indian agriculture'.[22]

Voelcker came to India and published his findings in 1893. It was he who emphasized on the need to use science for the development of famine affected Indian agricultural conditions. Voelcker laid great importance on the necessity of practical enquiry of agricultural conditions and practices that prevailed in India. However, the important theme which runs through the tracts of Voelcker is the fact that he also saw the use of science and various scientific experiments for the development of agricultural conditions and practices. In his own words, Voelcker saw that 'by the happy combination of science and practice, the work of experiment may proceed in a definite and useful direction'.[23]

At a conference of the Directors of Public Instruction held at Simla on 20 September 1905, Lord Curzon said that 'agriculture in India is the first and capital interest of this huge Continent, and agriculture, like every other money-earning interest, must rest upon education'.[24] The same view was again echoed by Lord Minto in his letter to John Morley, where he emphasized the link between agricultural science and practical agriculture.[25] Lord Minto was consistent in his endeavour to make practical use of scientific knowledge into agricultural practice. Thus, in one telegram to the Secretary of State, Minto stated, 'It is generally acknowledged that we are greatly behind other countries in the application of science to agriculture, more especially in view of the immense importance of agriculture to India, and now that funds are available, we consider it one of our first duties to remove this reproach.'[26]

It was this organic connection between existing agricultural practice and use of science for enhancement of better productivity that characterized the colonial ruler's idea of the development of agriculture, modernization, etc. The colonial authority viewed imparting knowledge of modern science as a corollary to the developmental project of agriculture. Everybody appears to have been preoccupied with the agenda to raise production/productivity. Application of scientific knowledge to elevate productivity took up a frontal position in the whole of the agricultural discourse. Thus, the agricultural discourse could well be identified as productionist, following the obsession, as found in the imagination of the colonial rulers, to raise production and productivity with the use of modern science.

Introduction of modern science into agricultural practice, as understood by the Bengali intelligentsia, could occur only through the impartation of agricultural science education. In this context, it could be recalled that several scholars like Deepak Kumar, Zaheer Baber, Pratik Chakrabarti have already noted down the beginning of science education in colonial India by the last quarter of the nineteenth century. The same time period also witnessed a number of initiatives leading towards an introduction of modern science into other branches of activity.[27] In striking contrast, agriculture as a subject of study was never given prominence and the colonial government, in the main, ignored any scientific investigation of agricultural practice. It was only towards the end of the nineteenth century that the colonial government focused on the possibility to impart agricultural education at different levels of academic curriculum in Bengal.

After realizing the need to impart modern science into agricultural production in colonial society, the British, for the first time, established the Agricultural and Horticultural Society of India (AHSI) in Bengal in the first half of the nineteenth century. The society, as it was hoped, acted as a medium to spread modern agricultural knowledge in colonial Bengal. Towards the end of the nineteenth century, an agricultural class was opened at Shibpur Engineering College.[28] But it was a short-lived venture and could not sustain in the long run. After the collapse of the Shibpur experience, Bhagalpore Agriculture College was set up in 1908.[29] This also did not have a long life as it ceased to exist in 1927.[30] Thus, the issue of formal agricultural education in the beginning of the twentieth century remained only as a subject of wide discussion in the form of reports, articles, editorials and speeches, involving both the colonizers and the colonized. Little was done to actually institute a learning centre of agricultural science education in colonial Bengal.[31] In a situation, where the issue of ceremonial agricultural science education could not gain currency, the colonial historical actors paid importance to the dissemination of the knowledge of modern agricultural science through setting up of experimental farms.

Experimenting on Experimental Farms

The setting up of experimental farms,[32] where a great deal of dissemination of knowledge took place, seems to confirm that the colonial authority was willing to assume some responsibility to bring about scientific investigation into agricultural practice. Experimental farms in different places came into being in the middle of the nineteenth century. Saidapet farm, Nagpur farm, etc., are examples of British initiatives to promote modern science in agriculture. One of the early such initiatives came in 1837 with the establishment of Saidapet farm.[33] However, the case of Bengal was unique as the experimental farms started coming up only after the middle of the nineteenth century. An unidentified correspondent from Bengali periodical called *Abad* noted the number of experimental farms that existed in Bengal in the first quarter of the twentieth century.[34]

At this point, not much is known about the working of these farms. Farms were not always maintained by the colonial government. In fact, the role of the Bengali population could be considered to be important for the promotion of experimental farms in different areas of Bengal. Sometimes these farms were maintained by the agricultural experts. In some cases, the eminent public intellectuals were engaged in the culture of supporting experimental farms to advance the knowledge of modern science and technology to the cultivating classes of Bengal. In this connection, one might recall Paikpara nursery, one of the oldest nurseries of colonial Bengal.[35] Although it was named as a nursery, it became a vehicle of disseminating knowledge of modern agriculture.[36] Bolpur Surul Krisi Khetro was another important venture, which not only aimed at extending the horizon of modern agricultural knowledge among the cultivators, but was also associated with the rural development plans of Rabindranath Tagore.[37] Similarly, Ranghat Mallick farm was another example which was maintained not by the colonial officials, but by Roy Bahadur Kumudnath Mallick.[38]

It is important to note here that even some princely states had agricultural experimental farms in different parts of Bengal. In this connection one might recall the agricultural farm of Cooch Behar.[39] This agricultural farm came under the jurisdiction of Cooch Behar rulers and was established in the last quarter of the nineteenth century. At this point, the model behind the establishment of the farm is not known. But then the farm did not run successfully, as is evident from the address of the Maharaja.[40] However, the farm did start working in a steady manner from the middle of the twentieth century as has been noted in many annual administration reports. The farm, as envisioned by the promoters, was supposed to be a vehicle for disseminating knowledge on modern agricultural conditions and methods to the cultivating classes. To that end, the improved varieties of seeds, manures and even sometimes implements were distributed and sold to the peasantry. Agricultural officers were recruited in the farm and they undertook extensive tour programmes in the rural areas, observing and explaining to the ryots, the need to embark on modern scientific development in agricultural production. Demonstration work by the agricultural officers in the field of the local growers in different subdivisions was considered important in the annual administration report of the state.[41]

It seems that these non-governmental experimental farms did not receive much state patronage. In most of the cases, some agricultural experts were the key figures in the running of the experimental farms. It has been said earlier that the zamindars did not care much about capital investment in agriculture. But in some exceptional cases, benefaction of the zamindars became a reality in order to meet with the expenses of the farms. Ranaghat farm was one of those which were run by the zamindars.[42] It could be argued that both the colonizers and the colonized Bengali population engaged either into putting up experimental farms or into a discourse of the necessity of instituting experimental farms. In the absence of any structured course of agricultural science education, at this initial juncture, the agricultural discourses of both the colonizers and the colonized wanted a practical oriented agricultural education. As a consequence, experimental farms were perceived as a site of knowledge. Although there were sporadic attempts, both by the colonizers and by the Bengali population, to institute a formal agricultural curriculum, it seems that the driving force was to construct a *hate hetre* (practical) model for undertaking agriculture.[43]

But the framing of an agricultural education curriculum was relatively a later engagement of the colonial authority. One may well question the late intervention on the part of the colonial government in formulating the educational curriculum in agriculture. Did they consider agriculture to be a subject that did not require a full-scale theoretical training and hence there was a thrust on experimental farms as places to disseminate knowledge on practical agriculture? The experience of other presidencies seems not to suggest the same, as the colonial government took up the case of promoting agricultural education from the middle of the nineteenth century. It is possible to conclude, from the Bengal scenario, that the colonial authority did not pay equal attention to agricultural education in Bengal, compared to the other parts of colonial India.[44]

Did the experimental farms meet the expectations? There appears to be a lack of historical evidence to address the issue. At this stage, it is yet not known as to whether there was any change in agricultural practices of the cultivating classes due to the dissemination project of the experimental farms. However, one could assume that some experimental farms could have gained some currency among the local ryots. But owing to varied economic, political and social reasons the diffusionist model of the experimental farms in colonial Bengal might not have ushered a general change in the agricultural practices of the cultivating class. In other words, it could be argued that these experimental farms remained at the level of experiments only and subsequently, it did not lead to a great change in the practices of the cultivating classes in colonial Bengal. In fact, there were instances where the idea of the experimental farm as an institution to disseminate the knowledge of modern agriculture to the cultivating classes came under attack by a section of the Bengali intelligentsia that took part actively in the agricultural discourse in colonial Bengal.[45] Thus, a thorough investigation is necessary to explore the effects of the experimental farms in colonial Indian agriculture.

Engagement of the Bengali Population

What was the role of the Bengali population in this agricultural discourse? Was it only the colonizers who initiated a discourse in agriculture? It would be historically fallacious if one attributes to the colonizers the sole responsibility for bringing about modern science into agricultural activity or at least initiating an agricultural discourse. Bengal witnessed the growth of an urban middle class from the second quarter of the nineteenth century. This new section could not entirely depend upon the income from the land. They constituted a part of the urban intelligentsia that was primarily salaried in nature.[46] It was almost from the final quarter of the nineteenth century that a section of the Bengali intelligentsia unanimously started articulating their concerns about the improvement of Indian agriculture. It was not the first time that the urban literati had dealt with the theme of agriculture. At an earlier instant, they had focused more on the question of rent, peasant rights, etc.[47] But now on, there was a paradigm shift and in place of agrarian relations, the intelligentsia articulated concerns on methods, tools, techniques, etc., of agricultural production. The focal point of the discourse was to raise productivity and improve agriculture by the application of modern science, thus taking a productionist approach.

How did they contribute to the agricultural discourse? What was the primary mode of expression for the contemporary urban literati that initiated a productionist discourse in agriculture from the last quarter of the nineteenth century? Their habitual contribution to this debate was expressed primarily in the contemporary Bengali periodicals, tracts, etc. Periodicals dealing with the issues of agriculture, trade and commerce like *Kajer Loka, Grihastha, Grihastha Mangal, Krisak, Pallibasi, Mahajan Bandhu, Khadya Utpadan, Krisi Sampad* and *Krisitatva* are important sources to understand the contribution of the Bengali literati in the agricultural discourse. Some general periodicals like *Basumati, Somprakash* and *Bangadarshan* also commented on the various issues of agriculture. These periodicals paid importance to a number of themes like the use of chemical fertilizers, new varieties of seeds, new agricultural implements and the linkages between agriculture and industry. But the most dominant theme of all these periodicals were the questions related to the introduction of modern science into agricultural production and the possibility of agricultural education in colonial Bengal. Studying the engagement of the Bengali intelligentsia as a mere response to the colonial discourse on agriculture would be an oversimplification that locates the colonized population at the receiving end of modern science. On the contrary, the spontaneous insistence of the Bengali literati, for varied economic, political, and cultural reasons, for the introduction of modern science into agricultural production, as evident from the review of the source materials, does not situate them at the receiving end. The demand, as voiced by contemporary Bengali historical actors, suggests an internally determined character of the development of a discourse on modern science into agricultural production in colonial Bengal.

Bengali Literati's Discourses

Examining a Paradoxical Case

The current of Hindu revivalism was evident in the agricultural writings of the literati. The prosperous picture of Indian agriculture became a subject to which the Bengali historical actors returned throughout the period under study. Thus, on one instance, nationalist leader Chittaranjan Das commented,

> We have not got poor within a moment and all of a sudden—slowly our body has become a skeleton. When the foreigners first came to our country, they got surprised to see our abundance of gold and silver. What was the source of that gold and silver? Bangladesh does not have a gold-silver mine, then we have to say that we earned huge money through the help of agriculture and trade.[48]

It is not entirely certain as to what was meant by the term 'foreigner'. The current of this sort never ceased to exist in the writings of the intelligentsia. In some cases, the term 'foreigner' was used to denote the rulers ranging from the Mughals to the British. The reference to ancient Indian agriculture occurred even before the twentieth century.

> In ancient India, agriculture had more admiration than commerce. Trade was not necessary in those days in India, because trade is only required to alter impoverished condition. . . . There was no scarcity in anything in ancient India. . . . In India there was plentiful exercise of agricultural education . . . agriculture used to fulfil all the shortage of the Indians. . . .[49]

Another independent correspondent of *Krisak* compared contemporary impoverished conditions of India in the following way:

> Famine has become chronic in this country . . . *Sujola sufola* Indian land has turned out to be a cremation ground . . . now India does not have those days. Dressed like a pauper, India is now geared up for begging at door to door. At one point of time, India used to send abroad foodgrain after giving enough food to her own children.[50]

The frequent reference to agricultural saints like Janak, Balaram, Khanar Bachan was probably done with an intention to challenge the discourse of the colonizers and to construct a Bengali identity of its own. A correspondent in the periodical called *Gramer Dak* romanticized the old agricultural traditions of ancient India, writing, 'Agriculture, unlike today, was not considered as contemptible in old times . . . Avatar Balarama, King Janak used to cultivate on their own, *Purana* is giving the witness to that.'[51]

It appears from the source materials that this stream of thought never ceased to exist. Probably this was done with a deliberate attempt to construct a separate identity of the colonized culture in the face of the colonial critique of Indian agriculture as 'primitive'. Hindu revivalism, as witnessed in the discourse of agriculture, was not an exceptional trend as we have seen in many cases that the Hindu revivalists in the early twentieth century offered a 'scientific explanation and interpretation' of the old scriptures. They were not opposed to science, but they were more interested to trace back the origin of science to the past traditions.[52]

It is interesting to note that the same Bengali literati, who romanticized the old tradition of Indian agriculture, recognized the impoverished conditions of contemporary agriculture. The one question on which the most significant degree of agreement was evident was the need for the development and modernizing of agriculture along the lines of modern science. In his presidential speech, delivered at the special session on agricultural literature at North Bengal Literary Conference, Ashutosh Lahiri defined 'agricultural improvement':

By the term agricultural improvement we understand the followings: (1) increase in production of the produced and ensuring quality of the produced commodity; (2) proliferation of the cultivation of new commercial crops; (3) development of Indian agricultural practice; (4) industrial utilization of agricultural produced; (5) enhancement of agricultural literature.[53]

The knowledge of science was seen as a vehicle to touch the pinnacle among all countries as India was viewed as a natural agrarian country. This realization was vivid. In the last half of the nineteenth century, one finds the realization to promote modern science into agricultural production.[54] The issue of national wealth through adopting a modern agriculture did not slip the attention of other periodicals as well. On one occasion, *Grihastha* commented on the theme in an articulate fashion:

If the energy, we have spoiled in many other works, could have been utilized/invested in the agricultural production, the country could have been benefited, economic development of the country would have been achieved. Agriculture is the treasure of the country, agriculture is the means for the development of the country, agriculture is the means to protect the property/resources of the country and agriculture is the *mokkha* (self-realization) of the country. Agriculture has elevated America and Japan in the higher position of the world.[55]

The reference to the centrality of agriculture to the national economy is striking. In a country where majority of the population were involved in agricultural activity, imparting modern knowledge of agriculture was realized as a key factor in the whole of national production. Thus, the foremost agricultural discourse viewed the use of modern scientific knowledge as an important aspect to increase the production and productivity in the context of colonial Bengal. It could be argued that the intelligentsia, which initiated an agricultural discourse in Bengal, saw the issue of agricultural development as primarily an economic project. In some instances, one would find a detailed account noting the development of agriculture in advanced countries. Many such articles were published in *Krisi Sampada*.[56] The explicit linkage between agricultural development and the use of modern science became an area which was discussed by all the historical actors of Bengal:

India is an agriculture based country. Superiority of agriculture in India is evident than the other countries owing to the eminence of water, air and soil. But Americans, showing their excellent technique and skill in agriculture, have been able to make it as one of the chief ways to earn money. Agriculture in India has been going on frequently in that old tradition. . . . When it is seen that the inhabitants of America and Europe, by adopting unique way, have been able to increase wealth of the country, when I see that the people of this India—inhabitants of Bombay and Madras, are benefitted by accepting new methods, then why is not that procedure adopted in Bengal?[57]

The Bengali literati realized the necessity to improve agricultural conditions through imparting modern knowledge, coming from the European territory. But it is interesting to note that in the same discourse the historical actors of colonial Bengal invoked a rich tradition of Indian agriculture to construct an identity, distinct from the colonizers. Invocation and subsequent glorification of the past agricultural conditions and the relentless call to modernize agricultural conditions along the lines of modern science, put the Bengali historical actors in a paradoxical situation throughout late nineteenth and early twentieth centuries.[58]

Conclusion

We have seen the context in which this initial debate took place between different historical actors to promote modern science in agricultural production. It was the commercial purpose that prompted the colonial power to develop a discourse on agriculture. Although the colonial power initially had relied on the zamindars for the development of agricultural conditions, it later on changed its views and saw itself as an agency responsible for bringing about change in agricultural production from the end of the nineteenth century. Initial effort of the colonial government to establish experimental farms for disseminating knowledge on modern agriculture could not probably usher a general change in agricultural production.

The last quarter of the same century also saw a growing body of urban intelligentsia which initiated a discourse on agriculture that identified the application of modern science as the driving force to modernize agricultural conditions. Thus, locating the Indian actors at the receiving end of modern scientific knowledge, coming from the advanced capitalist countries, seems to be problematic, as the case of modern science into agricultural production and an initiation of an agricultural discourse was internally determined, where the Bengali literati took up a leading role. Like the colonizers, the dominant Bengali view advocated the use of modern science and technology in agricultural production and this was thought to be important to address the economic problems of the common masses of colonial Bengal. In fact, it would not be wrong to identify both these discourses as productionist, where there was a constant thought to develop the forces of production by imparting modern knowledge of agricultural science and technology. The two apparently divergent discourses converged on some of the issues like reduction of poverty, improvement of Indian agricultural conditions, etc.

It is seen from the study that the Bengali literati, interested in agricultural discourse, invoked a rich tradition of agriculture from the ancient past. The tone of Hindu revivalism was added, probably in an attempt to construct a cultural identity of its own, separate from that of the colonizers. Thus, in a subtle way, the Bengali literati objected to the colonial idea of terming Indian agriculture as 'primitive and backward'. However, it seems to be interesting to note that the same intelligentsia also realized the impoverished conditions of contemporary agriculture. Impartation of modern knowledge on agriculture was considered important to address the problem of agriculture. The realization of assimilation and dissemination of modern agricultural knowledge, coming from the home territory of the colonizers, to improve production, productivity and thereby, national economy, frequently occurred in the thoughts of the Bengali literati. Thus, the historical actors of Bengal were put in a paradox of invoking a rich past and developing agricultural production along the lines of modern science.

This essay, thus, challenges the problematic reading of science as only a hegemonic force in the context of the history of science in colonial India, where the historians have discussed the use of science by the Indians as a political, cultural and social project. The concrete reference to the economy and production by the historical actors were not touched upon in this reading of science. The study, thus, shows that the Bengali actors referred to the concrete issues of economy, production and productivity and realized the importance of using the knowledge of modern science to improve agricultural production.

Notes and References

1. For a detailed study of the relation between modern science and industrial revolution, see Margaret C. Jacob and Larry Stewart, *Practical Matter: Newton's Science in the Service of Industry and Empire*, London, 2004.
2. The works of Deepak Kumar, M.S. Randhawa, Carl Pray and Prakash Kumar have tended to focus on the institutionalization of agricultural knowledge. For an overview, see Deepak Kumar, *Science and the Raj: 1857–1905*, New Delhi, 1995; Deepak Kumar, 'Science in Agriculture: A Study in Victorian India', *Asian Agri-History*, vol. 1, no. 2, 1997, pp. 77–103; M.S. Randhawa, *A History of Agriculture in India,* vol. III, New Delhi, 1983; C.E. Pray, 'The Impact of Agricultural Research in British India', *The Journal of Economic History*, vol. 44, no. 2, 1984, pp. 429–40; Prakash Kumar, 'Scientific Experiments in British India: Scientists, Indigo Planters and the State, 1890–1930', *Indian Economic and Social History Review*, vol. 38, 2001, pp. 249–70.
3. For this view, mention could be made of R.J. Henry and Satpal Sangwan. However, both of their works give a schematic overview of technology transfer in the realm of agricultural production. See R.J. Henry, 'Technology Transfer and Its Constraints: Early Writings from Agricultural Development in Colonial India', in *Technology and the Raj: Western Technology and Technical Transfer to India 1700–1947*, ed. R.M. MacLeod and D. Kumar, New Delhi, 1995, pp. 51–77; Satpal Sangwan, 'Level of Agricultural Technology in India (1757–1857)', *Asian Agri-History*, vol. 11, no. 1, 2007, pp. 5–25.
4. Vera Anstey, *The Economic Development of India*, London, 1929, p. 98.
5. Rajani Palme Dutt, *India Today*, 1940, p. 232.
6. M.N. Roy, 'India in Transition', in *Selected Works of M.N. Roy*, ed. Sibnarayan Ray, 1988, pp. 224–5.
7. Quoted in E.C. Schrottky, *The Principles of Rational Agriculture Applied to India and Its Staple Products*, Bombay, 1876, p. 8.
8. Ibid., p. 3.
9. Extract from *Report of the Indian Famine Commission*, Part II, Chap. IV, Secs. I and II, Home, Public, nos. 88–102, July 1881, Pt. A (acquired from Professor Deepak Kumar's collections).
10. *Curzon, Speeches*, vol. 1, Calcutta, 1900, p. 123 (Deepak Kumar's collections).
11. Irfan Habib, 'Colonization of the Indian Economy, 1757–1900', *Social Scientist*, vol. 3, no. 8, March 1975, pp. 23–53.
12. Economic historians have discussed this theme. For instance, see David Washbrook, 'The Commercialization of Agriculture in Colonial India: Production, Subsistence and Reproduction in the "Dry South", *c.*1870–1930', *Modern Asian Studies*, vol. 28, no. 1, 1994, pp. 129–64.
13. *The Tea Cyclopaedia*, E.C Lazarus of Indian Tea Gazette, Calcutta, 1881, pp. 15–16 (Deepak Kumar's collections).
14. *Kesari*, 15 May 1906, NNR, Bombay, 1906 (Deepak Kumar's collections).
15. Mike Davis gives probably the finest account of the famines. He accounts the unprecedented sufferings of the people in the last quarter of the nineteenth

century, a phenomenon which was identified by him, as a result of the European rule. See Mike Davis, *Late Victorian Holocausts: El Nino Famines and the Making of the Third World*, 2000, London.
16. To know more about famines in India, see B.M. Bhatia, *Famines in India: A Study in Some Aspects of the Economic History of India, 1860–1965*, Bombay, 1967.
17. There is a vast body of literature that has been written on the subject of the peasant struggle in colonial India. For instance, see Blair B. Kling, *The Blue Mutiny: The Indigo Disturbances in Bengal, 1859–1862*, Philadelphia, 1966; Ranajit Guha, '*Neel Darpan*: The Image of a Peasant Revolt in a Liberal Mirror', *Journal of Peasant Studies*, vol. 2, no. 1, 1974, pp. 1–46; Kalyan Kumar Sen Gupta, *Pabna Disturbances and the Politics of Rent, 1873–1885*, Calcutta, 1974.
18. Letter to W.R. Robertson by F. Nightingale, 5 May 1880, *Nightingale Papers*, IOR, MSS Eur B. 263 (Deepak Kumar's collections).
19. Quoted in E.C. Schrottky, *The Principles of Rational Agriculture Applied to India and its Staple Products*, Bombay, 1876, p. 8.
20. Ibid.
21. Famine Commission put forth this view in vol. II, sec. II in their report. Later this few lines were quoted by E.C. Buck in the seventh meeting of the Agricultural Conference, held on 13 October 1890. See Revenue Agriculture, nos. 7–11, May 1891, Pt. A, File no. 26 (Deepak Kumar's collections).
22. J.A. Voelcker, *Report on the Improvement of Indian Agriculture*, London, 1893, p. III.
23. Ibid., p. 296.
24. C.S. Raghunath Rao, ed., *Notable Speeches of Lord Curzon*, Madras, 1905, p. 404. XVI. NAI 954.084 C948N (Deepak Kumar's collections).
25. Letter from Lord Minto to John Morley, 26 July 1906, Revenue Agriculture, Agriculture, no. 7, August 1906, Pt. A (Deepak Kumar's collections).
26. Telegram, 19 February 1906, Revenue Agriculture, Agriculture, no. 2, August 1906, Pt. A (Deepak Kumar's collections).
27. Colonial rulers took initiatives for a number of subjects which included botany, forestry, geology, medical science, engineering and vocational education for varied economic and political reasons from the middle of the nineteenth century. It could be argued that the middle of the nineteenth century marked the foundation of science education in colonial India. See Kumar, *Science and the Raj*.
28. For an overview of the agricultural classes at Shibpur Engineering College, see N.G. Mukerjee, 'A scheme for the establishment of Agricultural Classes in connection with the Civil Engineering College at Shibpur', *Proceedings of the Government of Bengal for the Month of April, 1898*, Revenue, Agriculture, File 5-A/1 1, nos. 2–3, West Bengal State Archives.
29. 'Bhagalpore Agricultural College', *Krisak*, vol. 9, no. 5, Bhadra 1315 BS (1908), pp. 103–4.
30. 'Agricultural Schools in India', *Krisi Sampada*, vol. 14, no. 5, Bhadra 1330 BS (1923), p. 97.

31. It is interesting to compare the case of agricultural education in colleges in Bengal with those in other presidencies, namely Bombay and Madras. A detailed study is required to locate the gap that existed between Saidapet College and Coimbatore College on the one hand and Shibpur and Bhagalpore College on the other. For a study of agricultural education in college-level curriculum of Madras Presidency, see V. Chandrana, 'Institutionalisation of Agricultural Science in Colonial India: A Study of Madras Presidency, 1835–1928', unpublished Ph.D. thesis, Jawaharlal Nehru University, New Delhi, 2003, pp. 87–100.

32. The Bengali equivalent which was used to denote experimental farm was *krisi khetra*. Alternatively, contemporary Bengali historical actors also used the term called *adarsa krisi khetra*. Colonial government appeared to have an understanding of a clear distinction between these two concepts. Thus on occasion we find a resolution of the Government of India:

    For this reason it wishes that the record of the experimental area should be kept entirely distinct from the report on Model or Demonstration farms. Upon these latter the positive results of experiments may be usefully tried with a view to financial profit, and discredit should attach to any financial failure upon them not justified by very exceptional or unforeseen causes, such as a drought, cattle-murrain, etc. Any continued failure should lead to their contraction or discontinuance. It is only on these conditions that the Government of India can believe in the utility of expending public money upon agricultural experiments.

    See Resolution of the Government of India, dated 28 December 1885, Revenue Agriculture, Agriculture, nos. 12–14, February 1886, Pt. A, File no. 77 of 1885 (Deepak Kumar's collections). Voelcker commented in an unmistakable clarity:

    I wish at the outset to clearly distinguish between Farms which exist for the purpose of demonstration and those which are intended for pure experiment. The former are intended to show to cultivators the result of a practice found by experiment to be successful, and, therefore, they ought to be remunerative. In this way I should have no objection to their being called 'Model' Farms. But Farms at which the object is to put different methods to the test stand on a different footing altogether. The object at these latter is to find out *which* of several practices may be the best, and this of necessity involves doing a great deal that is of unremunerative character.

    See Voelcker, *Report on the Improvement of Indian Agriculture*, p. 338. The understanding of the Bengali literati on experimental farm was not always confusing. Thus, on some occasions, we would find that an unidentified correspondent of *Krisi Samachar*, which was published from Bengal Agricultural Department, Dhaka and was a short-lived project, made a distinction of different kinds of farms that existed in colonial India with an unmistakable clarity. See 'Work of Governmental Agricultural Farms', *Krisi Samachar*, vol. 4, no. 1, March 1924, pp. 18–19.

33. For a detailed study of the Saidapet farm, see V. Chandrana, 'Institutionalisation

68 *Arnab Roy*

of Agricultural Science in Colonial India: A Study of Madras Presidency, 1835–1928'.
34. This may not be an exclusive list of the experimental farms. However, the author mentioned some of the farms as listed below: Chinsura farm, Burdwan farm, Bankura farm, Baharampur farm, Sosoj farm at 24 Paragana, Jessore farm, Siuri farm, Khulna farm, Dacca farm, Kumilla farm, Maymensing farm, Barisal farm, Faridpore farm, Kishorgunj farm, Dhanbari farm at Tangail, Jaidebpore farm, Brahambaria farm, Karatia farm, Rangpore cattle farm, Rangpore demonstration farm, Burirhat tobacco farm, Rajshahi agricultural farm, Bagura farm, Pabna agricultural farm and Kalimpong farm. See 'Government Agricultural Farm in Bengal', *Abad*, vol. 1, no. 1, Baisakh 1333 BS (1926), p. 30.
35. *Krisi Sampada*, vol. 8, no. 2, 1324 BS (1917).
36. *Krisitatva*, one of the oldest periodicals that published articles related to the issues of agriculture came out from Paikpara nursery. This periodical from 1870s to 1880s published a number of articles covering different aspects of modern agricultural practice.
37. 'Agricultural farm of Surul', *Gramer Dak*, vol. 1, no. 4, Poush and Magh 1334–5 BS (1927–8).
38. 'Tour of Ranaghat', *Gramer Dak*, vol. 1, no. 5, Phalgun and Chaitra 1334–5 BS (1927–8).
39. Cooch Behar used to be a separate native state during the colonial times. Contemporary historical actors considered Cooch Behar to be a part of the Bengal province, although it had its separate administration under the raja of Cooch Behar. The similarities of cultural tradition between different parts of Bengal and Cooch Behar might have played an important role for the historical actors to include Cooch Behar within the Bengal province. Thus, Campbell writes that Bengal 'includes Bengal Proper, Bihar, Orissa, Chotta Nagpur, and the Native States of Cooch Bihar, Hill Tripura, Sikkim and twenty-six Tributary States of Orissa and Chotta Nagpur'. See A.C. Campbell, *Glimpses of Bengal: A Comprehensive, Archaeological, Biographical and Pictorial History of Bengal, Behar and Orissa*, Calcutta, 1907, p. 17.
40. See *The Annual Administration Report of the Cooch Behar State for the Year 1893–94*, no. 1380, North Bengal State Library.
41. To know about the activities of the farm, see *The Annual Administration Report of the Cooch Behar State for the Year 1934–35*, no. 816; *1935–36*, no. 744; *1938–39*, no. 4683; *1939–40*, no. 5296, North Bengal State Library.
42. 'Tour of Ranaghat', *Gramer Dak*, vol. 1, no. 5, Phalgun and Chaitra 1334–5 BS (1927–8).
43. There are examples of the demand by the Bengali literati to institute a formal course on agricultural education in colonial Bengal. The course on agriculture at Shibpur Engineering College was short-lived. But the dominant tendency was to design a practical oriented course for the students of the primary school. Moreover, the intelligentsia viewed the setting up of experimental farms as a vehicle of dissemination of modern knowledge of agriculture to the cultivating class. The term *hate hetre* became a popular expression in some of the periodicals

which dealt with the issues of agriculture. The term is used to connote practical agriculture.

44. The negligence of the colonial government to promote agricultural education was evident in Bengal, especially in the nineteenth century. It was probably to compensate the lack of infrastructural facility to institute agricultural education in Bengal that the provincial government of Bengal ushered scholarships in European colleges for a handful of Indian students. See *Report on the Administration of Bengal, 1879–80*, no. 159, Calcutta, 1880, pp. 27–8, Bihar State Archives (Deepak Kumar's collections). However, the project of scholarship could not also run for a long time and in 1887, it was discontinued. See Letter from P. Nolan, Secretary to Government of Bengal to the Secretary to Government of India, 26 March 1887, Government of Bengal, Revenue, Agriculture, nos. 76–80, March 1887, Pt. A, Bihar State Archives (Deepak Kumar's collections).

45. One such view came in the session on agricultural literature in the 11th North Bengal Literary Conference, held in 1335 BS (1928). Ashutosh Lahiri, a periodical contributor in the agricultural discourse, was the president of the session. In his presidential address, he said:

[T]he effort which is going on to promote development in agriculture through the establishment of experimental farms or demonstration farm to show the specimen of scientific agriculture or cattle farm in different places, we don't think that our peasants might get some special benefit from them. Will it be enough to get a potato of five *ser* after spending five thousands rupees?

See 'Agricultural talk', *Krisi Sampada*, vol. 19, no. 5, 1335 BS (1928).

46. To understand the emergence and nature of this newly emerged intelligentsia, see Kalyan Kumar Sen Gupta, 'Bengali Intelligentsia and the Politics of Rent, 1873–1885', *Social Scientist*, vol. 3, no. 2, September 1974, pp. 27–34.

47. Ibid.

48. His address at the conference of Bengal province was published in the form of a book which was called *Banglar Katha* (Annals of Bengal). Some parts of the book were reprinted in *Krisi Sampada*. See 'Annals of Bengal', *Krisi Sampada*, vol. 8, no. 2, 1324 BS (1917).

49. Rajendra Lal Singh, 'Agriculture and Trade—Which is Great?', *Krisitatva*, vol. 4, no. 6, 1288 BS (1881), p. 104.

50. Surendranath Mitra, 'Indian agriculture', *Krisak*, vol. 2, no. 8, Agrahayan 1308 BS (1901), p. 177. The term *Sujola sufola* requires clarification. This expression was first used by Bankim Chandra in his poem *Bandemataram*. This phase, as it is understood from its superficial meaning, was used to denote the land of India which was always full of water/river and blessed with good grown crops.

51. *Gramer Dak*, vol. 1, no. 2, 1334–5 BS (1927–8).

52. S.P. Udayakumar, *'Presenting' the Past: Anxious History and Ancient Future in Hindutva India*, 2005, p. 26, Westport, Connecticut; Meera Nanda, *Prophets Facing Backward: Postmodernism, Science, and Hindu Nationalism*, 2004, p. 65. New Brunswick.

53. 'Agricultural Literature Conference', *Krisi Sampada*, vol. 8, nos. 5–6, 1324 BS (1917).

54. 'Development of Indian Agriculture', *Krisitatva*, vol. 6, no. 11, 1884, pp. 204–10.
55. Iswarchandra Guha, 'Agriculture', *Grihastha*, vol. 4, no. 12, Aswin 1320 BS (1913), p. 918.
56. For instance, see Swarnakumar Mitra, 'Agriculture in the high schools of America', *Krisi Sampada*, vol. 7, nos. 10–11, 1323 BS (1916).
57. The extract of the speech by N.N. Banerjee was published in *Krisak*. See 'Agriculture in Bengal', *Krisak*, vol. 8, no. 9, Poush 1314 BS (1907), pp. 206–7.
58. Can this typical understanding of the Bengali literati be identified as reactionary modernism? The term 'reactionary modernism', coined by Jeffrey Herf, was used to denote enthusiasm for modern technology on the one hand and a wholesale rejection of the enlightenment on the other hand. See Jeffrey Herf, *Reactionary Modernism: Technology, Culture and Politics in Weimer and the Third Reich*, Cambridge, 1984.

4

Modernizing Agriculture in the Colonial Era: A View from Some Hindi Periodicals 1880–1940

Sandipan Baksi

A significant part of the study of specific aspects of science under British rule has focused on themes related to industrialization. It has also drawn the attention of disciplines other than history when exploring issues such as the divergence between the so-called Nehruvian and the Gandhian view of science and industrialization. The physical and chemical sciences and their development have been the subject of scholarly attention especially in the late colonial period. Two other themes that have more recently attracted interest and scholarly attention have been medicine and environment. In striking contrast though, agriculture and agricultural sciences have received much less attention. A noteworthy exception is Deepak Kumar's *Science and the Raj: A Study of British India*, which includes developments in agricultural science in the colonial era as an integral part of the larger account.[1] Some interesting issues in relation to the diffusion of modern agricultural knowledge, such as through agricultural fairs, have been discussed in some detail by Gyan Prakash in his book *Another Reason: Science and the Imagination of Modern India*.[2] Mention must also be made of the descriptive and chronological history of Indian agriculture in four volumes by Randhawa, which is the source of much useful technical information.[3] Two more recent writings that stand out, as exceptions to the general rule, are Bipasha Raha's study of the attitudes of the Bengali literati to agrarian issues[4] and Smritikumar Sarkar's study of technology and rural change in eastern India.[5] The major aspect of the historian's interest in agriculture though has stemmed from the perspective of economic and social history (where the social does not include science), where the relationship between social and economic conditions and

---

* This paper has made extensive use of essays and articles from Hindi periodicals from the late nineteenth and early twentieth centuries. The English translations of the quotations from the periodicals are all mine.

agricultural production and productivity has attracted substantial interest.[6] In much of these writings, the study of the history of agricultural science and its development in the colonial era has not been significantly in focus.

One may argue that agricultural science in the colonial era merits greater interest than has been evinced so far for the following reasons. The first is that agriculture was one of the major productive sectors of concern to colonialism. Agriculture was a major source of revenue for the colonial state throughout the period of their rule. The need to guarantee revenue was also the motivating factor for many irrigation works, which were among the first noteworthy technological interventions, along with the railways, of British rule. In the nineteenth and twentieth centuries, various actions of the colonial administration led to the introduction (and in several instances, imposition) of new varieties and new crops in agriculture as part of the larger phenomenon of the development and penetration of markets in all aspects of agriculture under colonial rule.

The second reason is that unlike the case (for the most part) of industrial goods or later modern industrial production, modern agricultural technology was not introduced on virgin soil by colonialism, particularly in the case of India. New techniques, new methods of production, new agronomic practices and the new knowledge associated with them, would have had to compete with existing techniques, methods and practices. To gain even a foothold, modern practices and related knowledge generation and/or diffusion, would have to confront what was already being practised. This reality becomes all the more stark in light of the fact that the introduction of the new took place in a context that was also shaped by well-established structures of socio-economic relations alongside deeply entrenched methods and techniques of production. Of course, the advent of colonial rule significantly changed the socio-economic context of agricultural production, through the eighteenth and the first half of the nineteenth centuries, with the institution of zamindari being the dominant feature in some parts of India, the ryotwari system being the dominant feature of other regions. In both settings though, the introduction of new science and new techniques in agriculture would have to contend with the perceptions of the landlord classes (whether of the zamindari or the ryotwari type) that dominated the countryside. Equally significant was the other important characteristic of Indian agriculture in the colonial era—the persistent poverty of the mass of the Indian peasantry. The bulk of the peasantry was more than likely to face considerable obstacles in any attempt to modernize their techniques of production or exploit the benefits accruing from the knowledge and application of modern agricultural science. This was due to their persistent insecurity of tenure, the exploitative levels of rent and their heavy, and often perpetual, indebtedness due to the usurious rates of interest for credit.

Indeed, as some economists have noted, innovation in agriculture is strongly adaptive in nature. In the case of introduction of manufactured goods (and the later introduction of manufacturing in the colony itself) that were produced earlier by craft techniques, the former simply overwhelms the latter in relatively short period. In the case of theoretical knowledge, as in the fields of astronomy, physics or chemistry, and to a partial extent in medicine, the displacement of the old by the new was, in the first instance, perhaps more cultural and social, calling into question established world views, before they really impinged on practice. However, the adaptive nature of the modernization of agriculture makes the context a critical aspect of the adoption of the new. Change, in the form of the entry of new techniques and knowledge, must in the first instance confront the applicability (a term whose meaning may be subject to much ambiguity) of the new in the existing agro-ecological as well as socio-economic setting. Even when it does so, it is unlikely that the new could be adopted without modification or alteration in some form. At the same time, new techniques and knowledge could also, in turn, modify the existing conditions of production and the social and economic setting. In the case of agriculture, therefore, we need to go beyond any simple-minded model of displacement and resistance.

One may expect, therefore, that the Indian response to the introduction of new techniques in agriculture and to new agricultural knowledge to reflect these specific features, and the subsequent assimilation of new knowledge and techniques, takes place at different levels, as Deepak Kumar has well described. However, as Kumar notes, an important feature of this response is the attitude and perceptions of different sections of Indian society to the increasing impact of science and technology under colonial rule.[7] The debate and discussions over the meaning and the utility of the new sciences that were introduced, the various attempts to diffuse such knowledge, and an increasing attempt to determine how Indian society could benefit from this new knowledge used along 'national lines and under national control',[8] all constituted a significant aspect of the Indian response. Significantly, the discourse of the Indian response was reflected prominently in the print medium, which itself was perhaps in the formative stages in many Indian languages. The response must also be charted in the active publications of books, pamphlets and regular writings in journals and periodicals. The latter, in particular, provide an interesting view of the evolutionary character of the response and the rising trend of assimilation that follows the initial phase.

While the study of the Indian response and subsequent assimilation of new agricultural science and techniques is a much larger subject, this essay restricts itself to examining the discourse on agriculture science and the modernization of agriculture within a limited period. In particular, it focuses on the discourse relating to the new agricultural science in some key Hindi periodicals. Two points may be made here with respect to our particular choice of the sources. One may expect that the response to agricultural science may be more interestingly traced in the vernacular press rather than the English language, as the former may be expected to have had greater reach outside the major urban centres compared to the latter. The vernacular press would perhaps also be more likely (though this need not be taken to be a certainty) to draw interest and correspondence on the conditions of production of agriculture and be closer to ground realities than the writings in English. The other interesting aspect is that there is indeed a close interrelationship between the rise of the vernacular press and the development of engagement with modern science and technology in different sections of society, a feature which we will not be discussing in any detail here, but which is of much relevance to our study. Examining the writings in the vernacular press would perhaps also help move the larger debate on science in the colonial era away from the tiresome preoccupation in attempting to determine whether the new science was the 'imposition' of a foreign knowledge system or, alternatively, whether science was the preoccupation of an emerging middle class, or an English-speaking babu class in more popular accounts.

More specifically, we examine writings on agriculture from a few significant Hindi periodicals, among whom particular mention should be made of *Sarsudhanidhi*, *Saraswati*, *Vigyan*, *Maryada* and *Madhuri*. Of these, the first was one of the very few periodicals in Hindi that published articles on science and scientific subjects in the late nineteenth century, which is often referred to as the Bharatendu Yug[9] in the history of modern Hindi literature, signifying the influence of Bharatendu Harishchandra on the language and literature of the period.[10] This period has also been understood as the formative years of modern Khari Boli Hindi and the Nagari script, which witnessed a remarkable rise in the quantum of Hindi writing, especially in the form of a plethora of literary periodicals. These developments in the language and literature witnessed a consolidation in the early twentieth century. *Saraswati*, under the editorship of Mahavir Prasad Dwivedi, distinctly contributed to these developments. Consequently, this period in the history of modern Hindi literature is often referred to as the Dwivedi Yug,[11] reflecting his influence.[12] *Saraswati*, and other periodicals like *Maryada* and *Madhuri* that followed, began publishing articles on science and scientific subjects with fair regularity. It was also the period when the first popular Hindi science periodical, *Vigyan*, began publication in 1915.[13] These periodicals contributed in a significant way in the formation of what has been termed as the Hindi public sphere.[14] These periodicals also depicted, as we shall witness in our study, among other tendencies, an influence of a growing nationalist sentiment on one hand and a persistent strain of revivalism of an 'ancient glorious past' on the other within which this sense of nationalism was articulated.

There were also other journals and periodicals devoted specifically to agriculture, to which we have not yet gained access. However, the journals under discussion not only provided a medium for the discussion of science in all its varied aspects, but also in their writing drew a close connection between the articulation of science and science-related issues in Hindi and the larger development of science itself.[15] This is also a matter of independent study, which however we will not enter here.

Disaffection with the State of Agriculture

An examination of the general writings on agriculture in the Hindi periodicals of the period reflects an acute and persistent sense of dissatisfaction with the then state of affairs. The concerns range from the prevailing social and economic conditions to the dismal levels of agricultural production and productivity. The acute distress in agriculture, and the difficult conditions of subsistence of the cultivating classes,[16] seems to have been the most striking and disturbing phenomenon in the experience of all commentators. Various forms of this distress were frequently expressed by the writers in the periodicals of the era. Bharatendu Harishchandra, for instance, compared the condition of cultivators to that of the beggars.[17] Such expressions of the dismal condition of the Indian farmers appear consistently across the periodicals, well into the twentieth century. A discussion in *Madhuri*, appearing almost half a century after the comment by Bharatendu, echoes the same anguish about the acute impoverishment that defined the life of Indian cultivators, which it bemoaned, was unparalleled in the world.[18] Another manifestation of the perennial distress in Indian agriculture, which found frequent expression in the Hindi periodicals, was the perpetual indebtedness of cultivators. While many of the discussions articulated the problem of indebtedness, there were also some more informed articles which even tried to quantify its real burden. For instance, an article published in *Madhuri* in 1931 proclaimed that the total indebtedness of Indian farmers amounted to Rs.6,000 crore. It further concludes that such a huge amount of debt would never allow any progress for Indian agriculture and the cultivators.[19]

The writings, on the one hand, are extremely critical of the exploitative character of the relationship of the cultivators with the landlord and the moneylender, as well as the revenue policies of the government. On the other hand, there is a persistent dissatisfaction with the poor state of agricultural production and productivity. In fact, disquiet about the declining productivity of agricultural land and the need to reverse that trend was an early concern reflected in the Hindi periodicals of the late nineteenth century, a concern that was increasingly articulated over the years. An article published in 1880 noted, 'If we consider seriously, it becomes obvious that day by day the golden land is losing its fertility. It does not produce as much as in the past.'[20] This concern of declining fertility was often associated with the ever increasing pressure on agricultural production, trying, thus, to provide a scientific explanation for the phenomenon. This pressure, due to multiple socio-economic reasons, including increasing internal and external trade, growing agricultural rent and its stringent collection, and usurious interest payment demanded by moneylenders among others, forced the cultivator to continuously try to produce on whatever land he had access to, and in turn did not allow land to be left fallow for a season or two.[21] The over-exploitation of land, not allowing it the time to rest and revitalize, had an adverse effect on the fertility.

This theme appears consistently in the Hindi writings of the period. Writing more than four decades after the article quoted earlier, Gopal Damodar Tamaskar repeats the argument of growing pressure on agricultural land, which in turn does not allow it the time to naturally regenerate its capacities, leading, therefore, to a decline in its productivity.[22] Tamaskar goes on to briefly describe the natural ways of revitalization and questions the basis of 'expecting good production year after year, in the absence of the time for rest and rejuvenation of land'.[23]

The persistent deficit in food availability for a significant section of the Indian population was often argued to be due to inadequate agricultural production. Dayashankar Dubey's 1920 article describes this deficit in some detail. Based on statistical information from various government reports, he calculates the average quantity of food required every year to feed India's population, including the feed needed for livestock, as well as the amount of grain required as seed for the next agricultural season. He then calculates the total food production for every year starting from 1911 to 1918. On the basis of these calculations, he concluded that there was a perennial deficit in agricultural production which needed to be catered to in order to ensure food availability.[24] The solution to the question of food scarcity appears quite clear to the writer: 'There are only two prime means to compensate for this loss— firstly to reduce the amount of food export and secondly to enhance the domestic foodgrain production.'[25]

The alarming recurrence of devastating famines was of course the most striking form of agrarian distress that naturally drew the attention of the Hindi commentators, and strengthened the call for raising food production. There were a number of such observations in the pages of these periodicals in the early decades of the twentieth century that mention the extent of devastation caused by frequent famines in India. Many of these articles appeared to tend to the view that the phenomenon of recurrent famines was of relatively recent origin, beginning in the nineteenth century, and worsening by the second half. The writings on famines often discussed the perceived causes for such recurrences and the ways to deal with the menace. Such discussions appeared to reflect a growing awareness of the 'need to scientifically develop agriculture as well as socially develop the cultivators to break the vicious cycle of famines'.[26]

Interestingly, the comparison with agriculture in other parts of the world, reinforcing the disaffection with its state at home, was a theme that appeared in the writings of the twentieth century and became increasingly frequent with its advancement. The comparison was not confined to the 'Western' nations, but extended also to countries like Japan, Russia and Turkey, which in the twentieth century witnessed substantial development in agriculture. In an article titled 'Krishi kee Dasha', in *Vishal Bharat*, one reads:

There is nothing that the Western countries have not already achieved. They have transformed infertile lands into productive [lands]. They now grow two crops in areas where there used to be only one. With the help of research they have given a new form to the older things. Today they rule nature. This is the story of the developed world, but we should be inspired even by those who till yesterday were residing in darkness, and were dependent only on traditional methods, like Turkey and Russia.[27]

These comparisons, in terms of agricultural production and productivity, clearly portrayed not only the recognition of the need to modernize Indian agriculture but also a sense of embarrassment at the prevailing conditions in Indian agriculture, even with the advance of the twentieth century. Unsurprisingly, the concerns regarding agricultural production over time were increasingly viewed through the lens of nationalist sentiment. Such discussions in the writings surveyed tend to draw on one or the other (and occasionally both) of the two attitudes. One class of commentary tended to hark back to a halcyon past, when agriculture in India was supposed to have been well developed, producing enough to feed its people, and which did not survive into the present because of subsequent misrule and neglect.[28] Such writings occasionally even asserted that the agricultural practices of ancient India were an inspiration to many other countries. For instance, in one essay, the author Gangaprasad Agnihotri, with typical rhetorical flourish, invokes the 'respect' accorded by 'ancient seers and sages' to agriculture, 'which was then followed by China and Japan, and is now being followed by the Western nations'.[29] The article blames 'the ignorance of the Indian kings and leaders, who do not realize that agriculture is the foundation of their kingdom, has led to its dismal state in the current times'.[30] The neglect of agriculture by the colonial as well as Indian rulers was often held responsible for its deterioration; however, the most common attribution of the cause of the 'downfall' was of course to the establishment of the British rule.[31] It perhaps comes as no surprise that with the progress of the twentieth century, these cultural arguments developed sharp political overtones. There are frequent references to British rule as the principal cause for the decline of Indian agriculture. These comments lay the blame for the decline in agriculture, on the decreasing respect for agricultural work under British rule, especially under the influence of the modern education system ushered in by them. This education system, it was argued, turned people away from agriculture, towards clerical jobs. The insensitivity of the modern educated pundits towards agriculture was argued to be the cause of the slump in productivity. The following quote, although coming from an article somewhat later in the twentieth century, succinctly sums up this line of argument:

There are many reasons for the present degeneration of Indian agriculture. But of all the reasons the most fundamental is the feeling among the present day intellectuals, that agriculture is the worst profession, which does not require any knowledge or science. Our Aryan intellectuals never had such disrespect for agriculture, as we witness today.[32]

The other line of argument linked concerns about lagging agricultural production and productivity to the question of national development. The emphasis in such writings was primarily on the question of the economic uplift of the bulk of the rural population and consequently, the development of agriculture within the larger context of the development of India.[33] Phrases which characterize the importance of agriculture for a national agenda, such as 'India is primarily an agricultural country' and 'more than 70 per cent of our population is dependent on agriculture', indicating a need to develop agriculture to achieve any significant progress on the path of national development, appeared frequently in the writings of the later part of the first half of the twentieth century.

It is important to appreciate that the two forms of expressions of nationalist sentiment cannot often be discretely distinguished. They were often interrelated. Interestingly, the relation between the two forms of articulation also portrays a degree of ambivalence. At times the development of agriculture is argued as the path to revive lost glory; however, one often comes across more pragmatic arguments, which, while acknowledging the 'glorious past' of traditional agriculture, accept the inferiority of the customary methods of cultivation in the modern era.[34] They recommend adoption of modern methods of cultivation for the development of the country. The current conditions of agriculture, such an argument contended, were no longer amenable to traditional methods of cultivation, and a shift to modern ways was essential for progress.[35] Thus, on many occasions, while the need to develop agricultural production and productivity was couched in revivalist terms, the perceived means for the attainment of that goal were nevertheless resolutely modern.

The Role of Science and Technology in the Modernization of Agriculture

The persistent disaffection, bordering almost on outrage, with the state of agriculture, as it prevailed in the late nineteenth century and through the first half of the twentieth century, was clearly associated with an urge for modernization of agriculture. Such concern, in an obvious sense, followed from questions relating to the decline in productivity and how this state of affairs was to be ameliorated or even radically improved. It was striking that the discussions on these themes in the periodicals surveyed were clear about the fact that the advance of Indian agriculture required both modernization of social and economic conditions of agricultural production, including agrarian and land relations, as well as the application of modern science and technology. Remarkably, the transformation of social and economic conditions of agriculture as the key factor in the process of its modernization comes out very starkly in these writings. In fact, the periodicals suggest an overriding focus on this aspect, as compared to the application of science and technology, for any improvement in agricultural production and productivity. Also evident in some of the writings was an understanding of the interplay between these two aspects. Given the focus of our study, we will at this point devote attention only to the perceived role of modern agricultural science and technology in improving agriculture, as was understood in these writings. Such writings which dealt with the role of scientific techniques and methods to enhance agricultural production and productivity were rarely to be found in the Hindi periodicals of the nineteenth century. However, with the advance of the twentieth century, with the establishment of *Saraswati* in 1900, and other later periodicals like *Vigyan*, *Madhuri* and *Maryada*, we witness a significant increase in such writings. As we shall see further, these writings concerned themselves not just with the importance of modern agricultural science and improved techniques and methods of agricultural production, but were also articulate about the role of appropriate socio-economic conditions for their proper application and adoption.

How to Improve Agricultural Productivity

The increased exploitation of land due to ever increasing pressure for production was perceived as the primary reason for declining productivity right from the last decades of the nineteenth century and this concern persisted thereafter. With the advance of the twentieth century, we witness a strengthening argument for the use of modern science and technology to deal with this concern. Gopal Damodar Tamaskar, whom we have already encountered in our discussion on declining productivity of land, after portraying it as an obvious outcome of the continuous exploitation of the land, goes on to list the ways to deal with the problem. He claims that, 'There are 3–4 ways to practically deal with this plight. To begin with we have to learn to apply fertilizers every year and to use the modern implements of cultivation. There should be proper arrangement of irrigation and good quality seeds should be sown.'[36]

We witness ample instances of discussions which went further into the nuances of the exact ways in which the soil loses its nutrients due to continuous usage. In the absence of the possibility of a natural process of revival, it was largely agreed upon that the land needed to be externally supplied with nutrients.[37] Thus, a general perception that comes out clearly from these twentieth-century Hindi periodicals is that the improvement of land, whether quantitative or qualitative, required the utilization of modern science and technology.

Countries which had already achieved some success in developing their agriculture, like Britain, Germany, America, Japan and Soviet Russia, were looked upon as obvious evidences of the potential of agricultural science and technology. Numerous articles discussed how these countries utilized modern science to improve the different steps in the production process.[38] While there was recognition of the fact that a number of factors were responsible for the improvements experienced in these countries, there was evident, at the same time, a secular belief that agricultural science and technology had been one of the most foundational elements of the story, as came out very succinctly in the discussion from *Saraswati*:

Indians cannot compete with Europeans in agricultural work. Indians are lagging even in the only profession left for them. The West is leading even in agriculture. The reason for this lead is the use of science. Their cultivation processes is much more developed than ours.[39]

The article concluded by stating the importance of adopting modern methods of cultivation, 'transcending the traditional means, as has been successfully done by the Europeans, in order to go beyond mere feeding the families'.[40] There was also visible, on occasions, a perception of loss and a sense of embarrassment at not being able to utilize science and technology to modernize agriculture. Narayansingh in his article 'Bharatvarsh kee Daridrata' explains two problems with Indian agriculture. The first problem, he states, is the law of diminishing returns—*Kramagat Hwas Niyam*—which, he says, operates in agriculture in every part of the world.[41] The second issue, however, he asserts, was specific to Indian agriculture. It was the inability to transcend the traditional methods of cultivation and utilize modern science and technology.

We do not know the scientific methods of cultivation. It is a matter of immense shame and sorrow that despite being the residents of a principally agriculture based nation; we are unaware of the methods that can increase the value of agriculture. We lack the intellectual capability to conceptualize and bring about new techniques of cultivation, new implements and machines. We are stuck with the same old implements and manures and outdated processes, which have not seen any development since thousands of years. After all, in the absence of any education in science and technology, how are we supposed to bring about any development?[42]

The other illustration of the success of modern agricultural science that apparently impressed the Hindi-speaking elite was the experimental farm. There were noticeable references to these farms as well as the agricultural department, at least since the 1920s. Discussions of the kind of research performed in these farms, the results of their experiments, as well as suggestions on how their work should expand and reach out to the cultivators can be found in Hindi language writings on agriculture.[43] There was a growing need to devise ways to replicate such results on the fields, by making them accessible to the cultivators.

Our land has become less fertile. Even a produce of 10 maunds of grains per acre is regarded as very good. Officials of the agricultural department have produced double and triple of this amount in similar lands, by using modern implements, fertilizers and proper irrigation, etc. It would be great if all our cultivators are able to do the same and become capable of producing at least double the amount.[44]

Surely, there were overt disagreements on the extent of use of modern agricultural science and techniques, and their suitable forms. For instance, an article in *Saraswati* criticized the agricultural farms for their ignorance of the ground realities of the cultivators and the impracticality of their approach. It argued for gradual improvements in the methods that were already in use, rather than pushing for improvements that require many new implements and substantive amounts of capital. It criticized the agricultural officials 'for living in a dream world, in ignorance of the real condition of the cultivators, devoid of modern tools, money and knowledge'.[45] However, these disagreements were rarely about the need for improvement per se or about the efficiency of modern science and technology. Neither was there any dissension over the role of the latter to bring about the desired improvements. The differences were generally pertained to their suitability in the given socio-economic or geographical context.

Interestingly, with the advance of the twentieth century, the argument for the need for intensive production, with higher levels of productivity, appeared to gain strength. We witness expressions of a realization, as of yet rarely expressed, that there are definite limits to the enhancement of production by means of extension of arable land through irrigation. This further bolstered the perceived importance of intensive cultivation with the help of science and technology.

The Indian farmer, unlike the trend in China and Japan, will extend his fields for extra production, but will not do intensive farming in small farms by increasing its productivity. 35.2 per cent land in India is not feasible for agriculture, 34.2 per cent land is under cultivation and rest 30.6 per cent is arable but not cultivated. The problem with further extension is that all good quality land has already come under the plough. Only bad quality land remains which in turn reduces productivity. Although such land superficially looks feasible but the cost of ploughing, irrigation and investing in such land is prohibitive. Intensive farming, like Japan, would therefore be more beneficial.[46]

Apart from these general writings, which discussed the need for science and technology for enhancing agricultural production and productivity and portrayed a growing consensus about its importance for modernizing agriculture, other writings dealt specifically with the various scientific aspects of agriculture. Such essays can be broadly classified into three groups. The first discusses how science and technology could be applied to different steps of the agricultural production process. They describe various elements that constitute *Vaigyanik Krishi* (scientific agriculture), as perceived by the commentators, and how the introduction of such elements impacted agricultural production and productivity. A March 1929 note in *Madhuri* discusses, among other things, the importance of knowing the composition of the soil in scientific terms and the specific requirements of the crop. This relationship, it states, was important as it was the soil which supplied most of the nutrients for the growth of the plant. A scientific examination could, therefore, help to realize the appropriateness of the soil as well as to know the materials with which the soil needed to be replenished for the cultivation of a particular kind of crop. Corrective steps accordingly could boost the health of the crop.[47]

Similarly, we find a number of discussions on the importance of manures and fertilizers, describing their intrinsic chemical constituents, the appropriate ways to improve the quality of traditional fertilizers, the positive impact of their application on productivity, recommendations that were at times even substantiated by quantitative estimates from the experimental farms. The use of alternate types of fertilizers, like *haddi kee khad* (fertilizers prepared from bones), *hari khad* (green manuring) and *kritrim/vaigyanik/rasayanik khad* (chemical fertilizers) in order to benefit specific kinds of soils or to nourish some particular crop were periodically discussed. Sripalsingh's article, 'Kritrim Khadein', in the *Madhuri* of December 1928, for example, did very simple calculations, to depict the relative advantage, in terms of efficiency, wastage, cost, land requirements, immediacy of impact, etc., of chemical fertilizers over dung-based manures.[48]

Discussions of a similar kind could be found for preparation of soil, on the importance of developed quality seeds, on irrigation by canals and tubewells, for use of improved agricultural implements, like the iron plough. They demonstrated the importance and the advantages associated with such usage. These discussions pointed at a general agreement on the potential advantages of the modern and scientific methods and innovations, and a growing desire to incorporate them in cultivation. Notably, we also witness a sense of sustainability, which finds expression, for instance, in the support for green manuring. Shankar Rao Joshi's piece in *Madhuri* of April 1928 can be regarded as a representative case. It discusses the importance of organic matter and humus, which is essential for maintaining the fertility of soil. It expressed the need for maintaining a balance between intensive production, by means of excessive and deep ploughing, and future sustainability of production. Excessive ploughing, it claimed, led to rapid exhaustion of the humus of soil, which in turn severely reduced its fertility.[49] Interestingly, the article made these claims based on scientific experiments and reports. The solution, it concluded, lies in resisting excessive ploughing, as well as in using a blend of chemical fertilizers and organic manures. It also strongly recommended green manuring, which could rejuvenate the humus of the soil. The process, it claims, may lead to the loss of a year of agricultural production, as the land had to be kept fallow for the year of application. However, it said, the long-term positive effects on productivity could in turn compensate for the losses in two to three years, and at the same time help in maintaining the fertility of the soil.[50] In fact, the question of deep iron ploughs versus shallow wooden ploughs was often repeated in these writings, but there appears to be an agreement on the superior utility of the former. Shankar Rao Joshi, the author of the article referred to previously, in another of his essays in *Madhuri* discusses the advantages of iron ploughs, thus indicating that it is excessive and regular deep ploughing that he was arguing against, in view of the potential long-term harm to soil fertility, and not deep ploughs per se.

The roots of soil go deep inside if the soil is loose. Hence it is important that we plough deep. Our country made ploughs cannot do this work, because they do not go deeper that 5–6 inches and there are some portions of the land that invariably gets left out. The result is that those parts of land remain unploughed and weeds are not totally destroyed. Masten, Ransom, Kirloskar and such iron ploughs allow better loosening of soil in lesser labour effort and cost. These ploughs go 8–9 inch deep in the soil. The wooden ploughs merely slash the land but the iron ploughs cut it. Apart from that, due to their special make, iron ploughs turn the soil and also break the lumps. It does not leave out any portion of the land. Moreover the different parts of iron ploughs are available separately and if required even the rustic/illiterate cultivator can replace faulty parts and re-assemble the plough. The iron ploughs properly serve all purposes of ploughing.[51]

The second group of articles consists of more technical and practical discussions which enter into the details of different steps or inputs required in the process of cultivation. Some discussions focus on one particular step or input required, for instance ploughing or *jotna*, while others discuss multiple steps or inputs required. The discussions would generally be detailed descriptions with a prescriptive feel about them. They analyse the local methods and prescribe how scientific improvements can be made through incremental efforts. They would, at times, also explain the scientific principles at work behind the local methods. Seldom do we find articles that totally reject the current methods of cultivation.

One striking characteristic of these discussions is the practicality associated with them. For instance, if the discussion is on fertilizers or *khad*, it would first describe the current methods and then prescribe the systematic improvements, like collecting animal urine along with dung in a deep pit that can be done through the traditional ways for preparation. They also recommend additional inputs, viz., 'Sulphate of Potash', which can enhance the efficiency of traditional manures.[52] At times, they read almost like extension documents for the practical use of cultivators.[53] The recommendations reflect a blend of experiential and scientific knowledge. It must be mentioned that *Vigyan*, an Allahabad-based popular science periodical which began publication in 1915, definitely stands out in these kinds of discussions, both in quantitative and qualitative terms. They would have a series of articles on any aspect of cultivation with detailed explanation of that aspect, its scientific relevance, the classifications within it and their specific uses, scientific ways to implement them and various ways in which cultivators can themselves practically make arrangements to bring about the improvements.[54]

The third group of articles focuses on crop-specific discussions. These would describe different operations in the cultivation of some particular crop, prescribing the scientific and systematic ways to execute them. Here again, one witnesses the same blend of experiential and scientific knowledge. For instance, an article on cultivation of grapes, prescribed, among other things, a fertilizer blend as: *Sadey gobar ka khad, aadhpav sodanatret, pavbhar super-phosphate* (Manure of decayed dung mixed with 125 g. of nitrate of soda and 250 g. of super-phosphate).[55]

Another illustration is the discussion on *Neel kee Kheti* (cultivation of indigo) by Murarilal Bhargav. The article describes the experiential knowledge of the cultivators—when the wind blows from the west, it gives a better crop with richer colour content. It then goes on to explain the apparent scientificity behind this experiential knowledge. It explains that indigo being a leguminous crop absorbs nitrogen from the air. 'Western winds', it explained, 'help in easy absorption because eastern winds have lot of water content, which prevents easy extraction of nitrogen. Moreover the eastern winds enhance the moisture content of the plants and wash the chemical content off the leaves.'[56] Clearly, the scientific validity of this explanation is arguable. Nevertheless, the attempt at providing a scientific explanation to experiential perceptions is interesting.

At times, such articles would even give a brief history of these crops. Significant overlaps, of course, can be found between these groups of articles. For instance, an article which describes the proper way of making compost may go on to describe the advantages to productivity and increase in production that it can bring about. Similarly, a discussion on the cultivation of wheat may go on to describe the proper blend of fertilizers for the crop and then further present evidence of the rise in production due to such application. Such writings reflect an earnest effort to disseminate modern knowledge by underlining its commensurability with contemporary practices. They clearly constitute an effort towards extension of scientific knowledge and modern techniques in a form that can be practically adapted by the cultivators at large. There appears to be a large consensus among Hindi commentators on the utilization of modern science for increase in agricultural production and productivity. These writings appear to be a part of the attempt by the Hindi literati to extend the new knowledge in a form and a language that appeal to a much larger audience, who would find practical utility in this knowledge. There is also an effort to portray this new knowledge as being synchronous with prevailing methods and techniques, which could be applied through small incremental steps, thus arguing for its adoption in contemporary practice. It is striking that modern science never appears to be an external, 'Western', imposition in these discussions, nor does its adoption need to be particularly justified.

It is also striking that this consensus prevails along with a number of discussions that try to situate the utility of agricultural science and technology in the socio-economic context.

Agricultural Science and Technology in the Socio-Economic Context

The general realization that scientific means cannot be adopted in agriculture without overcoming some significant obstacles is exemplified by the words of a commentator in the *Saraswati* of April 1920:

> There are innumerable problems in the use of scientific means. Indian farmer is very poor and the land holdings are extremely small and dispersed at different places. They therefore cannot draw much benefit from modern fertilizers and implements. Even the irrigation facilities are poor. Their oxen are too weak to pull the modern heavy ploughs. They are often in the grip of greedy moneylenders, paying usurious interest rates to their doom. Even the land rent extracted by the zamindar is huge. They also have to be bribed in some or the other form. A major portion of what they earn through their labour is uselessly lost. They (the cultivators) are trapped in darkness. It is a complex problem.[57]

The first aspect of the socio-economic context that drew the attention of these writers was land relations. The role of land relations in the adoption of modern technique in agriculture finds frequent reference by the first two decades of the twentieth century. Importantly, this was also the time when science writing in Hindi was getting established, with *Saraswati*, and to some extent even *Maryada*, coming out with regular expositions on scientific subjects. It was also the period when *Vigyan* started publication.

An article in the *Vigyan* of August 1917 discussing the need for utilizing iron ploughs, quality seeds and proper irrigation for improved productivity asserts that 'both the cultivator and the zamindar complain about the problem of productivity, but the blame for these shortcomings lies actually with them as can be inferred from the high productivity attained in the agricultural farms.'[58] It gives the example of the Kanpur agricultural farm that has witnessed a boost in production just by the means of proper ploughing. But it is evident to the writer that it is the zamindar who is primarily to blame because:

> The cultivator does not even have the surety of cultivating it (the same land) for the next year. As per the tenancy act of United Provinces, the peasants get security of tenancy after they cultivate for 12 continuous years, but for the first 11 years he does not have any hope/prospect/belief of retaining the land for the consecutive year. If the belief is established that he will be secured for even those 11 years, he will put fertilizers in the land, plough it properly and ensure its productivity.[59]

Clearly, without security of tenure, the cultivator would hardly consider maintaining or improving the productivity of the land through appropriate measures.[60] Such arguments persist in the literature which point to security of tenure as the most fundamental prerequisite for the effectiveness of all other steps of scientific agriculture.[61] With such security of tenure that gave the peasants some confidence, it was argued that the use of improved methods of production could become popular and yield good results.

We witness this perception being expressed in the periodicals, at least till the early 1930s. It is important to appreciate that this insecurity was not just confined to the fear of eviction. Another cause of insecurity was the variability of the rent payable by them. The zamindar had an overwhelming power to decide the amount of rent to be collected, which often led to sudden increases in their demand on any excuse whatsoever.[62] It was widely perceived among the literati that such arbitrary raises by the zamindars were leading to the tenants losing their land. In the face of such insecurity, the probability of cultivators retaining the fruits of their efforts was negligible, which made them lethargic to change and thus, non-enterprising.[63]

A similar insecurity prevailed with respect to the tax demanded by the government, as is succinctly expressed in a 1914 article, discussing the rent levied on the cultivator. It says that for the cultivator, both ryotwari and zamindari were equally penalizing because the rent payable by them remained high in any case and was always subject to arbitrary increases.[64] In such a setting, where any improvement by the farmer was, in all probability, likely to attract increased rent, it was understood as irrational on part of the cultivator to try and improve agriculture by the application of scientific methods. It was often argued that the combination of the lack of ownership rights in land and the high rent that were subject to arbitrary increases by the government or the zamindar were the fundamental obstacles to the application of improved or new techniques by the farmers. This impediment, it was at times claimed, would render ineffective other efforts, like the spreading of education among the cultivators.[65] Prannath Vidyalankar makes a striking comparison of the situation of Indian cultivators with that of the Irish peasantry. Just as in the case of Ireland, Vidyalankar is clear that peasants would never put any effort for extra production, as the gains from such efforts would invariably be usurped by the landlords. The author discards all cultural arguments which portrayed the Indian cultivator as lethargic and unenthusiastic to learn modern and scientific ways of cultivation. These, he sharply observed, were mere symptoms which pointed to the deeper malaise of zamindari and high rent. Vidyalankar thus concludes with the radical assertion that the eradication of such land relations and providing ownership rights to the peasantry was the primary solution to the problems of Indian agriculture as well as society.[66]

It is useful to appreciate the nuance in this interlinkage between land relations and the modernization of agriculture. It was the quantum of rent that determined the part of the produce that was retained by the peasant and that which was usurped by the zamindar and the colonial government respectively. The rent was, therefore, instrumental in deciding the investment on the application of new techniques. Although both the aspects, the security of tenure and a moderate quantum of rent, were perceived as important factors for encouraging the use of improved techniques, we observe that with the advance of the twentieth century, it was the latter argument that gradually began to gain ground. By the 1930s, it appears that the government tax rate, and not the insecurity of tenure, is regarded as the fundamental hindrance to investment in improving agricultural production, the question of tenure receding into the background.

Interestingly, by the 1930s, the writings in *Saraswati*, *Madhuri* and *Maryada* show a shift in the definition of the land question itself. The question of land fragmentation appeared to gain prominence. The problem of land fragmentation, it was frequently argued, was an important impediment to the modernization of agriculture. From the perspective of application of modern techniques, fragmented land presented many practical problems, for instance, 'the non-feasibility of the usage of various modern implements', 'the impracticality of extending irrigation facilities', 'the difficulties in collection of livestock excreta to be used as manure, since the animals were not located at the fields, and the huge amounts of wastage in carrying fertilizers to different places', and 'the impossibility of trying modern methods or new crops as every small patch of land used by a peasant was surrounded by land possessed by different cultivators' and so on.[67] These were problems of cost implications which mitigated the possibility of application of science and technology. The efficiency thus attained, per unit of effort, both in terms of time and money, was judged to be too low to encourage such usage. This indicated a clear understanding of the advantages due to economies of scale,[68] although we do not witness specific mention of the concept in the writings on agriculture. Land consolidation, using the mechanism of cooperatives or through legal intervention, was perceived as the solution to the problem of land fragmentation.

The issue of credit found frequent expression in the context of modernization of agriculture. The usurious rates that were charged by moneylenders and the power that they wielded in the Indian villages kept the cultivators trapped in an unending cycle of debt, which in turn made them repeatedly approach the moneylender. The moneylenders had a strong control over the economy of the village and it was impossible for the cultivators to escape him. It was evident to these writers that in the prevailing conditions, the benefits of higher productivity would automatically accrue to the moneylender as higher interest payment, even though it appears superficially that higher productivity would provide greater income to the cultivator. It was therefore felt necessary to ensure easy credit, and that too in such a fashion that the profits that ensued due to the improvements, made possible by credit availability, were retained by the cultivators. Scientific agriculture, it was argued, could be effective only when the cultivator is successfully brought out of the clutches of the moneylenders.

No matter how much ever scientific means are utilized in agriculture, it will not be effective in improving the conditions of the cultivators, until the means of freedom from sahukars and mahajans are not realized, and it is ensured that the profits of agricultural development reached the cultivator.[69]

In the writings of the twentieth century, the government was seen as the primary agency to facilitate the application of improved techniques to agriculture. This perception appeared to persist among the Hindi essayists. To begin with, there seemed to be an almost total agreement that promoting the infrastructure for irrigation, through canals and other mechanisms, like wells, at places where canal irrigation was not possible, was the duty of the state, and that it had been so historically.

One part of government support is to arrange water for the cultivators. More government support is expected in the execution of this work. From the times of Mahabharata, Hindu kings have got reservoirs constructed and wells dug for the irrigation of farms. Even the Muslim kings significantly associated themselves with this work, and the Yamuna canal is an example of their initiative.[70]

At times, the cooperation of zamindar was also solicited in order to aid irrigational facilities, for example, to dig wells and to set pumps and lifts, but here also, the primary responsibility seemed to fall on the government, as was made clear by this writer: 'It is the responsibility of the government and the zamindar, especially the government, that it provides all the assistance to the cultivator to construct wells.'[71]

Absolute responsibility of the government was also expected in the realm of scientific research and training and extension work. At times, although in much lesser frequency, such expectations were raised on the zamindars too. For instance, the February 1927 issue of *Saraswati* published an article, 'Bharatiya Krishi', which claimed that given its ignorance and poverty, the application of modern methods in agriculture was not easily achievable for the cultivator. Thus, the first step towards a scientific agriculture should be initiated by the educated zamindars. It then went on to suggest a series of concrete steps that could be taken by the zamindars, like 'establishing some experimental farms where cultivation is done using the modern methods, which then produces encouraging results', 'organizing agricultural fairs and exhibitions in different regions', 'distributing prizes to those who have taken initiatives to modernize their method of cultivations'.[72]

Mention should be particularly made of an interesting discussion in *Saraswati*, under the title 'Zamindar aur Kisan', dated June 1931. The discussion was actually excerpts from the speech of Rai Bahadur Thakur Raghunath Singh, the chairman of an assembly of zamindars in Barabanki district. The primary purpose of the gathering, it appeared, was to discuss and improve the organization of zamindars, but the talk went on to deliberate upon the need for improving the relationship between zamindars and *kisans*. Here we witness zamindars themselves discussing the need for facilitating agricultural modernization, through opening of agricultural farms, etc., in view of the 'lack of knowledge and financial abilities among the peasantry, rendering them incapable of bringing in improvements in their old methods'.[73] The article also talked about various initiatives, like opening modern agricultural farms and gardens in each village and provisioning better quality seeds and spreading knowledge about modern profitable ways of cultivation that the zamindars should undertake in order to facilitate scientific agriculture. It also talked about the need for other general improvements, like providing for health and education of cultivators, facilitating irrigational facilities, etc.[74]

Notwithstanding these rare and apologist references to the benevolent role of the zamindar, the initiative to popularize scientific agriculture was largely expected to come from the government. Interestingly, by the 1920s, we witness frequent demands being raised on the government. At times, there was strong criticism of the government for its failure to disseminate knowledge on scientific agriculture.[75] Expectations from the government also included providing initial support for the establishment of credit cooperatives, as well as cooperatives to purchase and maintain implements. The government was also held responsible for ensuring availability of scientifically treated seeds, pesticides and fertilizers, providing scientific education through agricultural schools and colleges, facilitating research for improving productivity, etc. It was understood, although perhaps only barely, that adoption of any new method had an inherent risk associated with it and it required an institution like the government to mitigate that risk. It was at times argued that this was all the more true for Indian cultivators whose lives were absolutely dependent on agriculture. In the context of extension of quality seeds to cultivators by the government, Tamaskar reflects the same feeling:

It is not enough to give seeds. It is important for people to know the appropriate soil for the seed, what fertilizers are required, what other points are to be taken in consideration for its usage. It has often been observed that seeds were given to the cultivators without proper knowledge. This prevents their benefits to reach the optimum levels. At times seeds from other countries were distributed without proper experimentation in this country, leading to destruction of the crop and huge losses for the cultivator. This failure made him averse to any new seeds.[76]

The spread of education was often depicted as a means to instil confidence and enthusiasm in the cultivator, helping him to gain consciousness about his own position and his exploitation by other classes, like the moneylenders and the zamindars. Apart from that, there also appears a strong argument on the need for propagation of education on aspects of scientific agriculture itself, so as to encourage and assist the proper and widespread application of the new knowledge in agriculture. The perception, as reflected in a number of discussions, was that the lack of education among the cultivators prevented the proper utilization of available scientific resources. It was an enduring argument that seemed to have prevailed throughout the decades of the twentieth century before independence. The argument often drew on examples from various countries, such as Denmark, England and Soviet Russia, to provide evidence of how education and extension programmes were instrumental in allowing successful utilization of science and technology in agriculture. Scientific knowledge and innovation, it was also argued, was of no help if kept confined to the experimental farms. The importance of taking that knowledge and innovation to cultivators and applying it to the field in a proper fashion, found a persistent echo in the periodicals of the era.[77] This, in turn, necessitated education of a practical kind through a healthy network of schools and colleges, as well as extension programmes for cultivators. It was persistently perceived that the requirement was to have proper extension programmes in order to disperse basic knowledge about scientific agriculture.[78]

Conclusion

It is evident that a strong sense of dissatisfaction with the state of Indian agriculture, particularly in terms of production and productivity, was felt among the Hindi literati by the late nineteenth century. The predominant concern that led to this realization was the acute distress in agriculture and the difficult conditions of survival of the cultivating classes. Various forms of this distress that found frequent articulation in the Hindi periodicals of the era included the recurring famines, the persistent shortage of food and the general impoverishment and perennial indebtedness in agriculture. The writings of the Hindi literati clearly reflect a growing perception among these sections of the need to increase agricultural production and productivity in order to improve the conditions of agriculture. Such a perception strengthened in the early decades of twentieth century, by which time, writings on the need for modernization of agriculture started appearing regularly in the Hindi periodicals.

We also saw an increasing tendency in the twentieth-century writings to compare the conditions of Indian agriculture with the state of agriculture in other more economically advanced countries. Such comparisons reinforced, on the one hand, the sense of disaffection with the dismal conditions of agriculture at home, and on the other hand, they further strengthened the desire to modernize agriculture. Another important tendency in these writings, when arguing for modernization of agriculture, was the shift in emphasis from the specific terms of improving the welfare of the cultivating classes per se to a more diffuse notion of the relationship between the development of agriculture and national development. This shift in emphasis was also accompanied on occasion by the development of agriculture as part of a revivalist endeavour to recover a mythical glorious past of prosperous agriculture. However, notwithstanding these different emphases, whether of a revivalist tenor or not, there was a general consensus on the need for the modernization of agriculture. It is striking that the discussions in these periodicals on the development of agriculture, paid close attention to both, the reform or transformation of the social and economic conditions of agricultural production and the application of new methods, practices and techniques. It is distinctly evident from our study that the socio-economic conditions had a significant nexus, in the view of the Hindi writers, with the need for improvement in agricultural production and productivity. At the same time, the commentators also appeared to be aware of and deeply impressed by the achievements of modern agricultural science and technology. Nevertheless, appropriate socio-economic conditions were perceived as a prerequisite for the proper adoption and application of modern science and technology to Indian agriculture.

What interests were represented by these commentators and writers from among the Hindi literati on the question of modernization of agriculture? While it will take us beyond the scope of this essay to explore this in any detail, some points may be noted in this regard. The general anti-zamindar sentiment is palpable at the beginning of the twentieth century, but it is a sentiment that gradually appeared to decline with the passage of the first few decades. The question of rent receded, and the focus moved gradually to the quantum of government revenue. Land relations also recede from focus, to be replaced by an increasing emphasis on land consolidation through cooperatives or state regulation. Above all, there was a striking silence on the nexus of caste and agricultural production, even though the larger issue of social reform does indeed find place in these periodicals (though not a direct subject of our study).

These considerations suggested that the discourse on the modernization of agricultural science and technology that we have traced originates from those who represent the interests of a section of the cultivator class that sees a conflict of interests with the zamindars, but which is not disposed towards sustained opposition to them. Nor do these voices emerge from any radical articulation of the interests of the bulk of the peasantry of the region, given the absence of any reference to social or agrarian movements. Instead, in later years, they tended to incline towards the idea of *gram sudhar* or village improvement, including education, provision of training and extension, provision of sanitation, improvement of health and social reform, but without any focus on land relations or even radical improvement in agricultural productivity.[79] While these considerations provide a preliminary view of the social (and class) origins of the views that we find articulated, a more detailed and accurate portrayal must await further study. One may also note that it would be imprudent to take away the autonomy of the Hindi literati in their understanding of agriculture and rural India. A further study must clarify the nature of their intellectual engagement with agriculture and agrarian issues and its relation to the wider class perceptions that it may have reflected.

Notes and References

1. Deepak Kumar, *Science and the Raj: 1857–1905*, Delhi, 1997.
2. Gyan Prakash, *Another Reason: Science and the Imagination of Modern India*, Princeton, N.J., 1999.
3. M.S. Randhawa, *A History of Agriculture in India*, vol. III, New Delhi, 1983.
4. Bipasha Raha, *The Plough and the Pen: Peasantry, Agriculture and the Literati in Colonial Bengal*, Delhi, 2012.
5. Smritikumar Sarkar, *Technology and Rural Change in Eastern India, 1830–1980*, New Delhi, 2014.
6. For instance, see G. Blyn, *Agricultural Trends in India, 1891–1947: Output, Availability, and Productivity*, Philadelphia, P.A., 1966; B.B. Chaudhuri, 'Agrarian Economy and Agrarian Relations in Bengal: 1859–1885', in *The History of Bengal: 1757–1905*, ed. N.K. Sinha, Calcutta, 1967; T. Kurosaki, 'Agriculture in India and Pakistan, 1900–95: Productivity and Crop Mix', *Economic and Political Weekly*, vol. 34, no. 52, 1999, pp. A160–8; A. Mody, 'Population Growth and Commercialization of Agriculture: India, 1890–1940', *Indian Economic and Social History Review*, vol. 19, nos. 3–4, 1982, pp. 237–66.
7. Kumar, *Science and the Raj*, pp. 192–201.
8. Ibid., p. 210.
9. See Ram Vilas Sharma, *Mahavir Prasad Dwivedi Aur Hindi Navjagaran*, Delhi, 1977.
10. See Vasudha Dalmia, *The Nationalization of Hindu Traditions: Bhāratendu Hariśchandra and Nineteenth-century Banaras*, Delhi, 1997.
11. See Sharma, *Mahavir Prasad Dwivedi Aur Hindi Navjagaran*.
12. See Sujata Sudhakar Mody, *Literature, Language, and Nation Formation: The Story of*

a Modern Hindi Journal 1900–1920, Ph.D. dissertation, South and Southeast Asian Studies, University of California, Berkeley, 2008.

13. *Vigyan*, which is in publication till date, was brought into publication by Vigyan Parishad (Vernacular Scientific Literary Society) established in 1913, in Prayag (Allahabad). The purpose of this institution was to make science literature available in Hindi, in order to popularize science among masses, on the one hand, and to enrich the Hindi language, on the other.
14. Dalmia, *The Nationalization of Hindu Traditions*; Mody, *Literature, Language, and Nation Formation*.
15. See Sandipan Baksi, *Attitudes to Modernizing Agriculture in the Colonial Era: A Study of Hindi Periodicals from 1876 to 1947*, unpublished M.Phil. dissertation, School of Habitat Studies, Tata Institute of Social Sciences, Mumbai, 2012; Pooja Mishra, *The Beginnings of Science Education and Science Popularization in Hindi (1860–1935)*, unpublished M.Phil. dissertation, Zakir Hussain Centre for Educational Studies, School of Social Sciences, Jawaharlal Nehru University, Delhi, 2010.
16. A clear-cut differentiation between different classes of cultivators seems to be absent in the discourse. Words like *kisan*, *krishak* and *kashtkaar* appear to have been used interchangeably.
17. 'The condition of those who cultivate is such that they appear like beggars, and those who are non-enterprising do not even get food to eat.' Bharatendu Harischandra, *Kavi Vachan Sudha*, March 1874, as quoted in Sharma, *Mahavir Prasad Dwivedi*, p. 13.
18. 'India is a land of cultivators. Agriculture is the business of 90 per cent people and they all feed themselves only by virtue of it (agriculture). But perhaps in no country in the world the condition of peasantry is as depressing as it is here (in this country)'. Pratapnarayan Mishra, 'Kheti aur Kala-Kaushal', *Madhuri*, Magh 305, Tulsi-Samvat, Year 7, Part 2, no. 1, March 1929, pp. 116–18.
19. Yugalkishoresingh Shastri, 'Bharat main Sahkar Andolan', *Madhuri*, Kartik, 308, Tulsi-Samvat, Year 10, Part 1, no. 4, November 1931, pp. 491–5.
20. 'Vaigyanik Krishi kee Avashyakta', *Sarsudhanidhi*, 4 October 1880.
21. Ibid.
22. Gopal Damodar Tamaskar, 'Gram-Sudhar key Kuch Prasnna', *Saraswati*, vol. 28, no. 1, January 1927, pp. 48–55.
23. Ibid.
24. The argued conclusion appears to agree with the actual trends of a falling per capita output of foodgrains, especially during the decade 1911–21. Agricultural production in this period grew at a rate lower than the population growth rate. See Blyn, *Agricultural Trends in India*; Mody, 'Population Growth and Commercialization of Agriculture'.
25. Dayashankar Dubey, 'Bharat main Aadha Pet Bhojan Paneywalon kee Sankhya', *Saraswati*, vol. 21, no. 2, February 1920, pp. 65–74.
26. Rudradatta Bhat, 'Durbhiksh aur Ussey bachney key Upay', *Maryada*, vol. 7, no. 6, April 1914, pp. 382–5.
27. Shiv Kumar Sharma, 'Krishi kee Dasha', *Vishal Bharat*, vol. 8, no. 6, Poush 1988, December 1931, pp. 697–704.

28. Rudradutt Bhat, 'Prachin Samay main Bharitya Krishako ke Samajik va Arthik Dasha', *Maryada*, vol. 7, no. 6, April 1914, pp. 372–6. This is a representative example of the glorification of Indian agriculture and the condition of cultivators in the ancient past.
29. Gangaprasad Agnihotri, 'Bharat main Krishi kee oor Upeksha', *Saraswati*, vol. 28, no. 6, June 1927, pp. 1459–61.
30. Ibid.
31. 'It has been known through the travelogues of foreigners and the contemporary writings that the condition of villages was good and the cultivators were prospering under the governance of Musalmaans ("Islamic Rule"). Neither did they (the cultivators) die of hunger nor was there such decline in their conditions. Such collapse is the gift of British rule.'
    Ishwariprasad, 'Bharitya Kisan', *Saraswati*, vol. 42, no. 4, April 1941, pp. 449–56.
32. Gangaprasad Agnihotri, 'Bharat main Krishi kee oor Upeksha', pp. 1459–61.
33. The principal reason for the decline in Indian economic condition is the alarming condition of the cultivators. Shastri, 'Bharat main Sahkar Andolan', pp. 491–5.
34. For instance,

    It is the law of nature that every thing/idea has a utility at a particular time and condition. It is not intelligent to follow the same thing at all times. The methods of cultivation of our ancestors and the manures that they use were useful at those times when everything that was produced in a land was also consumed there itself. However today when a major proportion of the produce is sent to other countries, and the population pressure too has increased, and at the same time the productivity of land has declined, it would be of no use to stick to the traditions.

    Harnarayan Bottham Sripalsingh, 'Kritrim Khadein', *Madhuri*, Marghshirsh 305, Tulsi-Samvat, Year 7, Part 1, no. 5, December 1928, pp. 937–40.
35. Ibid.
36. Tamaskar, 'Gram-Sudhar key Kuch Prasnna', *Saraswati*, vol. 28, no. 1, pp. 48–55.
37. For instance, see the series of articles on fertilizers. Pathik, 'Khad aur Khad Dalna', *Vigyan*, vol. 6, no. 1, October 1917, pp. 34–6. Or, the discussion on artificial fertilizers by Sripalsingh in '*Kritrim Khadein*', pp. 937–40.
38. For instance, see Krishna Sitaram Pendharkar, 'Gehu kee Kheti', *Maryada*, vol. 11, no. 6, April 1916, pp. 270–6; Vishnumanohar Shashikar, 'Mahayuddh aur Krishi Unnati', *Maryada*, vol. 16, no. 2, June 1918, pp. 81–4.
39. 'Bharatiya Krishi', *Saraswati*, vol. 28, no. 2, February 1927, pp. 276–7.
40. Ibid.
41. The author notes:

    India is primarily an agricultural nation and agriculture is always under the law of diminishing returns, which means that for some time land will give commensurate returns to more labour and capital, but if we continue to expand their quantum then a time will come when the law of diminishing returns will start operating, implying that the production will be less when compared to the cost. This is so because by that time land would have reached its fertility limits (*utpadakatva-seema*), and it is not possible for it to produce any further.

Narayansingh, 'Bharatvarsh kee Daridrata', *Maryada*, vol. 5, no. 3, January 1913, pp. 158–64.
42. Ibid.
43. For instance, see Narayan Dulichand Vyaas, 'Gobar kee Urvara-Shakti Badhaney key Upaya', *Madhuri*, Asarh 307, Tulsi-Samvat, Year 9, Part 2, no. 6, August 1931, pp. 884–8.
The article describes the experiments done by Pusa Institute on the use of *gobar* (cow dung) as a fertilizer, and quantitatively illustrates all the results of such experiments.
44. Dayashankar Dubey, 'Anaj kee Kami Dur Kaisey ho?' *Saraswati*, vol. 21, no. 4, April 1920, pp. 196–9.
45. Gopal Damodar Tamaskar, 'Gram-Sudhar key Kuch Prasnna', *Saraswati*, vol. 28, no. 2, February 1927, pp. 204–9.
46. Harikrishna Jaitlee, 'Bharatiya Kisnao kee Unnati key Sadhan', *Saraswati*, vol. 33, no. 4, April 1932, pp. 405–8.
47. Umashankar Nayak, 'Tarkari kee Kheti', *Madhuri*, Phalgun 305, Tulsi-Samvat, Year 7, Part 2, no. 2, April 1929, pp. 263–6.
48. Sripalsingh, 'Kritrim Khadein', pp. 937–40.
49. Shankar Rao Joshi, 'Bhoomi key Jaiv-Tatva kee Raksha', *Madhuri*, Chaitra 304, Tulsi-Samvat, Year 6, Part 2, no. 2, April 1928, pp. 261–2.
50. Ibid.
51. Shankar Rao Joshi, 'Jutai', *Madhuri*, Shravan 304, Tulsi-Samvat, Year 6, Part 1, no. 1, September 1927, pp. 176–7.
52. For instance, see the series of articles by Pathik, 'Khad aur Khad Daalna'.
53. Another type of periodical that seem to serve as extension documents are the official periodicals like *Hal*, published by the United Provinces Government and *Kisan*, by the Bihar Krishi Sabha.
54. For instance, see Sankarshan, 'Kheti ka Pran aur Uski Raksha', *Vigyan*, vol. 1, no. 1, April 1915, pp. 23–5. A series of articles follow by the same author under the same title. The first few articles deal with *khad* (manures and fertilizers) which is then followed by a series on *bij* (seeds), *jal* (water), *kheti key auzar* (implements) and so on.
55. Badrinarayan Joshi, 'Angoor kee Kheti', *Vigyan*, vol. 7, no. 2, May 1918, pp. 60–2.
56. Murarilal Bhargav, 'Neel ki Kheti', *Vigyan*, vol. 7, no. 4, July 1918, pp. 157–9.
57. Dayashankar Dubey, 'Anaj kee Kami Dur Kaisey ho?' pp. 196–9.
58. Pathik, 'Zameen kee Paidavaar main Kami', *Vigyan*, vol. 5, no. 5, August 1917, pp. 205–6.
59. Ibid.
60. Note:
There is no doubt that without scientific agriculture the condition of the peasantry in this country will not change. Western countries have grown in wealth by virtue of scientific agriculture. Countries where even ordinary grains were produced with difficulty are today producing 40-50 maunds of wheat per bigha. Hence if scientific method of agriculture is extensively used in this fertile country it will eliminate all our destitution. But this can happen only when cultivators are not devoid of their basic rights, and government accepts their rights over the land they cultivate.

'Sarkar Aur Kisan', *Saraswati,* vol. 27, no. 2, February 1926, pp. 236–7.
61. For instance, see Gopal Damodar Tamaskar, '*Gram-Sudhar key Kuch Prasnna*', vol. 28, no. 1, pp. 48–55.

    Here, the author discusses various prerequisites, most essentially capital and knowledge, for the successful adoption and application of agricultural science and technology, and the responsibility of the government to facilitate the availability of such prerequisites. However, he complements all such requirements with the fundamental necessity of the tenants having rights over land, independent of the will of the zamindar or maalguzaar. The argument maintains that it is most essential for the cultivator to believe that the benefits of his investment would accrue to him.

62. For instance, reviewing the amendments in the rent laws and the court cases thereafter, an article in the *Saraswati* on January 1927 states:

    The zamindar or talukdar can increase the rent even by Re.1 per Rs.100, giving the reason of price rise, which in turn is leading to the cultivators becoming rich. Thus they (the zamindars) can demand that the peasants give one-thirds of the increased income.

    'Kanoon lagan aur Kanun Kashtkari key Muqadmein', *Saraswati,* vol. 28, no. 1, January 1927, pp. 140–3.

63. Ibid.
64. K.D. Malviya, 'Humari Sarkar aur Kheti Ka Lagan', *Maryada,* vol. 7, no. 6, April 1914, pp. 376–80.
65. For instance, see Prannath Vidyalankar's monograph, strongly arguing for ownership rights of the cultivator over the land he cultivates. He claims,

    There would be no benefit of education of agriculture etc. till the time the most fundamental element of agricultural development is present on land. The question thus arises, what is that fundamental element on which the development of agriculture as well as the welfare and happiness of the cultivators depend? The absolute rights of the peasantry over land and no rent or tax whatsoever is that fundamental element on which the wheels of agricultural development can move. To achieve this fundamental element, the decimation of zamindars and talluqdars is necessary. Even government should not claim any ownership rights over land.

    Prannath Vidyalankar, 'Kisano key Adhikaar: Maalguzaari tatha Lagan ka Lena Paap hain', *Maanmandir*, Banaras, May 1921.

66. Ibid.
67. See Champaram Mishra, 'Kheto ka Sangathan aur Ekikaran', *Saraswati*, vol. 21, no. 3, March 192, pp. 158–61.
68. Ibid. For instance, see the following quote by Champaram Mishra:

    If there are ten farms at ten different places, the farmer will need to dig ten wells. When the fact is that he does not have the capacity to build even one well. As a result such huge cost implications scare him away from taking the effort to build any wells whatsoever. He cannot connect the well to a pump or engine, as no one can afford that kind of expenditure for such a small piece of land.

69. Shastri, 'Bharat main Sahkar Andolan'.

70. Matadeen Shukla, 'Hindustan main Krishako kee Dasha', *Maryada*, vol. 15, no. 6, April 1918, pp. 268–74.
71. 'Akal aur Praja', *Maryada*, vol. 7, no. 6, April 1914, pp. 341–3.
72. 'Bharatiya Krishi', *Saraswati*, vol. 28, no. 2, February 1927, pp. 276–7.
73. 'Zamindar aur Kisan', *Saraswati*, vol. 32, no. 6, June 1931, pp. 786–8.
74. Ibid.
75. 'In the last 150 years of British rule not a single scientific principle has been properly used in the execution of agricultural steps. It would not be wrong to claim that there are no means to gain knowledge about scientific ways to do cultivation, in this country'. Matadeen Shukla, 'Hindustan main Krishako kee Dasha'.
76. Tamaskar, 'Gram-Sudhar key Kuch Prasnna', vol. 28, no. 2, pp. 204–9.
77. In 'Krishi kee Dasha', *Vishal Bharat*, vol. 8, no. 6, Poush 1988, December 1931, pp. 697–704, Shiv Kumar Sharma writes,

    Although government has established experimental farms, but they have not been helpful to the extent they were expected to be. The principal reason for this is that the farm officials just stay in their farms and maintain their crops. But the duty of the officials should be to assist the use of scientific implements in the villages, which will in turn attract the cultivators towards them (modern implements).

78. 'How to cultivate in a better fashion, how should the implements be, what is the best possible way to use them, which seeds are to be used, how to produce and conserve such seeds, what is the proper way to plough, how can crop rotation work best and how to save and develop livestock. These things should be taught to them.'
    'Krishako Key Unnati key Upaya', *Saraswati*, vol. 38, no. 6, June 1937, p. 331.
79. For instance, see Premnarayan Mathur discussing the various classes of problems that *gram-sudhar* plans to focus upon, in 'Gramoddhar ka Prashna', *Saraswati*, vol. 39, no. 2, February 1938, pp. 149–52.

5

Missionaries as Agricultural Pioneers: Protestant Missionaries and Agricultural Improvement in Twentieth-Century India

Rajsekhar Basu

The idea of scientific agriculture, as it had developed in the non-Western world, is often defined in terms of scientific innovations which had characterized the social world of Europe and North America in the second half of the nineteenth century. It has been argued that much before the onset of the consortium of scientist and agricultural bureaucrats of the agro-industry lobby and farmers' organizations in modern times, such trends had been clearly visible in the nineteenth century, being often expressed in terms of interests for new disciplines, like economic botany and plant industry.[1] It is believed that such intellectual currents gained popularity in British India and this was evident in the teaching curriculums of the Universities of Calcutta and Madras, which during the last years of colonial domination tried to integrate knowledge systems drawn in from different Western scientific disciplines, like entomology, pathology, plant genetics and soil chemistry. While the research institutes for cereal crops, like rice and wheat, had been set up before the Indian Independence in 1947, there is very little discussion on how the processes favouring the dissemination of scientific agriculture in India was closely linked to the missionary involvement with agricultural productivity and the introduction of better breeds of cattle and poultry in different rural localities.

The debate on agriculture had emerged in the mid-nineteenth century when a group of European middle-class intellectuals were impressed with the 'rural sights and rural sounds' of India. There was a growing idea that Indian rural life revolved around the growing of crops and keeping of cattle. The Indian peasantry was valorized for its simplicity and for its inclination for simple village life. In fact, the great famine of Bengal and Orissa in 1866 led to a great deal of concern, which was expressed in the ideas favouring the creation of an agricultural department. But the project did not materialize, since many felt that the problem of low productivity could be checked through a greater spending on irrigation. Nonetheless, the formation of a separate Department of Agriculture was given serious thought by Lord Mayo in 1869. The Manchester Cotton Supply Association in their representation to the Secretary of State, favoured the establishment of a Department of Agriculture in each province so that the district officers could be more actively involved with matters connected to cotton cultivation. At the same time, there was also a growing realization that the prospects of Indian agriculture in the future years would also depend on the experimentations in the sphere of agricultural chemistry, a discipline in which England itself was in a backward state. In the 1880s, the government finally decided to incorporate these ideas in the agricultural branch which was attached to the Dehra Dun Forest School.[2] But the real opportunity for agricultural progress was witnessed in 1889 when Dr Voelcker of the Royal Agricultural Society was sent by the Secretary of State to advise the government as to how agricultural chemistry held the key to improvement in Indian agriculture.[3] In later years, the Government of India realized the importance of an agricultural research institute, an experimental farm and agricultural college, which was revealed in the Pusa Plan of the early 1910s. The idea of attaching an agricultural college to the research station emerged from the necessities of Bengal. The Agricultural College at Shibpur had not been successful and this was responsible for the emergence of Darbhanga in Bihar as one of the most important provincial agricultural research centres in India. By the early 1920s, the debate on agricultural improvement led to demands for the creation of an all-India agricultural organization to initiate a greater deal of connectivity between the imperial and the local governments.[4] At the same time, the Agricultural Adviser to the Government of India in his memorandum to the Royal Commission of Agriculture also emphasized that an imperial conference delving on the coordination of agricultural research within the British Empire could be held in London and this would definitely lead to recommendations applicable to agricultural conditions in India. The agricultural missionaries were successful in adopting many of these ideas in their programme of rural uplift and by the 1930s, they had edged out the government in terms of experimentation in the fields of agricultural research and in veterinary sciences.

In this chapter, I would like to argue that by the end of the nineteenth century, Christian overseas missions supported a shift from an all-out emphasis on proselytization to social service, which was directly linked to the human development. The expression of this change in the understanding of the Director Boards of the missions was revealed in their inclinations to establish agricultural missions to improve conditions of the rural population in different parts of Asia, Africa and Latin America.[5] While it has sometimes been argued that this shift could be explained in terms of a regenerative antidote to the spiritual impoverishment of a new industrial age, this really does not qualify for the explication of missionary involvement with agriculture and cattle breeding and the overall development of the rural population in British India. It could be argued that the missionary denominations which operated in British India witnessed an unprecedented wave of Christian mass conversion movements, mostly from the agricultural labouring groups which were despised in terms of their low-caste backgrounds and whose touch was considered as a major reason for defilement by the upper-caste Hindus. It was logical that these new Christian communities had to be provided with some degree of economic autonomy and education. In this situation, where the missionary agenda for social service was facing mounting competition from both government and non-government socio-political organizations, the missionaries decided to set up Christian villages, which could provide social security to their new converts as well as solve their problems related to poverty.[6] However, the missionary involvement with the rural world of India went far beyond the experiment in setting up Christian villages, with emphasis on vocational and industrial training. By the turn of the nineteenth century, agricultural missions were established in different parts of India to address the problems of endemic rural poverty. In fact, most of these missionary involvements remained largely European- or American-led programmes and there were very few instances where one could locate the independent voice of the native agencies.[7] Despite these criticisms, it needs to be stated that the broader social concerns on the part of the missionaries, which included education, provision of medical expertise and introduction of advanced technologies in the rural world, does bring out a narrative on how alien agencies, knowingly or inadvertently, helped to sustain as well as undo the fabric of exploitation which had been inherent within the policies of the colonial state. The missionaries did override their paternalistic attitudes towards their new converts and were usually appreciative of the specific needs of the indigent communities. This might possibly explain their intentions to identify both short- and long-term solutions to issues related to education, health, agricultural improvement and rural industries.[8]

The Beginnings

In the mission village schools, which had been established in different parts of India, more particularly after the outbreak of the devastating famine of 1860s and 1870s, there were some efforts to introduce basic agricultural training among the new converts, who were mostly drawn from the 'untouchable' communities. Despite these efforts, there had hardly been initiatives towards intensive specialized projects before the late nineteenth century. Prior to the missionary initiatives in improving the conditions of agriculture in India, there had been instances of few trained agriculturists being sent from Europe and the United States to Asia and Africa. The agricultural missionaries were fairly active in China, where an American missionary George Weissman Groff added an Agricultural Department to the Canton Christian College. In the beginning, most of these overseas agricultural missions were given leadership by the missionaries coming from the United States, which were far more advanced than Europe in matters related to agricultural education.[9]

The beginning of the agricultural missions in India is often connected to the establishment of the Allahabad Agricultural Institute in 1910 by Sam Higginbottom, an American Presbyterian missionary. Despite the opposition from the evangelical groups, the agricultural missions did receive some encouragement from the Church leaders. However, the fact remains that compared to the early decades of the twentieth century, when agricultural missions were favoured strongly by the mission boards, the situation in the late nineteenth century had been a lot different in India. Yet, some of the American missionaries who were actively involved with agricultural projects, from the early part of the twentieth century, introduced several institutions, ideas and techniques which influenced the patterns of agricultural development at the regional levels.[10]

In India, the agricultural missions from the very beginning were influenced by the logic of eradicating rural poverty.[11] The missionaries often tried to explain the incidence of poverty and malnutrition in the context of the exploitative caste system which had prevailed in India for centuries. The missionaries seemed to have been aware of the fact that the hierarchical social system had been responsible for their success in gaining 50 million converts, who were mostly categorized as 'untouchables'. Thus, it was quite logical on their part to come up with the assumption that Protestant agricultural missions had to undertake programmes which were sometimes in direct response to the daily needs of the converts who came from the lowest rungs of society.[12] Apart from the institution of caste, which acquired a central position in their argument, the social conditions in India were linked to wider economic issues centring around the organization of Indian villages, alongside issues of income disparity, agricultural productivity connected to the growing commercialization of agriculture, low crop yields, fragmentation of lands, problems related to animal husbandry and the growing incidence of rural indebtedness.[13]

Protestant agricultural missions began in India with the sole support of the funds provided by the overseas boards. The agricultural missionaries received very little support from the government in the late nineteenth and the early years of the twentieth centuries. It has to be understood that the government had its own involvement with agricultural research and development and was not willing to collaborate with any other non-governmental organization in this respect. The governmental interest in agricultural work basically remained centred around three sets of activities, notably research, education and demonstration. In each province and in many of the large native states, there was a Director of Agriculture who had been given the responsibility of undertaking steps for agricultural improvement in the area which was placed under their jurisdiction. The Director was assisted by specialists in genetics, plant pathology and entomology and in addition, there was the principal of the state agricultural college.[14] Depending on the size of the province, there were four to eight Deputy Directors who were entrusted with the work of demonstration. In the case of Madras Presidency, it was observed:

The Madras Presidency is divided into eight circles, and the Department (of Agriculture) seeks by research, propaganda, education, and financial assistances in every possible manner to improve conditions of village life. In this field of activity is found cooperative development which strengthens the cooperatives spirit; pumping and boring by the loan of boring plants; the distribution of improved agricultural implements which are kept in stock in various depots; publishing the best times for sowing or planting and information about the care of crops. . . .[15]

In fact, from the very beginning it was clear that the contribution of the agricultural missionaries in India would be foremost in the field of agricultural education. The missionaries emphasized on the need to acquaint the Indian agriculturists with Western advances in agricultural sciences. They were also interested in undertaking research and experimentation to determine the real value of the indigenous methods of cultivation and the strengths and weaknesses in the raising of native livestock and cultivating native crops. The missionaries started conducting experiments to improve the breed of Indian poultry and goats, upon which the Indian rural families depended heavily for their daily requirements of meat and milk. While some of the specialized vocational schools were opened by them for such purposes, a greater deal of work was done by the regular and normal schools where the emphasis was placed on agricultural and industrial trainees. Nonetheless, vocational schools were described in missionary documents as institutions which provided the greatest hope for the future of agricultural mission work.[16]

However, by the early 1930s, agricultural missionaries were involved with twenty-eight projects in different parts of the subcontinent. These projects varied in size and scope from agricultural colleges to lone efforts in a few villages. In the meeting of the International Missionary Council held at this time in Madras, it was observed:

Christian love cannot be indifferent to economic suffering either within or without the household of faith. . . . The comprehensive (rural) program includes better agriculture, better health, better recreation, better homes, better economic organization, the widening of the intellectual horizons, the enrichment of rural life through drama, music and the other forms of arts, the development of community spirit, as well as vitally important work of Christian preaching and teaching, and guidance in worship, fellowship and service.[17]

Rural Concerns and the Allahabad Experiment

The involvement of agricultural missionaries in rural localities of India is often equated with the successes of Sam Higginbottom and his colleagues at the Allahabad Agricultural Institute in bringing about rapid transformation in the agricultural educational curriculum. The beginning of the institute was quite unpretentious when one considers its importance in the later years. The Allahabad Agricultural Institute was established in 1911 as an experimental firm linked to the Allahabad Christian College. Sam Higginbottom with the help of some American and Indian staff members was able to gradually set it up as an autonomous institute. Higginbottom, before setting up the institute, had made an in-depth study of agricultural productivity in India and the weaknesses which posed obstacles to its higher rate of growth. While he appreciated the role of the Agricultural Department of British Government to increase productivity, he was quite vocal in his criticisms of the cultural background of the Indian agriculturists. In fact, this was evident in one of his early observations, wherein he had stated:

it takes so long for a foreigner trained in agriculture to get acquainted with the Indian conditions; the ignorance, the suspicion, the illiteracy and superstition of the Indian farmers so wide spread, that progress is necessarily slow. The illiterate Indian farmer has for centuries been fair game for anyone to exploit. It is difficult for him to believe that anyone is really trying to help him. When any improvement is being introduced he always imagines that some new trick is being played upon him. The government is establishing rural, middle and high agricultural schools but is compelled to go slowly because of the dearth of properly qualified teachers with the right attitude towards the villager. It is at this particular point that America can be of the greatest service to India.[18]

He had tried to compare the situation in India with the conditions prevailing in the American South where, the Afro-Americans and the poor whites lived in a poor economic environment, because there was no initiative to teach them modern farming techniques. It was believed that the situation could be improved if a few missionary institutions like Hampton or Tuskegee, staffed with trained Americans, were sent to India.[19] These institutions could train better some of the Indians in methods for earning larger profits than they usually received per acre. Others could be trained as demonstrators to undertake trips to the debt-laden, hopeless and despondent Indian agriculturists for disseminating the right kind of advice. Higginbottom believed that the demonstrators could be ideal teachers for the rural schools and they could encourage the rural classes to take to education.[20] Higginbottom admitted that though American missions had no business in India and that no legal claim could be made on the American Christians to send help to India, American missionary involvement was important in making the Indians realize the fullest measure of human freedom and also the training that had not been imparted to them.[21]

Higginbottom made a point-wise presentation as to why the missionaries had to emphasize on agriculture. He argued that agriculture was the main occupation in India and was likely to remain so in the future, because of the climate and the long growing season. Consequently, it was believed that improved agriculture was to be in the line of least resistance from a society bound by caste and that it was in consonance with the greatest wisdom. It was observed:

It is the simplest and most direct way to give India enough to eat and prevent famine. In the opinion of Higginbottom, improved agriculture taught the low caste convert to grow surplus food, which could be sold for meeting their daily monetary requirements. In other words, it helped them to learn the efforts through which they could support themselves and their families. But, it was asserted the fact that very few of the lower castes possessed land possibly could be used as major argument against mission agricultural training.[22]

Higginbottom argued that even illiterate lower caste non-Christians, who had worked on the mission farms for two or three years and had learnt the use of iron ploughs, harrows, rollers, seeding, mowing and threshing machinery and silage cutters, were in great demand at wages two and a half times as great as the average labour wage. In other words, it was stated that there would be no difficulty in getting keen labourers who wished to improve their own conditions by getting practical training from the missionaries which would entitle them to higher wages elsewhere.

It was strongly argued that what India needed was more technical and industrial education, all of which could have direct influence on the agricultural sector. The development of those industries was favoured which were related to and subsidiary to agriculture, viz., the making and repairing of modern farm implements and machinery; modern dairying; canning, preserving, and drying of fruits; sugar making; oil pressing; tanning and rope making. Higginbottom believed that the factor limiting the introduction of modern and efficient labour saving farm machinery into India was not money, but lack of men trained in the use of modern tools and in repairing them.[23] Indeed, it was felt that the aversion to technology was related to the weaknesses of rural and primary education in India. This form of educational instruction was criticized because it did not give a greater deal of emphasis on vocational or 'dollar' education. It was argued that if agriculturists could be trained, there would be an obvious interest in using better seeds, methods and implements. Furthermore, this would have a wider societal impact because the efforts of these agriculturists would be imitated by others, leading to a great deal of increase in the yield of crops and upgrade in the overall standard of living. Thus, technical training was seem as the quickest method which could reach the whole of India 'helpfully, naturally and economically'.[24]

The Allahabad Agricultural Institute was built on the patterns of the Rothampstead agricultural experiments stations in England. In Rothampstead, on one plot of land, for over seventy years, wheat was grown continually year after year, without using manure. The average yield per acre of wheat for the whole of India was less than the unmanured plot in Rothampstead. The missionary initiatives in Allahabad gained strength when Higginbottom, after graduating with a B.Sc. in agriculture, returned to India with a donation of $30,000. The donation had been collected by him from his friends to start an agricultural college. Thereafter, 275 acres of land were purchased for about $11,000. In order to secure this land for an agricultural college, the missionaries appealed to the government to put the land acquisition act in force. The land selected for the farm was rough and was very badly eroded and cut up into gullies. There were also a great many small, irregular-shaped fields. The land had been formed by the deposit carried by the river in flood time and it contained every kind of soil found in northern and central India. The Indian agriculturists could not plough this land except under the most favourable conditions. Higginbottom had reasons behind the choice of this land:

If I had chosen a good piece of rich, level land, irrigated from the canal, the Indian farmer would have said that anybody could farm and get a living on good and like that. I chose this poor land, eroded and full of pest plants difficult to eradicate, in order to show that the millions of acres of such land in Northern India could be redeemed and made profitable. Another reason for choosing this land was its location, so near the college and the city. Allahabad is the capital of the United Provinces which has a population of about fifty millions. In some time or other the leaders of the provinces come to the capital city. The farm being on the river bank, overlooked by the railway, and having two of the main roads into Allahabad to pass by it, is in a commanding position for a demonstration farm. Being so near the city provided a market for the dairy products and surplus vegetables. Furthermore, during the Hindu moth, Magh, from the middle of our January to the middle of February, Allahabad is the greatest pilgrim center on earth. On some of the big days of the

Mela, crowds of from two to four million pilgrims gather to bathe in the sacred waters of the two rivers which are seen, the Ganges and Jumma, and the river Saraswati, the river that can only be seen by the eye of faith, that is said to flow underground for hundreds of miles and joins the sacred Ganges at this hallowed spot. Where these three sacred waters unite great benefit is supposed to accrue to the one bathing under the right auspices during this month, Magh. Hundreds of thousands of these pilgrims each year walk past the Mission farm. Many stop to see our improved tools and implements, our sleek, well-fed cattle, our silos and sanitary barns. They carry the tidings to the most remote parts of the Indian Empire. We get many inquiries about the purchase of machinery from faraway places where these pilgrims have told of what they have seen.[25]

The most interesting initiative on part of the missionaries following the establishment on the Allahabad Agricultural Institute was to explore the possibilities of establishing a broadcasting station which would instruct the agriculturists in improved methods of cultivation, cattle production and poultry. The radio programmes were planned to meet the requirements of the villagers and in addition to agricultural dairying and health subjects, there were entertainment programmes of various sorts, including music. To make these broadcasts popular among the rural population, experiments were undertaken to produce a simple and inexpensive receiving set.[26]

However, the important projects were related mostly to the search for the scientific basis of food production and cattle breeding. The missionaries cooperated with the Department of Horticulture of the United Provinces to initiate research on the production of better quality of fruits. At the same time, there were collaborations with the Biology Department of Allahabad University to study the life history of beetles and moths which caused a lot of damage to wheat and flour. The Allahabad Agricultural Institute also worked with the Department of Animal Husbandry, United Provinces, to produce a fine herd of dairy buffalo and in improving the breeds of bees, goats, sheep and poultry. It also collaborated with the Agricultural Engineering Department of the Allahabad University to bring about a technological improvement in the methods of cultivation. Subsequently, there were also collaborations with the Agronomy Department of the Allahabad University to prepare reports on farm cost accounting and in bringing out the journal, *The Allahabad Farmer*. Interestingly, the missionaries also introduced short courses on fruit preservation for special students and on the rural aspects of church life for students of Theological College. The institute also set up rural extension centres to deal with the demands of the rural population.[27]

The missionaries also undertook projects to produce a herd of pure-bred Indian cattle and imported four strains, Jersey, Guernsey, Brown Swiss and Holstein, to produce a suitable cross for India.[28] Such experiments led to considerable increase in milk production and the major portion of it in the form of butter was transported to Calcutta. The remaining portion was sold in the local market for meeting the demands of the agricultural classes.[29]

The agricultural institute at Allahabad also made a valuable contribution by producing farm implements. The most noteworthy of them was the light plough which was suitable for the weaker cattle upon which the average farmers were forced to rely as beasts of burden in cultivation. In fact, this plough had been estimated to have tripled the productive capacity for its users as compared to the results which were obtained with the outmoded wooden plough which was in general use in India.[30] However, the most outstanding contribution of the institute had been the training programme, which it conducted for the agriculturists who wanted to be the enterprising section in the rural society. So considerable was the success in this field that only a small percentage of those seeking registration could be accommodated by the institute. Training was also given to women, particularly from 1936, when a new course in Home Economics was added to the curriculum.[31]

The Allahabad Institute in the early decades of the twentieth century also benefited from its connections with the princely state of Gwalior. Higginbottom, to raise funds for the institute, had himself worked as Director of Agriculture in the state of Gwalior from 1916 to 1919. Gwalior, which was among the largest of the native states, was similar in geographical size to the state of West Virginia in the United States and its Maharaja, Madhav Rao, provided resources for Higginbottom's initiatives in development. A research laboratory was established in the capital and experimental farms were set up in Ujjain. Model farming was also established in the districts. Books disseminating information related to improved methods of cultivation were translated into Hindi and distributed among the landowners. The development programme in Gwalior made the students and faculty of the institute conversant with the ground realities that prevailed in rural India.[32]

The Gwalior project did reveal some of the weaknesses of Higginbottom and his colleagues in introducing implements which were unsuited to the Indian conditions. For instance, the Scindia plough which had been developed by Donald Griffin of the Allahabad faculty and had been named after the royal family of Gwalior, though superior in many respects to the Indian plough, suffered from several defects. There were problems due to clogging under the beam and the grass trapped between the bar point and the share plate. This made it unacceptable to the rural masses.[33]

Nonetheless, the Gwalior programme of Higginbottom paved the way for collaboration between agricultural missionaries, foreign agricultural specialists and the British government in India. But in the 1920s, people like Frank Brayne became involved with development programmes in the Gurgaon district of Punjab. However, despite receiving the help and support of the government, these development programmes which heavily relied on enterprising individuals, failed to reach the desired levels. Interestingly, these failures did not prevent the Congress ministry in the United Provinces to imitate Higginbottom's approach towards rural uplift, though such initiatives remained short-lived because of the demands of the War and the decision of the national leadership to withdraw from the ministries.[34]

The Allahabad faculty was also known for introducing several fields of specialized study. Brewster Hayes, a Californian who had graduated from the Oregon State University, went to Allahabad in 1921 and became one of the first horticulturists to work in India. In his thirty-six years stint at Allahabad, there was extensive involvement with research and a standard text in the field was written by him. Hayes received a great deal of support from his two other colleagues—James Warner who specialized in dairying, and Mason Vaugh who specialized in agricultural engineering. Along with Higginbottom, Vaugh tried to persuade the educational institutions to recognize agricultural engineering as a valid academic discipline for which degree examinations could be held. Finally in 1942, the University of Allahabad decided to offer a degree in agricultural engineering, which was the first of its kind in South Asia.[35]

The involvement of these American specialists proved to be significant in more ways than one for the Allahabad Institute. The introduction of the steel plough instead of the wooden plough proved to be of much benefit to the agriculturists. First, it could be easily drawn by the farm animals, because its weight was comparatively light. Second, due to the fragmentation of the landholdings, the plough could be carried from one field to the other without much difficulty by the agriculturists. The steel plough gained its popularity from the use of the Urdu word *wah wah* and proved ideal for farming conditions in northern India. This plough, along with the *shabash* plough, enabled the agriculturists to till one acre of land in a single day. The demand for these ploughing instruments inspired Vaugh to open the Agricultural Development Society, a privately supported company which was integrated to the Allahabad Institute.[36]

Other Interventions: Small but Significant

The involvement of the missionaries with the agricultural sector was also seen in several other parts of India. The Allahabad experiment was reciprocated in the activities of the missionaries in the small Punjab town of Moga where, a comprehensive programme of village education and teacher training was devised. In the early years of the twentieth century, Ray Carter, a Presbyterian missionary, established a community middle school. After a few years, the training school for village teachers was established which served as the laboratory for rural projects. Interestingly, the dignity of labour was stressed from the very beginning and pupils were urged to participate in the construction of the school building. The Moga schools tried to make the villages more self-sufficient. They tried to initiate educational programmes with a definite 'rural bias', laying a great deal of emphasis on productive work in agriculture, spinning, weaving and all types of small industries. The educational programmes owed a lot to the three American faculties, William McKee, Arthur Harper and Irene Harper, who from the very beginning had been familiar with the progressive educational ideals of John Dewey, which were gaining popularity in the early twentieth-century United States.[37]

Interestingly, projects of nearly every imaginable type, ranging from experimental poultry to the study of the intricacies of banking, were initiated. The school children were given training in first aid and in the prevention and cure of common diseases. The children were also instructed to make clay models and were also given lessons in carpentry, needle work and arts. Such experimentations had new set of implications for the rural society in Punjab. The missionaries insisted that for improved cultivation, the consolidation of landholdings should be given top priority. Subsequently, the area under cultivation increased alongside yield per acre of land. The missionaries reported that the enlarged farms were worthy of irrigation and that there were less instances of litigations.

In the 1920s, the missionaries in Punjab formed agricultural societies whose activities were scattered over 1,500 villages. In fact, these societies reclaimed 20,000 acres of land and purchased an additional 13,700 acres for resettling their new converts, who were mostly drawn from the agricultural labouring classes. It was reported that improved agricultural methods and scientific innovations in dairying and poultry had led to profits over Rs.40,00,000 and one-third of the members of the societies had been redeemed from the debt contract.[38]

The early 1910s also witnessed missionary involvement with development work in western India. John Goheen, a graduate of Wooster College was sent by the Presbyterian Church to administer an industrial training school in Sangli which had been started a few years earlier. He quickly introduced an agricultural programme which was directed towards meeting the demands of the farmers in the locality. Since the training did not lead to a degree, the students were ineligible for employment in the government Agricultural Department.[39] Nonetheless, the training included a variety of occupational disciplines like smithery, masonry, carpentry, sewing, printing, auto-mechanics and agriculture. The Sangli graduates in most cases went back to their villages and tried to spread the message of Christ by setting up Christian homes, shops and farms.

The most outstanding feature of the work initiated by John Goheen at Sangli had been the success in raising the farmer's income through scientific experiments with poultry. His aim had always been to deploy methods within the scope of the average farmer's income. To popularize his ideas, John Goheen established the 'moveable' school which consisted of a truck carrying equipment which could produce electricity for lighting purposes and for the operation of a movie projector for showing samples of superior seeds, grains, cattle and poultry produce which were known as Christian calves and fowls. In fact, these demonstrations were largely made in the country market to attract the attention of the rural population.[40]

In eastern India, the Methodist missionaries, in order to involve themselves with the rural communities, became involved in the establishment of 'The Village of the Dawn' at Ushagram in Bengal. The model village was created by using modest materials and student labour. Sixty boys lived in Ushagram's one-room cottages enjoying few luxuries other than extra ventilation supplied by four large windows. Each of these cottages situated on a 60 ft. land had space for a flower and vegetable garden. In fact, it was from their gardens and private poultry that the demands of the village were fulfilled.

In this Indian counterpart of the American Boys Town, the youthful citizens were given instructions to operate their own savings banks, their private stores and a simple but adequate hospital. The experiment in Ushagram was successful because people from many of the neighbouring villages preferred to buy supplies from there. Thus, a market was found for much of the produce of Ushagram. In this model village school, the boys who were given education, that revealed a Christian influence, were made to understand the meanings of cooperation and gained valuable experience in village finance.[41]

In south India, the missionary involvement with agriculture was witnessed in both Katpadi in the erstwhile Madras Presidency and Martandapuram in the native state of Travancore. In the agricultural institute which was established in Katpadi, the missionaries were involved in experimentation with all varieties of stock and crops which were suited to the geography and climate of south India. The Katpadi experiment in producing better quality fowl was so successful that its assistance was sought by farmers in Burma, Ceylon, Arabia and Siam. It also received government recognition and this resulted in an agreement between the agricultural institute and the government for improvement of poultry stock in the neighbouring villages.

Further work at Katpadi included the introduction of breeding bulls in many villages. A missionary J.J. DeValois reported in the 1930, that in the past two decades these work had made great progress in the rural localities. The system of supplying bulls, which was defined as the 'Premia System', functioned under the government provision of earmarking $30 a year for each bull used. Such a policy helped to solve the problem of getting custodians for the animals. Out of the custodian's charge to the cow owner and the government grant, the bulls were properly fed and cared for. This whole project proved to be quite successful.[42]

The missionaries were also concerned with a project which was named as the 'Poor Man's Cow'. Under this project, the missionaries had supplied goats in many localities, which could provide cheap milk that was a necessary item in the diet of the villagers. The Government of Madras, recognizing the importance of this project, also provided a great deal of assistance to the missionaries.[43] Indeed, such services added to the training programme for the Katpadi students, made them realize the value of rural reconstruction work. In fact, such missionary involvement was replicated in small projects which were taken up by the missionaries in the Kistna district and Ramanathapuram in the Madras Presidency. The Vidyanagar Boy's School offered vocational training to local Mala and Madiga school students who had passed third standard in the village schools. The missionaries placed a great deal of emphasis on leather working and weaving, apart from training in the fields of farming, gardening, poultry raising and cattle farming.[44] The YMCA centre at Ramanathapuram imparted training to Christian converts who had been won over through mass movements. Here, the aim was the removal of illiteracy and the missionaries, while undertaking programmes for meeting this end, realized the importance of adding visual education, schemes for the improvement of agriculture, economic relief and revival of ancient rural institutions. All these were done to help the converts to resolve their problems locally.[45]

The most serious missionary involvement with rural development in south India was seen at Martandapuram in the native state of Travancore. The native state, in the last years of the nineteenth century, initiated a series of programmes to publicize the importance of scientific agriculture. The ruling classes introduced a programme for the teaching of elements of agriculture in primary school and also set up a Department of Agriculture, to bring about improvements in agriculture, cattle breeding and fisheries.[46] Missionary involvement with agricultural improvement in Travancore was influenced, to a great extent, by the government's efforts to introduce better breed of cattle and to employ veterinary *vaidyas* in the agriculturally prosperous territories.[47] All this definitely inspired D. Spencer Hatch who had been working under the Young Men's Christian Association (YMCA) in 1926, to establish a centre of rural reconstruction in Travancore. It provided training to extension personnel based on principles like those later used in the Government of India's programme for village workers. Hatch believed that demonstration was perhaps the best method for encouraging the villagers to make innovations and it received greater attention when it was presented in terms of self-help.[48]

Hatch and his staff trained hundreds of teenagers at Martandapuram, many of whom had been sponsored by private and government agencies. In fact, these students, after spending sometime in the villages, were sent for specialized training to Allahabad, Sangli, Moga and Etha. The Martandapuram graduates also worked at demonstration centres in different parts of southern India. They provided training for self-sustaining projects related to weaving, beekeeping and poultry raising. Hatch also organized programmes in the rural localities for disseminating ideas on cooperative credit, production and marketing facilities. The extension staff encouraged the villagers to improve economically and also in matters such as recreation, sanitation and citizenship. However, the scale of the operations faced frequent difficulties in view of financial stringencies and change of personnel. Yet, Hatch's programme of social education proved to be quite successful, since it helped the elderly villagers to resolve both their social and economic problems.[49]

Arthur H. Slater, a Presbyterian missionary initiated a much bigger project on rural development at Etah in northern India. Slater displayed the possibility of using poultry farming as a means of rural uplift. Etah had been at the centre of a mass movement, because of the conversion of 10,000 lower caste Hindus to Christianity. The new converts who had been scattered over 600 villages virtually, did not own land and had very little capital. In 1912, Slater who had graduated from the Ontario Agricultural College had begun his work at Etah amongst the Christian population. He felt that the development of the poultry industry would be the most viable means for securing the livelihood of the converts, since it required little land and capital. Moreover, the majority of the converts had been members of caste groups which had been associated with rearing of poultry stock.[50] However, there were problems related to the improvement of the quality of poultry. In 1912, the average weight of country chickens in their adulthood was 3lb. and they were sold at 16¢ in the market. Eggs weighed 1oz. and were sold for 6¢ a dozen. Slater tried to improve the quality of the stock by introducing the pure-bred fowls. He began his work with a gift of Black Minorca, Rhode Island Red, and White Leghorn breeds from the Bible class in Coatesville, Pennsylvania. In fact, within a few years, five breeding centres were set up at Etah for increasing the stock of the pure-bred fowls. Initially, when Slater distributed the eggs to the villagers, they were viewed as gifts and their economic importance was realized by them. But, when he started charging 1¢ per egg, the villagers took them and started hatching them for profits.[51] Indeed over a six-year period between 1912 and 1918, eggs from cross-bred hens averaging in weight 2oz. were sold for 18¢ a dozen. The weight of adult fowls had also doubled and their market price had reached $1 a bird. In 1915, Slater organized the first annual poultry show at Etah.[52] Subsequently, these shows became more frequent and were held about six times annually. In these shows, the entries numbered between 300 and 700. However, the grand show was held at Etah and the number of entries were as high as 2,000. These occasions were also utilized for both religious and secular purposes. The missionaries were not only involved with evangelism, there also seemed to be opportunities for them to discuss with government officials topics, viz., cooperatives, agro-industries and health and sanitation issues.[53]

By the 1920s, the poultry project had significantly improved the economic conditions of the region. In fact, the United Provinces Poultry Association reported to the Royal Commission on Agriculture that poultry farming had benefited more than 15,000 people. It has been pointed out that between 1931 and 1956, the poultry shows in Etah accounted for $55,000 for the villagers. Undoubtedly, this rural programme attracted the attention of the government officials and it encouraged the improvement progammes undertaken by the government in different parts of the province.

However, the poultry work at Etah also coincided with the project seeking to experiment with improved breeds of goats. Etah became India's first goat breeding and research centre with financial support of the government. The involvement of the government in this programme became evident in 1931, when the Imperial Council of Agricultural Research granted $11,633 to the missionaries for running a five-year research project. Subsequently, a splendidly equipped goat firm was established at Etah.[54] In 1936, it received another grant of $12,363 to carry out another five-year project.[55] Slater believed that research on goats was more practical, since the villagers would always find it difficult to obtain sufficient fodder for the cows. He favoured the gifts of pure-bred goats from America, despite the problems of eradication of the disease among the imported animals. Slater concentrated on improving the indigenous Jamnapari and Bar-bari strains. The Etah goats produced quality milk which was about three times the volume of the average goat. Indeed, this experiment was taken up by the government in the post-independence year, when improving the quality of goats became a fundamental part of the government's rural improvement programmes.[56]

By the 1930s, it was becoming increasingly visible that Christian missionary organizations were inclined in favour of rural development programmes. In fact, despite the Depression, which had an impact on missionary work in the 1930s, agricultural missionaries continued to be an important part of the missionary-sponsored welfare activities. Their service had two very important aspects. On the one hand, they were involved with institutions which tried to train agricultural leaders capable of grappling with the rural problems of the nation; on the other, there were also efforts to lift

Indian villages to higher standards in both economic and social terms. The Allahabad Agricultural Institute was an outstanding example of the first type of work and the model village of Ushagram, along with the outstanding contribution of community middle school of Moga, was an example of the second type of service.[57]

Conclusion

The contribution of the agricultural missions in India could be analysed from the point of view of a diverse range of activities which would include agricultural education, crop productivity, cattle improvement, cooperatives, village-based agro-industries and land reclamation. The agricultural education which was offered by the missionaries' schools, introduced syllabi which was better suited for the training of India's agricultural leaders. Indeed, these schools at the village level became the agencies through which the colonial government's rural uplift programme reached the villagers. At the same time, many of the ideas of the agricultural missionaries were adopted by the government officials while they were engaged with experimentation in cattle breeding. By the 1930s, rural reconstruction programme had definitely moved beyond the limits of the village schools and had become intricately linked to the training programmes which had been initiated by the agricultural colleges. Undoubtedly, the influence of the Allahabad Agricultural College was the foremost and many of its graduates had been entrusted with the responsibilities of discharging the duties of secretaries in the different rural cooperative credit societies which had been set up by the YMCA. Some of them had also been sent overseas to acquire training and on their return, they established agricultural schools of their own. However, there were Christian secondary schools which also were a part of the rural programme of the missionaries. The Anklesvar School produced students who went back to their own villages as teachers, labourers, artisans and tradesmen. Many parts of India were able to introduce vocational training as an important part of the rural uplift programmes. But, the most important element of agricultural missionary activity was the employability factor which was emphasized in the institutions run by them. These institutions, through education and training, produced a large number of field workers, who could be employed by the government as demonstrators. The other aspect which needs to be emphasized is the advice on the part of the agricultural missionaries to the Government of India to introduce demonstration trains in rural India. Indeed, these trains, made up of five or six cars, proved to be successful in imparting new sets of ideas to the rural masses on agriculture industry, cooperatives, public health, veterinary and several other issues. Lastly, the rural programme of the missionaries brought about a new awareness about banking particularly in the Christian mission settlements. By their active participation in the cooperative programmes which had been initiated by the governments in the provinces, the missionaries tried to integrate the principles of economics and ethics, so that the visitors could distinguish the cooperatives from the charitable institutions. However, the tenancy systems in Central and United Provinces proved to be a great obstacle to the missionaries. The missionary involvement with the cooperative programmes brought out the fact that this experiment would only be successful if the majority of the agricultural classes were owner-cultivators. Missionary involvement with rural uplift also did establish the fact that the fate of independent India would depend on village-based industries, rather than modern factories and in this sense, was sympathetic to the Gandhian vision of development.

Notes and References

1. Robert S. Anderson, 'Cultivating Science as Cultural Policy: A Contrast of Agricultural and Nuclear Science in India', *Pacific Affairs*, vol. 56, no. 1, Spring 1983, p. 40.
2. James MacKenna, *Agriculture in India*, Calcutta, 1915, pp. 2–3.
3. Ibid., p. 4.
4. In fact, from the mid-1910s, business associations, like the Indian Jute Mill Association, were offering suggestions that the government should involve itself with the selection and propagation of desirable varieties of seeds so that there could be better production of jute and cotton. It was also emphasized that the agricultural departments at both the imperial and provincial levels should have a better degree of understanding so that the benefits of agricultural policies would reach the small peasants and the agricultural labourers. For more details, see Revenue Agriculture, File nos. 1–8, A Proceedings, 1916, National Archives of India, New Delhi (hereafter, NAI); Foreign and Political, General, File no. 580-G/25, 1925, NAI.
5. Gary R. Hess, 'American Agricultural Missionaries and Efforts at Economic Improvement in India', *Agricultural History*, vol. 42, no. 1, January 1968, p. 23.
6. In fact, as early as 1858, the Christian missionaries were expressing the idea that there had to be definite efforts towards the improvement of the human and moral standards of their new converts. It was stated that severe famines in south India often forced the new converts to live on wild roots and berries and had also driven some of them to seek subsistence in some occupations, which had a depressing influence on their character and social mannerisms. In other words, it was stressed that in order to prevent these evil influences from casting a permanent shadow on them, it would be necessary to think of programmes and schemes which could provide economic security to them. For more details, see

Twenty-Fourth Report of the Annual Madura Mission for 1858, Madras, 1859, p. 6.
7. For more details, see Christopher Harding, 'The Christian Village Experiment in Punjab: School and Religious Re-formation', *South Asia: Journal of South Asia Studies*, n.s., vol. XXXI, no. 3, December 2008, p. 400.
8. I am indebted to the researches of Paul C. Byam for developing this idea in the context of India.
For more details, see Paul C. Byam, 'New Wine in a Very Old Bottle: Canadian Protestant Missionaries as Facilitators of Development in Central Angola, 1886–1991', Ph.D. dissertation, School of Graduate Study and Research, University of Ottawa, 1997, pp. 1–2.
9. Hess, 'American Agricultural Missionaries', p. 23.
10. Ibid., p. 24. Historians have pointed out that the agricultural missionaries in India were pioneers in the following ways: establishing the most distinctive agricultural college of the country, initiating the first large-scale programmes of rural development, introducing specialized fields, such as agricultural engineering, undertaking a series of scholarly publications on Indian village life, organizing training programmes for rural teachers and establishing local rehabilitation projects. Indeed many of these initiatives were heavily influenced by the thoughts of individual pioneers and the successes of the programmes depended on how and where they were implemented and how much the influence of the original ideas was adhered to by the missionaries and their native assistants at the ground level.
11. A missionary in the early years of the twentieth century had observed, 'To the stranger on his first visit, India is in many ways both a baffling and depressing land. It is not merely that poverty, squalor, physical malformation and malnutrition exceed these evils elsewhere. Rather the Sheer immensity of India and the complexity of its problems—political, cultural, economic, religious—in their intricate interrelationship seem too great for human management'. Henry P. Van Dusen, *For the Healing of the Nations*, New York, 1940, p. 103, cited in John Roland Same, 'The Program of the Rural Mission in India', Master of Religious Education thesis, Drakes University, Des Moines, Iowa, May 1952, p. 1.
12. Samuel B. Finlay, 'Protestant Agricultural Missions in India', Bachelor of Divinity thesis, Faculty of Theology, McMaster University, 1945, p. 1.
13. Ibid.
14. Ibid., pp. 14–15.
15. G.J. Lapp, *The Christian Church and Rural India*, YMCA Press, Calcutta, 1938, pp. 91–2.
16. Finlay, 'Protestant Agricultural Missions in India', p. 16.
17. Same, 'The Program of the Rural Mission in India', p. 14.
18. Sam Higginbottom, *The Gospel and the Plow or The Old Gospel and Modern Farming in Ancient India*, New York, 1929, pp. 42–3.
19. Ibid., p. 43.

20. Higginbottom, *The Gospel and the Plow*, p. 43. It was believed that demonstrators could teach the cultivators the advantages of book farming. The farmers would send their children to school and because they grew more crops it was possible to pay for their children's education.
21. Higginbottom, *The Gospel and the Plow*, p. 44. Higginbottom observed,

    The reason I advise that so many properly qualified Americans be sent out is not that India's own sons and daughters are not capable, but they have not had the chance for training in India which they need and which Americans has. Other things being equal, the greater the number of American helpers as a temporary measure, the quicker India will be able to manage her own affairs.

22. Ibid., p. 46.
23. Ibid., p. 49. Higginbottom observed,

    India has several million wells in areas where there never can be flow irrigation. At present the water is raised by bullocks, a slow and expensive method. The engineer who can overcome all the difficulties and give to India a cheap, durable, efficient and simple well-pumping outfit will do a great thing for India. We therefore wish to establish a strong agricultural engineering department to remedy this obvious lack.

24. Ibid., pp. 50–1. Higginbottom observed,

    By the present old fashioned and inefficient methods, India out of one of the richest soils on the earth has the smallest yield per acre or per man of any civilized country. So the rapid introduction of better farming is the most natural and easy method of giving to India the things of which she stands so sorely in need. This is the one sure way to rid India of the ever present nightmare, as well as the reality, of famine, and from the missionary standpoint the one sure way to get the self-supporting, self-propagating, self-governing church. Better farming for India means the introduction of modern machinery adapted to Indian conditions. The Indian farmer has gone about as far as anyone can go with implements made of bamboo tied together with weak string; to get bigger crops he must have better tools. The present tools and implements do not call out from the user any large degree of intelligence. It is for this reason that mission farms using Indian tools and methods have not made any substantial progress. But the Indian boy who learns to care for a tractor, or a threshing machine, or a silage cutter knows he was learned something that calls for more brains and effort. Modern machinery challenges the Indian farmer boy just as it has the American farmer boy.

25. Ibid., pp. 60–1.
26. Finlay, 'Protestant Agricultural Missions in India', p. 24.
27. Ruth Ure, *On This We Build in India* cited in Finlay, 'Protestant Agricultural Missions in India', p. 28; B.H. Hunnicutt and W.W. Reid, *The Story of Agricultural Missions*, Missionary Education Movement of the United States and Canada, New York, 1931, p. 28.
29. Finlay, 'Protestant Agricultural Missions in India', p. 25.

30. Milfred S. Hatch, *Serving India's Farmers*, cited in Finlay, 'Protestant Agricultural Missions in India', p. 25.
31. Gary R. Hess had pointed out that Allahabad environment produced leaders who were noted for their understanding of rural problems. One of the members of the American University field teams in India had remarked that in interviews of agricultural graduates, he could always identify those from Allahabad. They excelled in terms of their ability to discuss practical matters and for their sense of dedication to village improvement. The government colleges which had been increasing in numbers followed the Allahabad examples and gave greater emphasis to practical training. This sort of opinion was also expressed by the Royal Commission of Agriculture in India.
See Hess, 'American Agricultural Missionaries', p. 25; K.C. Naik, *Agricultural Education in India: Institutes and Organizations*, New Delhi, 1961, pp. 74–80; *Royal Commission on Agriculture in India, Evidence taken in the United Provinces, Government of India*, Calcutta, 1927, pp. 540–86.
32. Hess, 'American Agricultural Missionaries', p. 25.
33. In the late 1910s, the deliberations in the Imperial Legislative Council revealed that cooperative societies in central Punjab districts were increasing their stock of Meston ploughs for sale to agricultural groups. These societies were acting as subagents for the firms which supplied these implements and they also received a small commission from the banks for selling them. However, the government could not clarify whether these agricultural societies did receive any subsidy in lieu of the purchase and sale of agricultural implements, like in the European nations. The Government stated that advances to agriculturists were made under the provisions of land improvement and agriculturists loan acts for performing such acts which included the 'purchase of seed and cattle and any other purpose connected with agricultural objects'. For more details, see Revenue Agriculture, Land Revenue, File no. 98, 1918, NAI.
34. Ibid., p. 26; Frank Brayne, *Village Uplift in India*, Allahabad, 1927, pp. 1–183; Cedric Mayadas, *Between Us and Hunger*, London, 1954, p. 137.
35. Hess, 'American Agricultural Missionaries', p. 26.
36. For more details, see Hess, 'American Agricultural Missionaries'; Mason Vaugh, *A Review of the Work done by the Agricultural Engineers in India*, Allahabad, 1941, pp. 1–5.
37. Hess, 'American Agricultural Missionaries', p. 30.
38. Finlay, 'Protestant Agricultural Missions in India', p. 40. In 1928, the Royal Commission on Agriculture in India had observed,

> The results achieved may be said to be the provision of a large amount of capital at reasonable rates of interest, and the organization of a system of rural credit, which carefully fostered, may yet relieve the cultivator of that burden of usury which he has borne so patiently throughout the ages. Knowledge of the cooperative system is now widespread; thrift is being encouraged; training in the handling of money and in elementary banking principles is being given. Where the cooperative movement is strongly established there has been a general

lowering of the rate of interest charged by moneylenders; the hold of the moneylender has been loosened with the result that a marked change has been brought about in the outlook of this people.

For more details, see *Royal Commission on Agriculture in India*, cited in Samuel H. Finlay, 'Protestant Agricultural Missions in India', p. 41.

39. Gary R. Hess, 'American Agricultural Missionaries', p. 31. The school used the symbol of the Cross and the Saw which had been central to Christ's life and sacrifice and tried to link the common everyday tasks of life with His spiritual ideals. For more details, see H.W. Brown, *The Cross and the Saw in Village India*, cited in Finlay, 'Protestant Agricultural Missions in India', p. 41.
40. Lapp, *The Christian Church and Rural India*, p. 97.
41. Hunnicutt and Reid, *The Story of Agricultural Missions*, p. 37; Finlay, 'Protestant Agricultural Missions in India', p. 19.
42. Finlay, 'Protestant Agricultural Missions in India', p. 22.
43. Lapp, *The Christian Church and Rural India*, p. 96.
44. J.W. Pickett, *Christian Mass Movements in India*, Lucknow, 1933, p. 291.
45. Finlay, 'Protestant Agricultural Missions in India', p. 23.
46. In the early years of the twentieth century, the Government of Travancore introduced foreign breeds of chicken like the White Leghorn, White Wyandotte, Little Sussex and Rhode Island Red, for increasing the production of eggs. The YMCA at Martandam also gave special attention to the development of poultry farming, which was seen as an important sector of the cottage industry. Subsequently, a poultry breeders' cooperative society was organized for the cooperative marketing of eggs. A few consignments were also dispatched to Madras and sold at a good price.

For more details, see *Travancore Administration Report, 1928–9, Seventy-Third Annual Report*, Trivandrum, 1930, p. 118.
47. For more details, see *Travancore Administrative Report, 1930–1, Seventy-Fifth Annual Report*, Trivandrum, 1932, pp. 110–11.
48. Hatch's ideas were much different from other individuals', who had favoured change through their dealings with college graduates and not with illiterates bound by tradition and superstition. He believed that the extension workers had to meet and address the grievances of the villagers at their own level. See D. Spencer Hatch, *Up from Poverty in Rural India*, Bombay, 1933, pp. 42–3.
49. S. Thirumalai, *Post-war Agricultural Problems and Policies in India*, New York, 1954, pp. 234–5.
50. Arthur H. Slater, *The Animal Kingdom of Etah*, n.p., 1964, pp. 15–20. Finlay, 'Protestant Agricultural Missions in India', p. 30.
51. The plan of work in the villages was simple and effective. The eggs from purebred hens were sold for 2¢ each. The villagers after returning home set a number of brood hens, hatched the eggs and raised the chicks. When these reached a certain size they were examined and sometimes did fetch prices at the level of $1.75, if they met a certain standard. The chickens were usually advertised for sale throughout India and were sent to distant markets. For more details, see Arthur H. Slater, *The Animal Kingdom of Etah*.

52. Ibid.
53. Following World War I, the work at Etah gained importance. Branch farms were established throughout the Etah district and each of these farms started organizing poultry shows. The Etah poultry show annually received grants from the Government of the United Provinces and also donations from Nettie Fowler McCormick of Chicago. See Hess, 'American Agricultural Missionaries', p. 33.
54. Slater, *The Animal Kingdom of Etah*.
55. The work at Etah consisted of (1) selective breeding of milk goats with an aim to standardize a number of Indian breeds, (2) tests of milk production and progeny, (3) investigation of the causes of mortality and the preparation of remedies for diseases. The whole project was conducted in cooperation with the Imperial Institute of Veterinary Research, Mukhtesar, United Provinces. The local government had given financial aid in addition—making a grant of $14,000 in 1935 towards the extension of the work with goats and poultry. The Government of India following the Etah model had established eight goat farms of its own of recent date. See, Finlay, 'Protestant Agricultural Missions in India', pp. 30–40.
56. Hess, 'American Agricultural Missionaries', p. 33.
57. Sometimes, the native Christian leadership was also involved in such programmes. The National Missionary Society started its work at Bethlehem, a Christian village in Punjab. The National Missionary Society brought together a number of Christian families from neighbouring communities and settled them on the land donated by the Government of Punjab for that purpose. Schools for boys and girls and night classes for adults were organized. By 1936, the number of inhabitants had grown to 800, each one owning land of approximately 12 acres. See Finlay, 'Protestant Agricultural Missions in India', p. 36.

6

Transformation of Agricultural Practices: An Indigenous Experiment in Colonial Bengal

Bipasha Raha

In the last decade of the nineteenth century, Bengal witnessed the beginning of an attempt to introduce changes in the existing system of agricultural practices. It was Rammohun Roy who, for the first time, focussed the attention of the Bengali literati on the rural countryside by highlighting the adverse impact of British land policy. Subsequently, the peasantry came to occupy centre stage in all forms of literary expositions penned by them. Throughout the nineteenth century, the literati in Bengal continued to voice their concern at the deteriorating condition of the peasantry and degeneration of the villages. This could not be ignored as the nationalist movement gained momentum and a section of the literati strived to emerge as the representative of the masses. Till the 1880s, the Bengali literati's agrarian thinking was limited to issues related to peasantry and tenancy legislations. It was Rabindranath Tagore who gave a new dimension to the agrarian perception of the literati. Motivated by an urge to reconstruct the rural villages of Bengal, which he perceived to have decayed in course of the colonial rule,[1] Rabindranath Tagore, Bengal's greatest intellectual, evolved a programme of rural revitalization over a period of time that encompassed agrarian development.[2] Convinced of the fact that there could be no material change in the condition of peasantry, who constituted the bulk of the rustic dwellers, without transformation of existing agrarian practices, he applied himself to the task in real earnest. While rural resuscitation was his lifelong mission, modernization of agriculture was indispensable for it to attain any degree of success.

Interest in rural uplift was awakened during Tagore's prolonged stay in the countryside when he was entrusted with the work of looking after their zamindari estates by his father Debendranath Tagore in 1890.[3] During his stay in the countryside, he was struck by the decay that he witnessed there and the appalling condition of the villagers who were mostly peasants. He made certain discoveries. The colonial rulers had assessed land at a very high rate. The pressure of land revenue was great. He was aware of the exploitative nature of the zamindari system and the nature of tenancy rights. A large section of the peasantry was denied occupancy rights in land. But his main focus was agriculture. He was convinced of the fact that there could be no real improvement in the material condition of the peasantry without improvement in the existing system of agrarian practices.

He realized that agriculture was stagnant because of the use of age-old techniques. Peasants were unaware of the improved methods of farming already in use in parts of Europe. Consequently, agricultural productivity was low,[4] without any attempt at replenishment of soil fertility. With greater demand for food, land was cultivated every year. The fallowing practice was gradually abandoned. Law of diminishing returns was at play. Crop rotation and mechanization were unknown. Experimentation with new crops was non-existent, as were attempts at development of new and improved seeds. Profits from agricultural products did not show any significant increase. Agricultural wages tended to stagnate. The peasants were unaware of the practice of analysis of soil types to suit particular crops. Ryots were too reluctant to adopt new methods. Besides, the total land under fodder had also declined.[5]

Having diagnosed some of the ills besetting Indian agriculture, Tagore devised a work-plan over a period of time. He himself took an active interest in familiarizing himself with developments in this regard in the Western world. He sent his son Rathindranath, son-in-law Nagendranath Gangopadhyay and the son of a friend, Santosh Chandra Mazumdar to Illinois University in America to study agriculture and animal husbandry,[6] so that on their return they could be his active workers in the project that he was planning to undertake.[7] It was not a plan that he had envisaged from the beginning but to which he was to remain committed throughout the last half century of his life.

Shilaidaha: The First Phase

Tagore's agrarian experiments were carried out in three phases where he tried out his ideas. He hoped to find solutions to the problem of enhancement of agrarian yield and initiating material prosperity for cultivators. It was on the family estates of Shilaidaha that he first tried to familiarize ryots with the extent of agrarian development that had taken place in Europe and America. The poet always kept himself informed of all these developments. The need for adopting these changes was highlighted. Emphasis was on scientific agriculture and mechanization. This would save time, labour and expenses in the long run. However, its initial costs being high and beyond the reach of poor peasants with dwarf holdings and little or no capital, adoption could only be possible with introduction of collective agriculture in each village or *mandali*.[8] Tagore advocated the adoption of the principle of cooperation at a time when there were no government cooperatives in Bengal. The societies present were mainly credit societies. It has been rightly observed that:

It is now not necessary to explain to enlightened readers the virtues and advantages of co-operative enterprise. Its usefulness is now taken for granted. . . . But . . . when nobody bothered about the principle of co-operation or of their application to the rural problems, Gurudeva Rabindranath Tagore thought about them and devotedly worked in this field of study as a pioneer for the uplift of the countless men and women residing in remote villages scattered all over Bengal and wallowing in the mire of poverty, ignorance and superstition.[9]

Tagore imbibed the ideas of cooperation from Sir Horace Plunkett (1854–1932), the father of the Irish cooperative movement, whose work also impacted upon the development of the agricultural cooperative movement in Great Britain and the Commonwealth.[10] Tagore was also inspired by the Irish poet George Russell and his work *The National Being*. He too held that the cooperative idea had immense possibilities and could be applied to all kinds of economic enterprises.[11] When applied to agriculture, collective efforts could be highly remunerative:

Many labour-saving appliances have been invented in Europe but the smallness of our holdings and our lack of resources make them almost useless to us. If the farmers in a village or, better still, in a community unit combined and engaged in joint cultivation of their land on a co-operative basis, they could all profit by the use of these modern machines which would reduce expenditure and give larger yields. It is economical for them to buy even an expensive machine if all the sugar-cane in the village is crushed by it. They can afford to have a jute press in the village if the produce of all the fields and homesteads is brought to a common centre. There would be an improvement in animal husbandry if all the milkmen in the village combined. Similarly, weavers in the village can indent for improved power looms if they pool their resources and work on a co-operative basis.[12]

Union of fragmented lands, he suggested, accompanied by common use of tractors, ploughs, looms, etc., would facilitate productivity.[13]

It was a daunting task that the poet undertook. The ryots on his estate were not easily convinced. He wrote:

At one time I called the peasants together to discuss the question of combining agricultural land. From the verandah of the house where I lived in Shelidaha, one sees nothing but field upon field stretching out beyond the horizon. From early dawn one peasant after another comes with his plough and cattle, ploughs round and round his tiny plots and departs. How great is the waste of divided effort! And I have seen it daily with my own eyes! When I explained to the peasants the advantages of joining their land together and tilling them by machine plough, they readily agreed.[14]

Tagore advised the use of mechanical plough and harvester to take optimum advantage of the growing season. Both sowing and harvesting could be accomplished with less labour and time, thereby entailing minimization of production costs.[15] The poet considered the use of time-honoured ploughs to cultivate strips of land separated by ridges, comparable to filling up a 'bottomless pit'. He warned ryots about persisting poverty as long as they refused to consolidate their fragmented and scattered landholdings.[16] The meagre resources of individual ryots were not conducive to agricultural growth. Most peasants did not have sufficient farm animals. Besides, much of the energy of the plough animal was wasted as the direction of the plough changed frequently because of the twisted boundary lines of the plots.[17]

At Shilaidaha, Tagore experimented with new ideas.[18] Scientific techniques were introduced and the results of experiments were demonstrated to local cultivators as part of the extension programme. Practical training was imparted to them.[19] As an attempt to eliminate the fallowing year and increase agrarian yield, cultivation of vegetables was promoted. New vegetables were introduced in the locality, viz., cauliflower, maize and peas. Double or multiple cropping was advised. Cultivation of fruits, e.g. different varieties of melon on sandy soil, was taught. Soil experimentation was carried out and potato cultivation was undertaken on the estate.[20] Large quantities of potato seeds were procured from Nainital at a considerable expense. New kinds of fertilizers were experimented with to enhance soil fertility.[21] Seeds were also supplied by the state Agricultural Department which were distributed among the peasants. It was an uphill task to motivate peasants to adopt new techniques as most in the region were content with their single rice or jute cropping.[22] But these experiments, the first of its kind, undertaken on the zamindari estate, were significant in that some of the local enterprising ryots were inspired enough to try them out. One of the tenant farmers obtained a bumper crop using improved potato seeds which inspired others to grow the crop. Maize and paddy cultivation was experimented with. High-quality maize seeds were procured from America and paddy from Madras.

After his return from America, on the completion of his studies, the poet's son Rathindranath set up an agricultural farm for research on 80 bighas of land at Shilaidaha. Seeds of maize, clover and alfalfa were imported from America; discs, harrows and such modern implement suitable to Indian conditions were introduced. Even a small laboratory was fitted up for soil testing.[23] Farm workers attempted to show local ryots how the usual paddy could be combined with maize, sugar cane, cauliflower and peas of the Patna variety. Cultivation of *kankur* and *kalai* on sandy soil was promoted. Sugar cane of the *gandari* variety was brought from Dacca and cultivated. A sugar cane mill was set up. The focus was on experimentation with varieties of crops and fertilizers. Pump irrigation was also introduced. An agricultural circular was published from the estate giving details of crops and seasons and was distributed among ryots of the neighbouring regions.[24] Rathindranath even imported improved tractors from America. As part of the training programme, plots of land were entrusted to volunteers to experiment with varieties of flowers, groundnuts, peas and onions. Silk cultivation was introduced to promote silk weaving. The success of these experiments was not in the actual profit or loss incurred but in their demonstration effect. A peasantry unused to such innovative techniques was shaken out of its stupor and exposed to alternative possibilities. Agricultural work in Shilaidaha was sought to be based on cooperative efforts.

In 1907–8, the *mandali* system or rural self-government was introduced here.[25] A cess called *kalyanbritti* was introduced. It was first collected at the rate of 3*p*. per rupee of the revenue. An amount equal to that collected from peasants was contributed from the income of the zamindari.[26] For the first time on the estate of a zamindar, a *krishibank* or agricultural bank was established to provide soft productive loans to farmers and monitor spending.[27] Capital of the bank was built up with money borrowed from friends and local moneylenders. A cooperative grain bank or *dharmagola* was established to which each ryot contributed a part of his produce after the harvest as a safeguard against future calamities. Provision was made to provide ryots with seeds at nominal rates.

Patisar

After the partition of zamindari estates, Tagore had to abandon his work at Shilaidaha and concentrate on Patisar in Kaligram pargana in Rajshahi district, where rural reconstruction work had already begun. Here agriculture was not easy. Land was prone to waterlogging during the rains. After soil testing, Tagore advised peasants to concentrate on fruits, e.g. pineapples, date palms, etc., and potato cultivation. Tractors were used for the first time. As a result of the demonstration effect, some of the neighbouring farmers were induced to follow suit. New techniques were hence adopted.[28]

The *mandali* system was introduced in Kaligram. Over a hundred villages were brought within the scope. In 1905, a *krishibank* was established at Kaligram. Agriculture and cottage industry were financed. Borrowers were required to repay at 9 per cent interest. Agricultural loans were to be repaid after the harvest. Ryots were often given a remission of 3 per cent on the interest rate. Once his loan was repaid, the ryot was free to borrow again.[29] The *krishibank* sanctioned loans only on the recommendation of the *karmi sangha* entrusted with the charge of rural welfare in the region. Records prove that all these efforts succeeded considerably in reducing indebtedness in the region and substantially affected the business of local moneylenders causing resentment.[30]

Rural revitalization work was started in real earnest.[31] Agricultural improvement was accorded pivotal importance. Irrigation schemes were undertaken and many wells were bored. A cess, *hitoishibritti*, similar to *kalyanbritti*, was introduced to cover the expenses. Sub-infeudation of land by the peasants within the estate was forbidden. Ryots were not allowed to sublet their land and were afraid of ejection. Estate officials were instructed to deal fairly with ryots.[32] Remissions of rent were granted when inability to pay was proved. Grain stores were also established.[33] The degree of success attained in such work by 1915–16 was significant enough to acquire official recognition, 'It is clear that to poetical genius he adds practical and beneficial ideas of estate management, which should be an example to the local zamindars . . . .'.[34]

However, the work had to be abandoned soon as some of the workers aroused the suspicion of the British authorities regarding their links with the political movement. But these setbacks were not powerful enough to stop the litterateur. He was to take it up with the same spirit and enthusiasm in another place on a much larger scale.

The Final Push

The next and the most enduring phase began in Sriniketan. The poet had earlier bought some lands in Surul in Birbhum from the local zamindar. It is close to Santiniketan, where his father held some lands and where the poet himself had established a Brahmo-Vidyalaya. Rathindranath and Nagendranath were entrusted with the task of setting up an ideal agricultural farm and laboratory there. Santosh Majumdar joined them soon. These men brought with them the knowledge and expertise they had acquired while studying in America. Sriniketan served as the Agricultural Department of Santiniketan in the early days. The poet stated that the object of Sriniketan was to bring back life in its completeness into the villages, making them self-reliant and self-respectful, acquainted with the cultural tradition of their own country and competent to make efficient use of the modern resources for the improvement of their physical, intellectual and economic conditions. Both agriculture and dairy work continued simultaneously. The dairy first opened in Santiniketan was later shifted to Sriniketan in 1916.

The farm had about 50 bighas of land under cultivation with a large tank for irrigation. In the farm, besides paddy and other common crops grown in the locality, groundnut, onions, maize, linseed, cowpea and chillies, fibre crops, viz., jute, *san* (hemp) and *dhainchia* were grown and also different types of fruits and vegetables, viz., lime, jackfruit, tomatoes, *kachu*, radish, papaya, plantain, pineapple, gourd, brinjals, etc. Napier grass and *jowar* cultivation was undertaken to solve the perennial shortage of fodder. Experiments in growing a few varieties of wheat and cotton were also carried on. A Japanese expert Kashahara was invited to monitor these activities. Mulberry cultivation and the rearing of silkworms were begun with the help of government experts. In cooperation with the Department of Agriculture and the Government of Bengal, a sericultural farm was opened where systematic training was provided to a large number of apprentices. The main objective was to attempt the establishment of sericulture as a village industry.

The garden had about 6 bighas of land with a small well for irrigation. Efforts were made from the start to increase the area and the supply of water. Multiple varieties of Chinese, Japanese and other foreign fruits, vegetables and fodder crops were experimented with. Considerable success was attained. The village boys, under the guidance of the Village Work Department, started gardens in their home compounds and kept their families on fresh vegetables during the rainy season, thus building up the vitality that the poet considered to be the bulwark against attacks of malaria. Seeds and saplings were raised and supplied from the garden. The Japanese gardener demonstrated the conservation and use of night soil in his garden. It was considered the most practical, sanitary and profitable method of disposing of waste materials of the most dangerous kind.

A new dimension to agricultural experimentation and work was added once L.K. Elmhirst, an agricultural scientist, joined the poet's programme.[35] With a band of young volunteers, he embarked upon a training programme. In 1922, the Institute of Rural Reconstruction was established. Its aim and avowed objective was to win the friendship and affection of villagers and cultivators by taking a real interest in all that concerned their life and welfare, and by making a lively effort to assist them in solving their most pressing problems and to take the problems of the village to the classroom for study and discussion and to the experimental farm for solution. Its extension work gave crucial importance to agriculture.[36] Covering a wide radius, rural reconstruction work was begun in a large number of villages adjacent to the Institute. The knowledge and experience acquired in the classroom and the experimental farm was imparted to the locals to help them develop their resources, make them aware of better techniques of growing crops and vegetables and breeding of livestock and to make them realize the benefits of associated life, mutual aid and common endeavour.[37]

Land in the region was first analysed:

Sriniketan stands on what is by rail the nearest rising ground to Calcutta. The ocean of green paddy fields which covers so much of Bengal, finds its limit at our door, and behind us stretches a rapidly increasing area of red laterite desert, which today supports neither man nor beast, but which once was covered by thick forest and jungle.[38]

Undulating land in the western portion of the district consisting of undulating uplands was broken up by wedge-shaped depressions that received detritus from the highlands that hemmed them. It received plenty of water from the drainage of the slopes. Rice was grown in these depressions and in terraces of the slopes.[39] But this being in the lateritic zone, the sterile laterite soil did not permit the cultivation of rice in the crests of the ridges. Red soil formed from the decomposition of the solid laterite lying below covered the surface for a few feet in depth. However, rocks in large masses were found at frequent intervals. In the eastern side of the district, the land was low and of alluvial formation. The soil was mostly a light sandy loam that was enriched at places by detritus from the uplands and elsewhere by silt from the overflow of the local rivers. Generally, the rivers deposited sand when they flooded. Embankments were built in many cases to protect the cultivated area from the drifting routes of the rivers. *Aman* or winter rice was the most important crop that was grown on the major portion of the cultivable land. *Aus* rice was also cultivated to an extent. Other *bhadoi* crops were infrequently grown. Rabi crops were rarely cultivated except in small tracts in the Nalhati and Murarai *thanas*.

In the three years that Elmhirst stayed in Sriniketan, he laid the work of rural development on a solid footing. Agricultural experiments started under his direction, continued even after his departure. The farm, that gradually grew in size as more and more areas were acquired over the years, had threefold aims—experimentation, training and extension. Different types of crops and fertilizers and the system of crop rotation that would be viable, given the type of soil condition of the district, were experimented with. Special attention was paid to the problems of fodder and fodder shortage, deep ploughing, manuring, trenching, selection of seeds, conservation of moisture and the use of low-cost and affordable but improved techniques.[40] Different kinds of paddy were experimented with, viz., *basmanik, paramannashal, raghushal, nonarmashal, sindurmukhi, chapshal, badkalamkati-65, jingheshal* and *basmati*.[41] The maintenance of a seed bank was suggested. To encourage sugar cane cultivation on a profit basis, the *hadi* system of *gur* (jaggery) was promoted. Cultivation of tomatoes, soybeans, cauliflower, beetroot, beans and potatoes were tested, as also *motihari* tobacco, Agartala *kalai* and C0213 and C0421 sugar cane. Fruits, viz., papaya, banana, guava and pineapple were grown. To provide fodder, *kalai*, napier grass and *jowar* cultivation was promoted. Silo method of fodder preservation was adopted. Cultivation of camphor, cardamom, clove and cinnamon was started in 1931.[42] The farm demonstrated to the villagers improved methods of farming, especially the utilization of manures, conservation of moisture and rotation of crops and helped them by the distribution of better seeds. Elmhirst and his band of volunteers called the *Brati Balaks* sought to demonstrate the ills of faulty treatment of soil to the farmers. The need for proper fertilization to restore soil fertility was highlighted and organic fertilizers were also introduced.[43]

Elaborate classification of land and varieties of soil was made for facilitating profitable cultivation. *Do* land was identified as a rich soil area. Autumn rice, gram, *masuri*, peas, *khesari*, wheat linseed, *til*, sugar cane and the occasional cotton was prescribed for such soils. There were three sub-kinds of this soil. On the *awal* land, multiple cropping was possible. The winter *aus* crop could be followed by wheat, gram, *masuri*, linseed, *khesari*, peas or mustard. This again could be followed by *til*. In between harvesting and sowing of crops, land needed to be manured and ploughed. After reaping the *til*, it could be made ready for the next annual agricultural cycle. On the comparatively inferior kind of *do* land, i.e. *doem* land, same kind of crops could be grown. But since this land was not easily irrigated, a change in the crop pattern was suggested. After the *aus*, onion or garlic was grown best, followed by a crop of *kashta til*.[44] The third type, i.e. *soem do*, was similar to the second kind in soil composition but more difficult to irrigate. The crop pattern similar to that on the second could be followed with the expectation of a lower yield. On the *do* land sugar cane could also be grown.[45]

Small amounts of moisture could be retained by the *suna* lands. Same cropping pattern as on *do* land could be followed except sugar cane, which could not be cultivated on *suna* land. The cost of cultivation on such lands was greater and the yield lesser. *Suna* lands were subdivided into *suna korpa* or *awal suna*, *doem suna* and *soem suna*. Another category of land was *sali* lands, subdivided into *jol* or *awal*, *doem* and *soem*. On the moist muddy *jol* or *awal sali* land, three crops may be sown annually, viz., *aman* rice, *khesari* and *kashta til*. Transplanted in March–April, *aman* is harvested in November–December. *Khesari* could be sown among the paddy as it begins to ripen and is to be harvested in February–March, while *til* needs to be sown in middle March as it ripens by early May.[46] As the best *sali* land occupied a level lower than the *doemsali* land, it received the silt from the higher levels during the rains and is also easier to irrigate. *Doem sali* land could produce only two crops annually, viz., *aman* rice and *til*. The yield was however considerably less. Situated at a yet higher level, the *soem sali* land produced a small quantity of a single crop, viz., rice. It was not really conducive to agriculture. Even the rice produced was of poor quality.[47] *Jedanga* or *danga* land, which is difficult to irrigate, was to be found near the homesteads, as also in the open plains. It was a mono-cropping land, able to produce either *arhar, san* (hemp) or brinjal. There could be orchards of mango, jackfruit and other fruit trees. Silt carried by rivers, deposited along the banks, cover the *olan* land. Though fertile, it was prone to inundation. Cucurbitaceous plants, viz., watermelon, *kankur, lau* or gourd, bitter gourd (both *uchhe* and *karela*) and *khero* (a variety of gourd) could be profitably grown on this land. *Dihi tut* and *mahal tut* were the two kinds of *pat-jami* or mulberry lands. Situated near the village, the former was a highland that was immensely favourable to mulberry cultivation. The latter was highland in the open that was situated away from the village. On the banks of the Ganutia River, mulberry was grown in the alluvial soil. It was very suitable as it did not require manuring.[48] There was even a silk *kuthi* there. In the *jangal-bhumi, sal* trees were grown best. It could be sold as timber, while new shoots could grow from the roots of the old trees. Composed of clay that was continuously moist, *methel* land was good for growing betel. *Ghas* was reserved grassland. In the *kati ghas* land, grass was cut for fodder while the *charai ghas* land was pasture land. Found on the river banks, *sarbera* or sandy lands were home to the *sar* or reed that grew wild. It was used primarily for thatching and preparing shade for betel plants (*panerbaraj*). *Sarbera* lands, where there were small deposits of mud, could be used for pasture, as grass grew there sometimes. *Bastu* or homestead land was classified into *nijbastu* and *udbastu* lands. The former was the land where the house stood, while the latter was the land near the homestead. The latter was again subdivided into *tarkaribastu* and *saribastu*. Cucurbitaceous plants, viz., pumpkins and gourds, could be grown in the *tarkaribastu* or lands within the courtyard. Some chillies, banana, brinjals, *karela, uchhe* and *dingli* were grown on *saribastu* or lands lying about the enclosed part of the house. However, the greater portion of the latter remained fallow.[49] *Shabek patit* or *danga patit* was land that was always left waste. It is the highest land of all that was actually stiff clay or laterite. It was too expensive to irrigate.

The soil in the district was also classified. *Metel* soil was best suited for growing winter paddy, sugar cane, wheat, gram and *kalai*. It was a clayey soil that could retain moisture. *Etel* was brownish clay that became sticky when wet. Again it was prone to getting hard and cracking in long fissures on drying. It was a poor-quality soil and was capable of producing paddy only if irrigated. Even with irrigation it was impossible to cultivate rabi crops on this soil.[50] *Banga etel* is a reddish soil—sticky and tenacious. It contained limestone nodules. It becomes hard when dry but retains moisture for a longer period than any other soil. But this too is poor-quality soil that can produce paddy only after adequate manuring. *Pali* is deposit of silt in the riverbed. It is loose, friable and yellowish in colour. This is soil rich in content. It is suitable for sugar cane, wheat, gram, potato, cabbage and other vegetable

132   *Bipasha Raha*

cultivation. Rabi crops can be grown even without irrigation. *Reti* or *ret* is reddish, loose and friable alluvial soil that is best suited to cultivation of vegetables, wheat, barley, etc. Its moist quality allows cultivation of rabi crops without irrigation.[51] *Bindi* was sandy soil that improved with continuous cultivation. It was reddish, loose and friable but could not retain moisture. It was poor-quality soil that could produce low-quality paddy. Rabi crops could be cultivated only with adequate irrigation. *Doansh* is a mixture of clay and sand forming a blackish, loose and friable soil that was unable to retain moisture. It was rich in content and suited to the cultivation of all major kinds of crops. Rabi crops however required irrigation. *Bele* was whitish, loose and friable soil that was not moisture retentive. It was poor in mineral content and could only produce paddy and vegetables. Even with irrigation, rabi crops could not be grown. *Kankare* or gravelly soil was reddish, loose and friable laterite. It contained ferruginous concretions and was a poor soil. Only *bajra, kurthi, marua, gondli*, maize and peas could be grown. Rabi crops could be supported only if irrigated. Land was good for growing jackfruits. *Bastu* land could generally be used for rabi crops. It was blackish, friable rich soil. It could be easily manured with cow dung, ashes and other refuse from the village. It was unable to retain moisture. However, paddy, sugar cane, wheat, peas, linseed, tobacco, maize, *til* and *bajra* could be easily grown.

Rice was the predominant crop in the district and the farmers were heavily dependent on it.[52]

Elmhirst underscored the importance of attempts to control soil erosion that was a bane in some regions of the district. Large tracts of the district consisted primarily of a porous soil and rapid drainage. The need for artificial drainage was felt in years of scanty rainfall, especially for rice that was grown on terraced slopes. When there was sufficient and timely rainfall, artificial irrigation was not required. The farmers would divide their fields into many little plots and enclose each by a raised bank so that the rainwater could be retained. Each plot was thus a small reservoir.[53] The lower fields could be irrigated by letting water into them from the plots situated at the higher level. Well irrigation was not practised, except in the case of garden produce. Efforts were made at Sriniketan to promote well irrigation. A water engineer, Akhil Chakraborty, was invited from outside to drill a well. It was stressed that such ventures that entailed considerable expenses could easily be undertaken by farmers of each locality if they cooperated and combined their resources. The normal practice followed by farmers was to water their lands with water stored in the tanks. There were already several old and large tanks that pre-existed in several parts of the district, viz., Dantindighi near Dubrajpur and Raipur Sair and Lambadarpur Sair near Suri. Besides, there also existed innumerable small tanks. O'Malley estimated that each village had, on an average, 5 such small tank reservoirs.[54] He cites that in the village of

TABLE 6.1: The Normal Acreage of Principal Crops in the District (1910)

Name of crop	Normal acreage	Percentage on normal net cropped area
Winter rice	6,04,600	77
Sugar cane	9,000	1
Total *aghani* crops	6,13,600	78
Autumn rice	1,44,100	18
Total *bhadoi* crops	1,51,000	19
Wheat	5,000	1
Barley	300	–
Gram	7,000	1
Other rabi cereals and pulses	3,500	
Other rabi food crops	1,000	–
Linseed	800	–
Rapeseed and mustard	2,000	–
Til (rabi)	400	–
Other oilseeds	200	–
Other rabi non-food crops	700	–
Total rabi crops	20,900	3
Orchards and garden produce	25,000	3
Twice cropped area	23,900	3

Source: L.S.S. O'Malley, *Bengal District Gazetteers*, Birbhum, Calcutta; repr. 1998, p. 61.

Sankarpur as a case in point, there were 111 tanks occupying 167 acres. In fact, 46 of them were situated so close to each other that only footpaths on the top of the banks separated one from the other. But owing to the neglect of the zamindars, most of whom were absentees. A large number of these irrigation tanks had silted up and become unfit for the purpose. In fact, many of them had become so dry that they were let out for cultivation. The apathy of the village dwellers was also to be blamed. Tagore suggested restoration of these tanks to their former condition. Till his visit to Russia in 1930, he did not advise changes in the existing land system.[55] Distinguishing between 'good' and 'bad' zamindars, he appealed to the former to be active partners in rural development work. He considered it their responsibility as they were the 'natural leaders of rural society'.[56] Well aware of the adverse impact of absentee landlordism that had become so rampant in the countryside of Bengal in the post-Permanent Settlement decades, he advised zamindars to take up residence in their estates, at least for some months in the year. Outlining a plan of work, he appealed to them to take an interest, among others, in agrarian development, keep themselves informed of strides made in such practices and techniques elsewhere in the world and educate the ryots on their estates. It was suggested that a part of the rental be spent on constructive work. The general decay that he witnessed in the countryside was blamed on the disinterest of zamindars in rural affairs.[57] He suggested that zamindars take an interest in restoration of these tanks, many of which had been constructed by their predecessors. He himself attempted to set an example by arranging for the restoration of Bhubansagar in Bhubandanga, Santiniketan, constructed at the behest of the Sinhas of Raipur, the erstwhile owners of this region from whom Debendranath, and later Rabindranath himself, had bought their property in this region. In the village of Supur, situated on the bank of the Ajoy River, which formed the southern boundary of the district of Birbhum and had once been a part of the Sinha zamindars of Raipur, five tanks were reclaimed at the initiative of the Rural Institute.

Once the tanks were full, water could be let into the fields by farmers, through channels constructed in the banks. When the water was low, the cultivators could raise it by either using *acheni*, i.e. a swing basket, or by a mechanism called *dhuni*.[58] The *do* fields on the banks growing sugar cane, oilseeds, flax and vegetables could be irrigated by the *teura* which was a kind of Grecian lever.

The local farmers practised many of these techniques. The effort of Sriniketan was geared towards helping peasants to make optimum utilization of their meagre resources. This had a positive impact on the prolonged attempt made in the district, since the turn of the century, to improve the methods of cultivation by the organization of Suri Cattle and Produce Show, which was managed by a committee of local respectable men under the presidency of the Collector. A district agricultural association had also been started to experiment with new crops. To impart training in new and improved agrarian techniques, use of modern agricultural implements, use of approved seeds and manures, method of soil testing and proper marketing of the produce, apprentices were accepted in the farm at Sriniketan. Long- and short-term programmes were arranged for facilitating such efforts. Major and minor projects were offered to the students to work on. Preparation of soil before the start of the agricultural season, selection of approved seeds, use of adequate fertilizers, proper irrigational techniques, rotation and protection of crops were some of the necessary basics that were taught.[59] Trainees were also taught a preparatory course on agricultural economics, collection and analysis of relevant data and analysis and conservation of soil.[60] Each year educated urban agricultural enthusiasts, as also ryots from neighbouring regions, began to enrol themselves for training at Sriniketan. In order to make new ideas and practices acceptable to a class engaged for generations in the only pursuit they knew, Sriniketan struggled to raise the average yield of paddy in its farm to 9–10 maunds per bigha, i.e. one-third of an acre as compared to the yield of 5 maunds per bigha, which was the average of the region.

As in Shilaidaha and Patisar, Rabindranath evinced the same faith in the importance of collective agriculture at Sriniketan. He continued to enumerate the benefits of consolidation of small holdings and proliferation of cooperative joint farms. But his idea of cooperative farming was not the Russian *kolkhoz* or collective farms in lieu of personal proprietary right over land. His idea was such that even while retaining personal proprietorship over land, it would be possible to obtain the benefits of large-scale production, as consolidation of holdings would make establishment of bigger farms possible. Mechanization, scientific agriculture, use of tractors, etc., were more viable when resources were greater.

Sriniketan adopted an elaborate 'extension' programme. To serve the interest of poor and middle cultivators, who could not afford to remain absent for a prolonged period of time from their holdings, particularly during the agricultural season, efforts were made to familiarize them with new ideas without having to leave the village. Bolpur-Bandgora, Adityapur, Benuri, Bahiri and Goalpara, among others, were the villages brought within the scope of the work. Experimental farms were started in many of them. Dedicated band of volunteers and experts from Sriniketan were dispatched to carry on the experiments and guide willing ryots. Sriniketan provided ryots in the adopted villages with approved seeds, manures and saplings and stem cuttings at nominal or no cost depending upon their resources. Help was extended to make provisions for minimum irrigational opportunities. Attempts were made to promote lift irrigation. Produce shows and exhibitions were organized. Local agricultural produce, as also samples of experiments carried out in the Sriniketan farm and laboratory, were displayed. The efforts of local farmers were recognized. Agricultural manuals with detailed information on the new knowledge were published from Sriniketan and circulated locally. The entire cost was borne by the estate.[61] Introduction of the system of green manuring was given importance in the extension programmes. Along with crop rotation, farmers were advised on appropriate methods of fighting crop diseases and insects and pests. The farm superintendent supervised the farms where seeds and saplings from Sriniketan were used.

Equal emphasis was put on animal husbandry as an indispensable part of the agricultural process. Oxen and buffalo were used for agricultural purposes, sheep was reared for commercial purposes, and goats and pigs were for local consumption. Besides performing plough work, bullocks were used as beasts of burden, for drawing carts and carrying packs of grains and sundry other goods. They were also yoked in the oil mills. The local breed of cattle was poor in spite of all attempts made to improve their quality. The apathy of cultivators and graziers to the issue of breeding could be a partial explanation.

The Suri Cattle and Produce Show and a dairy farm, established in Suri, since their inception had endeavoured to attract the focus of agriculturalists to this issue albeit with limited success. Some Hissar and English bulls had been imported by the District Board and the Suri Cattle Show Committee for improvement of the local breed almost two decades prior to Elmhirst's arrival at Sriniketan. A veterinary dispensary had also been established at Suri.[62] There was increasing extension of cultivation in the district.[63] Consequently, there was growing difficulty of finding good pasturage for the cattle as grazing grounds had dwindled substantially in the eastern part of the district by the 1920s. Here, practically the only grazing lands were small plots of commons near the villages which yielded poor-quality grass in a very insignificant quantity.[64] These commons and the occasional herbage found in uncultivated and uncultivable land, on the banks of the tanks or the raised boundaries of the fields, and the stubble left in the rice fields provided all the grazing for plough cattle by the 1920s. This had to be supplemented by fodder consisting of rice straw. In the west, there was still some pasture land on the uplands. However, the *sal* forests, in which the cattle used to graze, had mostly been cut down for timber by the beginning of the century.

Realizing the need of the hour, Tagore insisted that dairy, poultry and breeding of goats and bees be included in the itinerary of Sriniketan. The dependence of Indian agriculture on good-quality cattle was an established fact. Nagendranath was advised to visit Ireland to study the advanced techniques of dairy farming in use there.[65] It was important to breed superior quality of cattle at Sriniketan. The villagers within the scope of Sriniketan were advised to undertake scientific methods of breeding so that they could have sturdy milk cows as also draught animals. The latter could be produced without the importation of foreign stock by scientific selection and by breeding from existing strains of Indian cattle. The herd in the dairy consisted of upcountry and cross-breed cows. A good Montgomery bull was attached to the herd and was also utilized by the neighbouring villages for the improvement of their cattle. Efforts were made to induce some of the villagers to get rid of their bulls which were very poor and to use the bulls of the Sriniketan dairy when required. The Dairy Department also selected the best types of local cows for breeding with their bulls at the farm in order to build up a good herd from the indigenous stock by proper selection and culling. Calves were sold to villagers from the Sriniketan dairy at nominal prices.

Tagore felt that in India, little or no attention was given to the poultry industry which he considered could be very profitable. Poultry work began in 1924. It attained significant success. The Poultry Department was placed under an expert and within a short time, the stock of white Leghorns and Rhode Island Reds multiplied rapidly. Chicks hatched in the incubator were reared successfully. Different breeds of hen were experimented with. Efforts were made to cross native hens with Leghorn cocks and thus build up a better breed which would meet the demands of the people. Crossing *Deshi* hens with Leghorn cocks showed marked improvement in the size and number of eggs, as well as in the size of the birds. Among the stock, there were imported Rhode Island Reds, White Leghorns and Chittagongs, all of which won prizes at the All-India Poultry Show. The results were demonstrated to local villagers. Many of the latter were enthused enough to join the project as apprentices. It was reported that they set up their own poultry farms on the completion of their training in their respective villages.[66] The Poultry Department was able to introduce White Leghorn cocks in some of the villages in which it succeeded in getting rid of the native cocks. This increased the laying capacity of the native hens. The superintendent of the Poultry Department gave lessons to poultry men in the villages on treatment of poultry diseases. All the chicks in the Sriniketan poultry were hatched in incubators and the villagers were also persuaded to bring their eggs for hatching there. A goatery was also opened later in 1933 but it never really took off and had to be closed down soon. The absence of demand for goats in the locality was blamed for the failure. The poet's attempt to start breeding of bees was also not successful.[67] The whole Sriniketan project was based on the cooperative system. Elmhirst shared the poet's faith in its applicability. He held that it should be for India to lead the way towards cooperation for a fuller and more abundant life, both spiritual and material, because the memory of such a life in the past was not yet dead, and the will 'to sacrifice material acquisition for the pursuit of high ideals and spiritual gain' was perhaps more alive in the soil of India than anywhere else in the wide world.[68] All kinds of constructive work was undertaken on the basis of cooperation. A number of credit, health, pisciculture, irrigation and other cooperative societies were established over the years to make the poet's dream of resuscitation of indigenous villages a reality. A central cooperative bank was started in 1927 that still continues to function. At a nominal rate of interest, ryots could avail themselves of loans for agricultural purposes. Buying and selling cooperative societies, as also credit societies, to facilitate the agricultural process were initiated in villages adjoining the Santiniketan and Sriniketan project.[69] In 1928, a *dharmagola* or cooperative grain store was opened. Peasants contributed a part of their surplus to build up the stock. The shortage in the occasional bad year could be met out of this. Besides, this represented the total savings of the village community and was maintained in a well-protected store which was easily accessible to all the people of the area, so that each could put his own savings in the bank, draw loans whenever required and pay back their dues with interest after their respective crises were overcome. The saving scheme was an indigenous enterprise and could claim substantial results to its credit.

Tagore laid great stress on organizing festivals, fairs and exhibitions as a part of his programme for rural recovery and underscoring the importance of agriculture as an integral part of village life. The Varsha Mangal festival was a welcome to the rains which was destined to bring prosperity. 'Briksharopana', now popularly known in India as 'Vana Mahatsova', was a plantation ceremony that offered a special impetus to the growth of vegetation in rural areas and initiated a determined onslaught against the expanding pace of soil erosion. 'Halakarshan', the ploughing ceremony was the most important cultural aspect of agriculture.[70] Organization of village fairs was important for their value in integrating the interests of villagers and in strengthening their contact with the outside world. It was also a powerful mass medium for dissemination of knowledge. These festivals and fairs were enthusiastically accepted by the locals, along with the inmates, and continue to be celebrated till this date and are a part of the annual calendar of both Santiniketan and Sriniketan. In organizing the village festivals, the poet insisted that these should be dependent on local resources devoid of any ostentation. Each ceremony was to commemorate one aspect of the total programme of rural reconstruction, as also be a point of community contact and action. They could be so organized as to create group integration and community cohesiveness, as well as provide opportunities for development of local leadership. These, in turn, could contribute to the production of surplus wealth in the community.[71]

Conclusion

It is to be argued that when Tagore began his work at Shilaidaha, Patisar and Sriniketan, he had no models before him to emulate. He was a pioneer in what he aimed to accomplish. The colonial rulers, primarily concerned with the issues of steady realization of revenue and maintenance of law and order in general, never concerned themselves with the quality of life of the peasants, as long as it did not impact upon the stability of the empire. While the projects at Shilaidaha and Patisar had to be abandoned midway for reasons mentioned earlier, Sriniketan, during the poet's lifetime, attempted something more stable and enduring. The importance of its contribution towards a comprehensive development of the life and welfare of the peasantry and the rural folk in general could not have, therefore, in that context, been overstated.[72] It is to be admitted that the Institute of Rural Reconstruction introduced an experimental programme and almost single-handedly carried on 'its lone struggle for rehabilitation of the villagers of a particular area'.[72] It was the poet's hope that the results achieved there would ensure widespread application of his programme. He himself stated that he could not take responsibility for the entire country. But he expected that he would be able to hold up a model before the rest of the country to follow. The role of Sriniketan was that of an educator. One may agree that the role of Sriniketan in experimenting with more advanced programmes of welfare, development, training and research, in concentrated form and in a limited geographical region, could be used as a model by contemporary educators and rural rejuvenators.[73]

The material and non-material achievements of Sriniketan in its area of work, however, need to be analysed. The adequate follow-up data regarding the development that had taken place in every operational area of Sriniketan is not available. Examination of the available limited data reveals that all the 85 villages which lay within the area of operation of Sriniketan did not bear the same impact of the Institute's work.[74] Standard of living did not rise significantly everywhere. The condition of the villages lying within the area of work hardly changed in all the centres and subcentres of activity. Agricultural production showed only marginal increase in pockets. Ryots continued to be poor, though the cooperative societies were able to provide some relief in the form of easy loans, buying and selling facilities and accessibility to basic medical treatment. Notable results were achieved in a small area and in a few villages. The economic returns there were such that the rising standard of living in the area was very noticeable. There was a new confidence that developed among the villagers. Based on a survey, it was reported that in the village of Laldaha, located 13 miles away from Sriniketan there was, by the sheer effort of the people, a 1,000 per cent rise in the per capita income of the people. There was considerable employment available for each individual in the village. The consumption pattern of its residents had been substantially augmented. A comparative survey of the living conditions in Laldaha, carried out in 1959, with the base line data of 1939, recorded and maintained by a worker of the area, revealed that there was great improvement in the standard of life in comparison to other villages which either lay outside the work area of Sriniketan or which were manned by less competent workers. Again in the cluster of villages represented by hamlets of Benuria, Islampur and Bahadurpur, the income per capita was Rs.35 as compared to Rs.16 of the neighbouring villages. Elsewhere, lack of financial resources and dearth of trained personnel partially accounted for the difficulties in realization of targets. It is admitted that any programme of rural reconstruction needed large financial support to be able to produce any tangible results.[75] In spite of all the appropriate methods being adopted, and the determination to realize a mission, the material standard of living of peasantry living on a barren piece of land, with little or no resources, could not be improved without access to financial support. The poet's resources were meagre. The income from the estate was gradually dwindling. The earnings from his literary creations could not support an enterprise of this magnitude. Endowments were small. Sriniketan had no access to any kind of state patronage. The workers who joined Sriniketan from time to time were often severely handicapped in that they were unaccustomed to the new approach envisaged by the poet. Besides, many did not stay long enough. This seriously affected the continuity of the programme and often caused frustration among the villagers. This, at times, resulted in opposition to the work. However, the poet's programme was neither fundamentally flawed nor even inaccurate in broad details.[76]

Besides, the real purpose of Sriniketan and what it aimed to achieve, and secured to an extent in certain areas of its work, 'was not so much the physical targets of external growth but non-material development of culture, initiative and group integration'. The aim was to 'create the power of a community to lay down its own goals, and to enable the same to reach these by its own efforts'.[77] The poet did not have any other end product in view. Elmhirst observed, 'Tagore's idea was always to illustrate a few basic principles so that objectives and methods which he would thus evolve could be used as tools for the benefit of the rest of the country.'[78]

The importance of the work lies in the fact that it was undertaken at a difficult period of the country's history. It was a path-finding effort and its demonstration effect was significant. It drew some attention of his countrymen to the importance of such constructive work.

Notes and References

1. Sasadhar Sinha, *Social Thinking of Rabindranath Tagore*, Calcutta, 1962, pp. 1–2.
2. Rabindranath Tagore, 'City and Village', *Towards Universal Man*, Calcutta, 1961, pp. 317–18.
3. The extensive zamindari estates of the Tagores were spread over large areas in the districts of Pabna, Rajshahi and Nadia in Bengal and Cuttack in Orissa in the last decade of the nineteenth century.
4. Binay Bhushan Chaudhuri, 'Agrarian Economy and Agrarian Relations in Bengal, 1859–1885', in *The History of Bengal (1757-1905)*, ed. N.K. Sinha, Calcutta, 1968, pp. 304–6.

5. Rabindranath Tagore, '*Bhumilakshmi*', *Palliprakriti*, Calcutta, Aswin 1325 BS, 1987, pp. 37–9.
6. Letter, 12 Kartik 1314 BS, in Tagore, *Palliprakriti*, p. 230.
7. Rathindranath Tagore wrote, 'He (Rabindranath) thought that in order to resuscitate rural life, agriculture, which is the basic economic resource of the people must be improved. He, therefore, desired that Santosh and I must go abroad to get technical training in agriculture and animal husbandry so that after our return we could help him.' Rathindranath Tagore, *On the Edges of Time*, Calcutta, 1958, p. 74.
8. Rabindranath Tagore, 'Presidential Address', Pabna Regional Conference, Falgun 1314 BS, *Palliprakriti*, p. 2.
9. Sudhi Ranjan Das, 'Preface', in *The Co-operative Principle*, Rabindranath Tagore, Calcutta, 1963.
10. The poet met Plunkett in England in 1920 where they discoursed upon the future of the cooperative movement in India. Rathindranath, who had accompanied his father on that visit, described what transpired between the two. Tagore, *On the Edges of Time*, p. 35.
11. Sinha, *Social Thinking of Rabindranath Tagore*, p. 15.
12. Rabindranath Tagore, 'Presidential Address', *Towards Universal Man*, p. 119.
13. Bipasha Raha, 'Economic Thought of Rabindranath Tagore', *The Calcutta Historical Journal*, vol. XXV, no. 1, January–June 2005; *The Plough and the Pen: Agriculture, Literati and the Peasantry in Colonial Bengal*, New Delhi, 2012; *Living a Dream: Rabindranath Tagore and Rural Resuscitation*, New Delhi, 2014.
14. Rabindranath Tagore, *Letters from Russia*, Calcutta, 1984, pp. 21–2.
15. Rabindranath Tagore, 'Co-operation', *Towards Universal Man*, p. 327.
16. Tagore wrote,

    As long as the landowner and the tiller go each his separate way, neither can thrive. All over the world, men are forming unions to mobilize strength. Whoever remains isolated in the modern is bound to remain enslaved and a hewer of wood. Unless we unite to build an embankment by our joint effort, the results of our labour will, like trickles of water, slide down hill-slopes to fill alien reservoirs. We shall then produce food for others and ourselves starve and not even know why this is so. We must therefore first bring together those whom we wish to starve.

    Tagore, 'Presidential Address', *Towards Universal Man*, p. 119.
17. Tagore was convinced, 'If each cultivator did not regard his small holdings as an independent unit, if all adjoining strips were reckoned as one, fewer ploughshares and bullocks could do the tilling and much wasteful labour would be eliminated. There would again be a great saving of energy and expense if after harvest the farmers collectively stored and marketed their produce.' Tagore, 'Co-operation', *Towards Universal Man*, pp. 325–6.
18. He was supported enthusiastically by his friends Dwijendralal Roy, the poet, who was an agricultural expert, and Acharya Jagadish Chandra Bose, the scientist. They helped with new ideas and expertise.
19. Sachindranath Adhikary, *Shilaidaha O Rabindranath*, Calcutta, 1974, p. 71.

142 *Bipasha Raha*

20. Tagore's letters to Jagadish Chandra Bose from Kumarkhali in Shilaidaha, 10 Asar 1306 BS, *Palliprakriti*, p. 211.
21. It was reported in official survey:

    Experiment with Nainital potatoes were made by Rabindranath Tagore in The Tagore estate of Shelaidah in the Kusthia subdivision. The crop was not satisfactory owing to the defective cultivation. One of Mr. Tagore's tenants however working under more favourable circumstances obtained a bumper crop from a portion of the same seed and—the experiment is said to have induced several neighbouring ryots to take to potato cultivation.

    Land Records of Agriculture, Bengal, 1899.
22. The poet lamented, 'I tried to encourage them to do so, but failed; those men who willingly sweated over paddy would make no effort to grow vegetables.' Rabindranath Tagore, 'The Striving for Swaraj', *Towards Universal Man*, p. 276.
23. Tagore, *On the Edges of Time*, p. 86.
24. Adhikari, *Shilaidaha O Rabindranath*, p. 424.
25. Rabindranath divided Birahimpur (Shilaidaha) pargana into five mandals, eachunder an *adhyaksha* or head. They were to establish village societies there, so that the villages could themselves look after their own welfare, maintain roads, provide drinking water, settle disputes through arbitration, establish schools and grain stores, etc. The *adhyakshas* were instructed to build houses for tenants and plant banana, pineapple, date palm and other saleable fruit trees.Prabhat Kumar Mukhopadhyay, *Rabindrajibani*, vol. 2, Calcutta, 1961, p. 187.
26. Tagore, *Letters from Russia*, pp. 148–9.
27. Letter no. 60 to Prasanta Chandra Mahalanobis, *Desh*, Asar 1382 BS.
28. Tagore, *Palliprakriti*, pp. 244–5.
29. Tagore borrowed from friends and acquaintances to build up the capital of the bank. He also later deposited Rs.1,80,000 of the Nobel Prize money into the bank.
30. Sudhir Sen, *Rabindranath Tagore on Rural Reconstruction*, Calcutta, 1978, pp. 97–8.
31. Efforts were made to extend medical facilities, elementary education and scope for solution of internal disputes through arbitration without recourse to government law courts. Arrangements were made for provision of drinking water.
32. L.S.S. O'Malley, *Bengal District Gazetteers, Rajshahi*, 1916.
33. Rathindranath Tagore, *Pitrismriti*, Calcutta, 1968, p. 248.
34. O'Malley, *Bengal District Gazetteers, Rajshahi*, p. 149.
35. L.K. Elmhirst, *Poet and Plowman*, Calcutta, 1975, pp. 61–2.
36. P.C. Lal, *Reconstruction and Education in Rural India*, London, 1932, p. 233.
37. *Visva-Bharati Bulletin*, no. 6, July 1928.
38. *Visva-Bharati Bulletin*, no. 11, December 1928.
39. O'Malley, *Bengal District Gazetteers, Birbhum*, 1910; repr., Calcutta, 1996, p. 55.
40. Lal, *Reconstruction and Education in Rural India*, p. 65.

41. *Visva-Bharati Bulletin*, no. 11, December 1946.
42. Satyadas Chakraborty, *Sriniketaner Gorar Katha*, Calcutta, 1985, p. 65.
43. Bipasha Raha, 'Economic Thought of Rabindranath Tagore', p. 30.
44. O'Malley, *Bengal District Gazetteers, Birbhum*, p. 56.
45. Ibid.
46. Ibid.
47. Ibid.
48. Ibid., p. 57.
49. Ibid.
50. Ibid., p. 59.
51. Ibid.
52. Ibid., p. 61. The statistics was prepared by the Agricultural Department.
53. Ibid., p. 60.
54. Ibid.
55. Tagore, *Letters from Russia*.
56. Rabindranath Tagore, '*Sabhapatir Abhibhasan*', Pabna Pradeshik Sammelani, *Palliprakriti*, pp. 6–8.
57. Ibid.
58. The *cheni* was a scoop made of matting with ropes attached to its four corners. It was worked by two men, each of whom held two of them. After dipping the scoop in the water they tilted its contents into the channel for irrigating the field. The *dhuni* or *drauni* consisted of a trough with a bend in the middle or towards the end. The two portions of the trough were of unequal length. The shorter end called the *ankra* was closed. The whole mechanism moved upon a pivot. To the end of the *ankra* was attached a rope that was fastened to one end of an elevated lever bearing a counterbalancing weight. The *ankra* was dipped into the tank and when filled the weight was released and dragged up the closed end. The water was poured through the open end of the trough into the irrigating channel. O'Malley, *Bengal Districts Gazetteers, Birbhum*, pp. 60–1.
59. Raha, 'Economic Thought of Rabindranath Tagore', p. 30.
60. Chakraborty, *Sriniketaner Gorar Katha*, p. 65.
61. Tagore, *Palliprakriti*, pp. 11–18.
62. O'Malley, *Bengal District Gazetteers, Birbhum*, pp. 62–3.
63. Ibid. According to the returns for 1907–8, the net cropped area was 6,50,900 acres. Current fallows accounted for 2,43,460 acres, cultivable wastes other than fallows occupied 90,000 acres, and the area not available for cultivation accounted for 1,36,920 acres.
64. O'Malley, *Bengal District Gazetteers, Birbhum*, pp. 62–3.
65. Tagore, *Palliprakriti*, pp. 233–4.
66. Kalimohan Ghosh, '*Sriniketaner Pallisamgathan*', *Bhandar*, repr. May 1975, p. 18.
67. It was found to be a difficult task to prevent bees from going away during the long dry season. *Visva-Bharati Bulletin*, no. 11, December 1946, p. 4.
68. Elmhirst, *Poet and Plowman*, pp. 42–50.
69. Ibid., p. 62.

70. Sugata Dasgupta, *A Poet and a Plan: Tagore's Experiments in Rural Reconstruction*, Calcutta, 1933, pp. 99–101.
71. Ibid.
72. Ibid., p. 103.
73. Ibid., p. 107.
74. Ibid.
75. Ibid.
76. Ibid.
77. Ibid., p. 113.
78. Elmhirst, *Poet and Plowman*, p. 62.

7

Barbaric Hoe and Civilized Plough: Tribes, Civilizational Discourse and Colonial Agriculture in the Khasi-Jaintia Hills of North-Eastern India

Sajal Nag

Among many other things that the British colonialists had discovered in India, was also the phenomenon of subsistence agriculture. From the reports and statements they left behind, it is evident that the British could not believe that such vast agricultural pursuits in India did not generate the kind of profit that it would in Europe. With their experience of agriculture that was producing commodities suited for world market, the British wanted that Indian agriculture should be completely commercialized so that it produced items suited for the international market, through which Indian agriculturists would be able to pay the spiralling revenue demands of the colonial government. They believed that the reason for the backward state of Indian agriculture was the 'indolence' of the indigenous farmers. Hence, they repeatedly increased the revenue to make them work harder, to be able to pay the enhanced revenue demand of the government and to go for extensive cultivation. The colonialist also found that massive 'arable lands' were lying 'waste' which had to be brought under cultivation to increase the revenue generation potential of the countryside, and if local farmers were not willing to cultivate those, migration from other provinces were to be encouraged, induced and settled, in these so-called wastelands. Hence the integrated discourse of backward agriculture, lazy natives and migration of stout farmers from neighbouring provinces, was put forward by the British throughout colonial India. The tribals were not spared from this discourse too. The hill tribes who practised swidden cultivation were also seen as lazy natives who performed less than their potential and produced so little that they could not meet the revenue demands of the state. In this essay, we study the colonial attempts to transform the tribal agriculture of the Khasis of north-eastern India from a subsistence level to commodity production, to save the expensive forests that the tribals were supposedly destroying through their farming practices, and to import and settle agriculturists from Naga Hills who were skilled in terrace cultivation in the area so that they could teach the new technology to the Khasi farmers.

Shifting Cultivation as Primitive Agriculture

Shifting cultivation was another domain that the Europeans discovered in the colonies. Colonialists from Europe had derived the notion of shifting cultivation as the earliest form of agriculture from the recent discourses on simple and complex societies that were being developed in the Occident.[1] Although shifting cultivation had always existed as a parallel form of agriculture, it was unknown to the outer world as there was either no interest or no knowledge of subsistence pursuits of people living in deep forests and margins of large state formations. Prior to the European arrival, the imperial states based on lowland agricultural systems had little control over mountainous zones in central-peninsular and eastern India in which societies based on shifting cultivation existed. The forested areas were often kept out of sustained administration.[2] With colonial rule, forest resources lying between imperial and local societies were to acquire a new significance since the trees would attract the interest of the state. The colonial period saw increasing interest of the state, both in expansion of wet rice frontiers into shifting cultivation areas and in the control over primary forest resources involved in the process of production within shifting cultivation.[3] The mercantile interest in timber until the nineteenth century and then the timber requirements of the World Wars put the matter of efficient control over all forest dwellers at the forefront of state policy.[4]

As knowledge of shifting cultivation had increased, models of early agriculture were developed. Shifting cultivation, known variously as slash and burn and swidden, was perceived to be a primitive mode of land-based subsistence pursuit, mainly because it was seen as such in Europe. In northeast India, such agricultural pursuits were known as *jhumming* and communities who practised it were called *Jhummias*. *Jhummia* has nothing to do with the recently coined word *Zomia* for the highlanders of South-East Asia.[5] There are conflicting theories about the origin of the word *jhumming*.[6] Swidden agriculture had been associated with some distinctive features, like, it is practised on very tropical soils, it represents an elementary agricultural technique which utilizes no tool other than the axe, it is marked by a low density of population and it involves a low level of consumption.[7] It is marked by lack of tillage, less labour input than other methods of cultivation and a concept of private land ownership.[8] It was also seen to represent a special stage in the evolution from hunting and food gathering to sedentary farming—a stage evidently consisting of such null traits as the unrelatedness to pastoral pursuits and the production of little that is of trading or commercial significance.[9] The most striking feature of the practice is that it is attended by serious deforestation and soil erosion.[10] It is however admitted that much of these features are unqualified generalizations.[11] Subsequently, the use of fire in shifting cultivation was recognized. The burnt vegetation is often seen as the only source of direct nutrient replenishment for the soil. In shifting cultivation systems, the fields that have yielded one or two crops are left to regenerate vegetation for a long time, unlike other intensive agricultural systems. The cultivable plot is only one of the sources from which food supply, either yam or manioc or seed crops like maize and millet are obtained. Gathering, hunting, domestication of animals and tending of tree crops and vegetable gardens are mixed in with the cereal crops for a more broad-based diet.[12]

Colonial Intervention in Tribal Agriculture

Agriculture and horticulture, the primary occupations of the tribes, received earliest attention of the British authorities but, of course, for different reasons. Like rest of their empire in India, the British wanted to commercialize the agriculture so that it could generate more revenue and also produce surplus for the world market. It was done through two methods—introducing foreign corps which were more commercially viable and increasing the productivity of agriculture. The British entered the north-eastern region in 1824 and took over parts of Assam in 1826 as a result of Treaty of Yandabo. It formally took over the neighbouring tribal Khasi hills in 1834–5. But even before that, it had already started interfering in the subsistence pursuits of the people. The most significant development was the introduction of foreign crops and fruits and their cultivation on a large scale. Potato and a few other agricultural crops of European consumption owed their introduction to the initiative of David Scott, the Agent to the Governor-General. The objective of Scott in introducing foreign crops and fruits originated in his plan to establish a sanatorium for the Europeans in the Khasi hills. He began horticultural experiments by growing European fruits in Nongkhlaw. The experiment proved a success.[13] Later, he established farms at Mairong, a village near Sohra (Cherrapunji) where he successfully cultivated turnips, beet, millet, maize and various cereals of European consumption, besides potatoes. Even wheat was found to be thriving well in the hill slopes.[14] First introduced by David Scott in 1830, W.J. Allen observed, 'potato had been the greatest boon that a British ruler had conferred upon the Cosseah (Khasi) people'.[15] Allen was correct as the cultivation spread rapidly throughout the Khasi hills and within two decades, it became one of the principal cash crops of the country. When J.P. Mills visited Cherrapunji in 1853, the production of this crop amounted to 30,000 maunds.[16] Two years later, Jenkins noticed that cultivation of potato was practised even in the interiors of Khasi hills, viz., Jeerun, Sohiong and Mawflang. The volume of export of potato in one year from Jaintia hills alone was 50,000 maunds.[17] In 1872, Chief Commissioner Bivar reported even more increase of production which he computed at 1,85,000 maunds a year. It is remarkable that the Khasis cultivated potato only for trade. They did not consume this tuber at that time. Consumption of potato was confined only to the 'Europeans and better class natives'.[18] The intensive cultivation of cash crops, such as potato and orange, especially the former, has been described by Mills later as an 'agricultural revolution'.[19] But it had its impact on the traditional land tenure system of the tribe, which from communal ownership gave way to the concept of private ownership. Land in the hills was held by the community or clan or as a whole. However, if a plot of land was cultivated continuously for some years by an individual, such land would become his property. Even in 1858, all foresaw the rise of private proprietorship in land in the Jaintia hills where the Dolois and others were misappropriating the *rajhali* land, which actually belonged to the government.[20] In the Khasi hills, the new system of landholding came into existence almost imperceptibly. Because of the large-scale cultivation of potatoes and oranges, pressure on land was increasing.

In contrast, the indigenous crops like rice and cotton did not show any increase in production. In 1872, Bivar noted that the cultivation of rice was insufficient for the consumption of the people. There was no improvement in the rice grown nor was any 'sensible increase in cultivation' noticed even though it was the principal crop in Jaintia hills.[21] In 1873, a total of over 27,000 maunds were produced in both these hills taken together. The cotton cultivation had also gone down. Mills reported in 1853 that 9,000 maunds were produced in the hills which dwindled to a mere 2,152 maunds in 1873 and this was in Jaintia hills only.[22] Lindsay noticed that a large variety of fruits, including oranges, were growing wildly in the hills.[23] Similarly, Mills reported the growth of *Tejpat*, pan or betel leaves and wild cinnamons, which grew in large amount and were exported in considerable quantities.[24] The government not only encouraged this trade but also tried to popularize fruits, vegetables and medicinal plants of foreign origin. Thus, Chinchona plants were brought from the Botanical Garden of Calcutta. Local intelligentsia like Jeevan Roy supported the government's endeavour.[25] To test the viability of growing foreign fruits in Khasi hills, the government established an orchard, a tree nursery and a public garden in Shillong in 1873.[26]

Shifting Cultivation as Primitive Unprofitable Agriculture

A new pattern of crops had come into being. The expansion of indigenous as well as crops of foreign origin was unprecedented. The mode of cultivation, however, remained primitive. *Jhumming*, the locally known practice of slash and burn cultivation, was still in vogue. While commenting on the state of agriculture, Carnegy observed that while the Jaintias cultivated their lowlands with plough and bullock, higher lands were cultivated by them and Khasis only by *jhumming*.[27] Speaking of their agricultural implements, he noticed that the Khasis used only hoe and declined the use of plough entirely. Even potato was cultivated by *jhumming*. The procedure was that when a particular land was selected for cultivation, the forests and trees covering it were slashed and burnt down and the ashes and wood were used as manure.[28] In wet fields, only dung was used for manure. For the recuperation of its fertility, land was left fallow for a period, which in the case of Jaintia hills extended even to twenty years. Carnegy noted that in the Jaintia hills with an area of 2,100 sq. mi., only 20 mi. of land were cultivated, while in the Khasi hills, with an extent of 4,450 sq. mi., only 30 mi. were under cultivation.[29] Hence, Carnegy held that the country had great potential, especially for cultivation. Water soil, iron, wood and cattle were available in plenty. He argued that if potato could be grown successfully, European cereals could also be cultivated with success. Of course, prior to it, certain measures like introduction of rotation of crops, proper drainage of the fields and change of mode of production was necessary. He recommended the substitution of plough for hoe. In other words, he suggested the abandonment of *jhumming* method of cultivation. He hoped that the Khasis would adopt the system once they were made aware of its benefit. In other words, the European agency might be employed to introduce the new system of agriculture. Meanwhile, the government might establish a public garden to test the agricultural capability of these hills.[30]

In 1886–7, despite the annual increase in potato production and export, the crop registered a decline, reportedly due to a disease because of the presence of fungus, and in the following year, the crop was reported to have rotted in the ground.[31] Owing to this disease, the exports continued to fall from 41,548 maunds in 1887–8 and 24,386 in 1889–90 and to 12,016 in 1890–1. In 1892–3, exports decreased to 10,776 maunds.[32] The matter alerted the administration. The chief commissioner of the province sought detailed information on the cause of the decline and also overall agricultural situation in the Khasi and Jaintia hills. This brought the question of the unproductiveness of shifting cultivation method to the agenda of official discussion again. The Director of Land Records and Agriculture, after an enquiry, reported that the potato crop was indeed affected by a disease named *Phytophthora infestans* in 1885–6. Due to the fall in the production of potato, the farmers reverted back to rice cultivation resulting in a considerable increase in the area of rice cultivation. In Jaintia hills, it was about 14,000 acres by 1898.[33] The varieties of rice found in the Khasi and Jaintia hills were divided into two main categories—one grown as a dry crop on highland and the other raised in valleys and hollows which were artificially irrigated from hill streams. The lowland rice was more productive than the one grown on highlands, the average returns per acre being 11.70 maunds of paddy per acre, as against an average of 9.40 maunds per acre. The Director commented,

the average outturn of both kinds is extremely poor as compared with that of any description of rice grown in the plains. The rice grown in the Khasi and Jaintia hills is also very inferior quality, the grain when cleaned being of a red colour and extremely coarse. In the Shillong Bazar it sells retails from a seer to two seers per rupee —cheaper than the cheapest kind of imported rice. It is not considered fit for food by the foreign population of the station and so far as I have been able to ascertain is eaten by Khasis only.[34]

The officer concluded,

bad quality and short out turn of rice grown in the district are probably both due to the primitive methods of cultivation employed as well as to inferiority of the seed used. In the Khasi hills the use of the plough is almost unknown and land is turned up with the hoe, both for low land and up land rice. On the steeper slopes the latter description of rice is also raised without the use of the hoe, by *jhumming*: the seeds being dibbled into holes made in temporary clearances in the forests. In the Jaintia hills a rude kind of plough is used in cultivating low lands . . . experiments have been made in his part of the district of substituting the plough and of introducing seeds of superior quality but the results have not justified the additional expenditure. It is very possible that seeds of varieties of rice grown in the plains would not be successful in the hills. There seems however to be no reason why the plough should not be used in low lands and much of the up land rice cultivation and it is probably only ignorance of its advantages which prevents the Khasis from employing it and from considerably extending the area under rice.[35]

The report has two important mentions. One, the missionaries working in the area were also taking initiatives to make the Khasis give up the shifting cultivation method and adopt settled rice agriculture. Most prominent being Revd C.L. Stephens of the Welsh Methodist Mission, based in the village of Khadsawprah. It states that Revd Stephens' 'experiments have already been made in his district of substituting plough for hand labour and of introducing seeds of superior quality' but he adds that 'the result have not justified the additional expenditure'.[36] The other mention being a variety of rice seed which had a returns of 15 maunds of paddy per acre, which was successfully grown at the elevation of 6,000 ft., were proposed to be introduced in the region. But it was doubted if the people would take to it in spite of 'the advantages which would be gained if the Khasi cultivators could produce rice of better quality for the consumption of the rapidly increasing foreign population of Shillong, which at present draws its food supplies almost entirely from outside the district are obvious . . .'.[37] The Sikkim variety of rice was to be experimentally planted in the Shillong farm. But the difficulty still of making the Khasi farmers adopt the plough remained. The Director of Agriculture commented,

the difficulties in the way of introducing agricultural improvement into the Khasis and Jaintia Hills are of course great but it may be hoped that they will not prove insuperable. Even now in their agriculture the Khasis are far in advance of all other tribes in this province, indeed, in some respects, such as the conservation and use of manure they are in advance of the majority of cultivators in the plains of Assam; and it is curious that among them should be found along with some enlightened methods of, indications of barbarism, such as the practice of *jhumming* and ignorance of the plough.[38]

The Director's report was based on the ground investigation by the Deputy Commissioner of the Khasi and Jaintia hills district that,

the practice of *jhumming* is to be regretted because the fertility of the soil is rapidly exhausted by it. But so far as these hills are concerned there is enough land to spare for the population for some time to come. It is possible that a model farm may be of use. But it must be remembered that unless the improved methods of cultivation can be shown to pay, shrewd conservative people like the Khasis will not adopt them.[39]

Renewed Effort to Introduce Settled Rice Culture

During the World War I period, there was renewal of endeavour to make the Khasi-Jaintias give up shifting cultivation and take up terrace cultivation this time. In the summer of 1914, all the deputy commissioners were instructed to institute enquiries in order to a ascertain whether there was a flatland on the north-western borders of the Khasis and Jaintia hills suitable for terraced cultivation and what steps, if any, can be taken to induce the villagers to substitute wet rice cultivation for *jhumming*.[40] They were also asked for opinion as to the desirability of endeavouring to induce the villages of the Bhoi and Nartiang circle in the Jaintia hills to abandon *jhum* cultivation by imposing a special tax on *jhumming* and substantially reducing the revenue now assessed on wet lands in those circles. This had become necessary in view of the climatic condition that had rendered it impossible to cultivate rice successfully on the southern scarf of the hills. It was also asked to enquire into the extent of damage that *jhumming* cultivation had done to the forests of the Khasi and Jaintia hills. The deputy commissioners were asked to make every effort to introduce permanent cultivation in the northern part of the district. Interestingly, the Commissioner proposed that the Angami Nagas could be imported from Naga hills and settled in Khasi hills to conduct the experiment.[41] It is interesting because Angami Nagas of the neighbouring Naga hills district were expert terrace cultivators and by now the colonial administration had come to recognize that the Angami Nagas had historically evolved a very developed and extensive technique of terrace cultivation in their district, due to which they were not only sufficient in food production, but were using rain water irrigation very effectively for rice production. The quality of rice that the Angami Nagas produced was much better and significant in returns. The proposition that Angami Nagas should be brought from Naga hills and settled in the Khasi-Jaintia hills to familiarize the Khasis with terrace cultivation and teach them the technique was interesting, as was the proposal to increase the land revenue rates, so that the revenue demand was so high that the Khasi farmers were forced to go for improved methods of cultivation to produce more, so as to be able to pay the increased revenue demand of the administration. This was an old nineteenth-century colonial policy in Assam being brought back again in the twentieth century.

The Ensuing Debate

In reply, the deputy commissioners reported that there was practically no flatland in the areas like Nonglang, Nongriangsi and Jirngam Sirdarships in the Jaintia hills that would be suitable for wet rice cultivation. There was some good *hali* (*sali* wet rice) cultivation in the vicinity of Malangkona and Aradonga. The villagers were not willing to learn how to terrace. Both the Deputy Commissioner and the Subdivisional Officer thought that there was very little land suitable for wet rice cultivation in the Jaintia hills which had not been converted into *halis*. The main difficulty was in getting water.

The local officers were not in favour of general prohibition of *jhumming* or of imposition of an enhanced rate of house tax. This was originally fixed at Rs.2 but in 1907–8, it was raised in 30 villages to Rs.3 and in 353 villages to Rs.2–Rs.8, leaving 333 villages at the old rate. The *rajhali* lands in nine circles in the Jowai subdivision were resettled at an all-round rate of 12*an*. per bigha, for a term of twenty years from 1 April 1912. The three circles of Nongphyllut, Narpuh and Bhoi were assessed at the old rate of 10*an*. per bigha. It was also pointed out that the Land and Revenue Regulation of Assam were not in force in the Khasi and Jaintia hills. According to the spirit of Rule 75 of the Settlement Rules of the Assam Land and Revenue Manual, the revenue free term of three years may be allowed. A sectional concession rates (8*an.* per bigha) may be allowed for the remaining term of twenty years. As per the encouragement and incentives to be given to the farmers for taking up permanent cultivation, it was stated that agricultural loans were ordinarily given for the purchase of seed, plough cattle and implements (vide rule 10 of the Rules). There was perhaps no objection to grant such advances on a liberal scale. In 1914–15, the government had placed a sum of Rs.60,000 as the disposal of the Commissioner for Agricultural Advances in Surma Valley and Hill Districts.[42] The commissioners also did not agree that the settlement of an Angami village close to it was practical, as far as the introduction of terrace cultivation in the district was concerned. If the demonstration for terrace cultivation by the Agricultural Department was necessary, there was an Agricultural Inspector for the Khasi, Jaintia and Garo hills who knew the art of terrace cultivation. The Director of Land Records and Agricultural was proposed to be asked to arrange for demonstrations in the Jowai subdivision.[43]

Commenting on the reports submitted by the Deputy Commissioner of Khasi-Jaintia hills and from Jowai's Subdivisional Officer, W.J. Reid, Commissioner, Surma Valley and Hills Districts, wrote to the Chief Secretary of the Chief Commissioner of Assam that:

the time has not come either for the general prohibition of *jhumming* or for the imposition of an enhanced rate of house-tax on those villages which follow this method of cultivation. Clearly however every encouragement should be given to people who are likely to go in for permanent cultivation. In such cases the revenue-free term of three years which us admissible under the rules should invariably be allowed, and this should be succeeded by a term during which special concession rates of revenue will be charged. At the same time advances should be given on a liberal scale for the purchase of seedlings, agricultural implements, and plough-cattle in places where those can be used.[44]

Although Reid agreed that most of the flatlands from Shillong to Jowai and on to the Koplili riverbank have been converted into rice fields, a lot of flatlands still remain for terrace cultivation. He felt that an Angami village would have doubled or trebled the area of this cultivation by terracing.[45]

On the issue of destruction of forests as a result of extensive *jhumming* cultivation, A.W. Dentith, Deputy Commissioner-in-Charge of forests in Khasi and Jaintia hills, gave a detailed report and recommended that the *sal* forests of the Nonglang, Nongriangsi, and Jirngam sirdarships; the forest tract about Mathan in the Mylliem State; the *sal* forests at Amiong Kmi in the Khyrim state and the forest tracts of the Bhoi and Mynriang circles in the Jowai subdivision should be inspected,[46] because they were unable to discover any *sal* forests that were worth reservation. In the Nonglang, Nongriangsi and Jirngam sirdarships, they found no *sal* trees whatsoever once the plains ended.

In the plains, the *sal* was almost entirely on the Kamrup side of the inter-district boundary. In the Nonglang sirdarship, the officer had reserved a small patch of good *sal* forest, less than 1 sq. mi. in extent, in the vicinity of Malangkona village, together with the neighbouring Hashram hill. The sirdar of Nonglang and the villagers were not unwilling that *jhumming* should be prohibited in these two areas, as there was plenty of wet rice cultivation in their neighbourhood and *jhumming* was, consequently, a secondary consideration. At Aradonga, in the Jirngam sirdarship, orders were passed that no *jhumming* would be permitted on the Rangmasi-Umlaru hill, about 3 sq. mi. in area.

It was also found that the practice of *jhumming* had been stopped in this area by the siem of Khyrim on the instruction of the colonial administrators. The officer also recommended that no further lease for the extraction of *sal* should be granted by the siem, until the forests were examined in detail by a forest officer, who could then advise whether after four years of working, these forests should be given a rest. Meanwhile, *jhumming* would continue to be prohibited in these *sal* hills. The two best forests in the Jowai Bhoi country, viz., the Mawrynjong forest, 3 sq. mi. in area, near Umswai and the Umshlong forest, near Amiong Okuri, contained very little *sal* and it was felt that they were not worth reservation.

The officer then reported that there was practically no flatland in the Nonglang, Nongriangsi and Jirngam sirdarships which would be suitable for wet rice cultivation. Generally speaking, the country was a congeries of precipitous but that was all. The villagers, when questioned, stated that they did not want to learn how to terrace. The administration, however, contemplated experimenting with terraced cultivation next year at Nongmaweit. These experiments would doubtless settle the question of terraced cultivation in these sirdarships for some time to come. If they were successful, the administration felt, no doubt they would be copied each year by an ever increasing number of Khasi and Garo cultivators. If however, they were unsuccessful, terraced cultivation in these sirdarships would be conspicuous by their absence.

On the subject of *jhumming* in the Jaintia hills, it was found that there was very little land suitable for wet cultivation which had not already been converted into *halis*. At Umswai, Umpanai and Mynser, there were excellent *halis* that were as productive as *dhankhets* in the plains. There were also excellent examples of terraced cultivation in the vicinity of Shillong-Myntang and Nongbah, though their returns was not as good as that of the Bhoi *halis*. Terraced fields, as a rule, were not as productive as *halis*, for the good surface soil was thrown away when the field was levelled and not replaced, while it took four years before the soil weathered and became oxidized. In the upper hills, every available piece of flatland suitable for wet rice cultivation was taken up; consequently, terraced cultivation was by no means uncommon in the Khasi and Jaintia hills.

In the lower Bhoi hills, however, very different conditions existed. Nomadic tribes, like the Mikirs (known as Karbis in contemporary north-east India) who were hereditary *jhummers* did not practise or have knowledge of any other mode of cultivation. The British considered that the Mikirs:

> were a conservative people, independent and suspicious of innovations, they had plenty of land available for *jhuming* (for the country is thinly populated) and the average Mikir, being indolent by nature and an opium eater by choice has a got unnatural aversion from the hard labour involved in converting a hill side into terraces, which require to be irrigated and manured. But the water-supply is often extremely scanty and he is too poor as a rule to keep cattle for ploughing. He is also ignorant of the use of a *khodali* (spade). On the other hand, but little labour is required to prepare a *jhum* on which several crops can be grown simultaneously: no manuring is required and the outturn is generally on a par with the outturn of *hali* fields. In these circumstances so long as there is land available for *jhuming* the average Mikir is not likely to undertake the trouble of terraced cultivation.[47]

In these circumstances, he was not in favour of adding 'thou shalt not *jhum*' to the Mikir Decalogue. The colonial officer came out with a punch line, 'the forests were made for man and not man for the forests'.[48] Still it was not considered that every Mikir was unprogressive, for though the Lalungs chiefly go in for *hali* cultivation, it was found that the Mikirs at Mawlongjarain and Jaikyndeng were about to start *halis*. It was consequently quite possible that some Lalungs and Mikirs might be induced, but not driven, to terrace; but, there must first be some successful demonstrations. Some intelligent headmen were found who were not averse to learning how to terrace in the vicinity of Jowai, Nartiang and Nongbah and could also send two or three Synteng experts, who, unlike the Angami Nagas, know the Mikir and Lalung dialects, to tell them how to terrace. He was however strongly opposed to the proposal that a special tax should be imposed on *jhumming*:

> The Mikir is poor and uncivilized but he must be allowed to live. In a year of scarcity, like the present, he is entirely dependent on his *jhum* crops and on jungle tubers. In ordinary years he must have some highland cultivation, for his great vegetation . . . he should be exempted from payment of revenue until he is able to convert land into permanent *salis*. If, at the same time, he is convinced by practical and successful demonstrations that *hali* and terraced cultivation is a good thing. I see no reason why progress in this direction should not be made. But such progress, with all deference to the *silvicultural* (and impatient) idealist, is bound to be slow.[49]

The administration, on the whole, was unanimous that *jhumming* was essentially wasteful and unscientific method of cultivation.[50] No doubt, when a magnificent forest was ruthlessly destroyed or when the hill slopes were extremely steep, as in the War (a subtribe of the Khasis) gorges, *jhumming* was extremely objectionable to the administration. But in the Bhoi areas, the hills were not considered so precipitous and were usually covered with bamboo or other not so valuable jungle. The Forest Department was for keeping trees till they rotted, rather than allowing *jhumming*, or letting the trees go for far less than the prescribed royalty; but in the Bhoi country forests, 'first class' trees were very few and far, and so, when such forests were discovered, their commercial exploitation was rendered impossible on account of their inaccessibility. *Jhumming* was not wholly unreasonable and with ordinary care it could be contained indefinitely, and land would again be ready for *jhumming*, ten or twenty years later.[51]

Arguing for Shifting Cultivation

The local administration in some parts of the Khasi and Jaintia hills also found that there were intelligible justifications for going for *jhumming*. In the Bhoi country, the system of cultivation by *jhumming* was popular for two reasons—one was that the people did not have enough flatlands which they could convert into *halis*, the other was that they preferred *jhumming* to wet cultivation, as they could raise several kinds of crops on the same land, without going to the expense and labour of having filed for each particular crop.[52] They could raise only paddy in wet cultivation, while in highland cultivation again, they could obtain not only paddy but *lil*, yams, chillies, maize, pumpkins and other vegetables, all at once. Cotton was the first crop raised on the land and lac trees were planted with it. The following year (1915), paddy was grown along with *lil*, maize, chillies, etc., and they also had the lac tress on which lac was reared. Paddy could be grown on the same land for two years in succession, so that with the cotton crop, the field was made use of for three years at a time. There was not much difference in the out-turn of the paddy crops from *jhums* as compared with those of *hali* lands. The out-turn of *jhums* was almost as much as that from wet rice fields.[53]

Besides, the flatlands in the valleys of Umsuwai, Umpaai and Mynser were partially all under cultivation; there was not much flatland in the Bhoi circle. So that unless terraced cultivation was adopted, there was practically no more room left for wet rice cultivation. Moreover, in Bhoi country terraced cultivation was not possible in the winter, as there was difficulty in getting water on the hilltops. In any case, it would require a deal of expense and labour in making terraced cultivation for the reason that the faces of the hills must be cleared of all the bamboo and tree jungles that they at present contained, as only then could the terraces be made and aqueducts constructed:

'The Mikir was too poor for one thing, and too lazy (being an opium eater) for the other, to undertake this job, and unless some one does it for him free of cost and gives him possession thereafter, it is not likely that any amount of inducement will possess him to take it up himself.'[54]

Another colonial official opined that tax on *jhumming* would be of little use. When asked what effect a tax would have on *jhumming*, several Mikirs unanimously informed that rather than lose their *jhums,* they would pay any tax imposed on them. If a tax was imposed in these hills, similar action would have to be taken in Nowgong and north Cachar hills also, for otherwise most of the Mikirs would migrate to these districts in order to escape the tax. It would be hard on the people to impose a tax, for the people had some highland field, even if not to grow paddy in, but to grow other crops, e.g. lac, cotton, maize, *til* and other vegetables. The Mikirs were great vegetarians and could subsist on vegetables when the paddy ran out. In lean years and years of scarcity, the Mikirs fell back on their vegetables.[55]

The only thing I can think of to encourage more wet cultivation among the Mikir is to exempt him from all revenue until such time as he is able to convert a land into a *hali* and make it into a permanent one I say permanent, because very often a man makes experiments at first. If he finds, because he has not enough of water, of for some other reason, that the *hali* can not be made into a permanent *hali* he gives it up. Many people are kept back from making experiments, for no sooner they do this the lands are measured up for *pilla* at once, and naturally, the people fear having to pay revenue when they are receiving nothing in return from the land. As an experimental measure we could engage 2 of 3 Synteags who are versed in terraced cultivation to go round the Bhoi country and show the people the mode of this kind of cultivation and at the same time, find out if it is possible to have this cultivation in Bhoi or not. It is no use getting an Angami Naga because he does not know the language.[56]

As regards *hali* lands in Nartiang, it was reported that most of the *hali* lands in this circle also had already been taken up. It was only where there were too many wild animals, elephants, pigs, deer, etc., and where the people could not keep them out of their field for want of guns, and where the crops were destroyed year after year, that the people relinquished the lands. The people in the higher region of these hills were so keen about wet rice cultivation that almost every available bit of land was taken up. The yearly increase in the number of wasteland applications would prove this. Table 7.1 shows the applications received since 1900–1.[57]

There was no wet cultivation on the southern slopes of these hills for the reason that the soil was too rocky for *hali* fields. Terraced cultivation was also not possible for the same reason. A good deal of the tree forests in the War

TABLE 7.1: Number of Wasteland Applications

Year	Number of applications
1900–1	20
1901–2	20
1902–3	55
1903–4	43
1904–5	32
1905–6	42
1906–7	41
1907–8	79
1908–9	47
1909–10	35
1910–11	99
1911–12	320
1912–13	314
1913–14	378
1914–15 (up to 9 March 1915)	413

Source: The Subdivisional Officer, Jowai to the Deputy Commissioner-in-Charge of Forests, Khasi and Jaintia hills, no. 125, 11 March 1915, Revenue Agriculture, A Progs., August 1915, no. 50, Assam Secretariat.

country was preserved for *pan* cultivation and it was only the small forests that were *jhummed* for millet, chillies, etc. The people generally bought all their rice from the plains.[58] As far as importing Angami Nagas and settling them in the Khasi hills was concerned, the officer felt that it would be of no use as the latter would not understand the language of the former.[59]

Shifting Cultivation as Destroyer of Forests

As far as the impact of *jhumming* on the forests was concerned, the colonial administrators in Khasi-Jaintia hills were unanimous that it was detrimental to forest resources. After completion of inspection of Jirang and Nongkhlaw, the local administrators reported that the best *sal* forests usually occurred in the forests of South into the Khasi hills, in the Sirdarships of Jirngarm, Nonglang and Nomonianesti. It was remarkable that unlike the country of Jirang and Nongkhlaw, this portion of the Khasi hills did not contain *sal* trees, meaning that they had been denuded as a result of *jhumming*. Between

Malangkona and Nonglang, which was to its south, there was one solitary *sal* tree standing.[60] The report said:

The whole country appears to have been completely *jhummed* over and now the hills sides contain only grass, bamboos and few trees, generally evergreens, in the *nullas* and deciduous on the hills. The principal trees seen to be *Makria* (Schima Wallichii) Castanopsis, Kamili (Careya Arborca) occasionally *sida* (Largest *memia Parvillora* and *Mullata*, a species of *Mararanga*) which is very common. There are of course other trees besides these, but none of these trees occur in any quantity; they are only isolated and scattered. The country gets barer as one goes south, until by Nongmaweit one gets large areas practically treeless covered with grass. I noticed some half-dozen pine trees between Nungmaweit, and Nonginiaw, so this must be about the lowest limit of their growth.[61]

The isolated trees standing in these bare areas in many cases had obviously been lopped, so presumably, the areas must have been *jhummed* at some time or the other. The old *jhums* were now covered in grass and generally got burnt off every year. Therefore, they always appeared to remain grass covered, and showed no signs of coming up in the tree forests. In the plains, there was generally a certain amount of *sal*, more or less along the foot of the hills.

However, in order to preserve the small patches of *sal* which did exist at Malang in the Khasi hills district, the deputy commissioner had issued orders to the sirdar of Nonglang (who was not at all unwilling to assist) that these patches of Hashram hill and Malang forest should not be *jhummed*. The people also did not show any desire to *jhum* there either, for at Malang they went in chiefly for wet rice cultivation, and *jhumming* was a secondary consideration.[62] From Malang, eastwards into the hills, which ran out into the Kamrup district forming the Gijang reserve, and marched down the Gijang River to the village of Gijang, there were no *sal* on the flatland just south of the Gijang River. The hills to the south, inside the Khasi hills district, did not appear to contain *sal*. On the hill above Aradonga, there was a fair amount of *sal* on the lowland along the Singra River. The *sal* forest extended up the hill towards Kamriangsi, and it was the only *sal* worth considering that was seen in this part of the Khasi hills. The deputy commissioner had agreed to preserve this block as far as possible and had issued orders that *jhumming* was not to be permitted. This caused very little hardship to the villagers as there was practically no *jhumming* now, except a little on the flatland between the hill and the Singra River.[63]

In the eastern Khasi and Jaintia hills at Nongpoh, there were two patches of *sal*—one some 4 mi. up the road to Shilling on the east of the road, where at the foot of the hills, there was a patch of *sal*. This, however, was comparatively small in area and did not extend far up the hills which had been *jhummed*. The other patch was on the west of the road about 1 mi. above Nongpoh. Marching from Nongpoli to Nongbir, a strip was found of *sal* along a stream, about 1 mi. east of the main road. This again was small in area and badly stocked. The hill sides above had been *jhummed* and there was nothing left worth preserving. From Nongbir to Mathan, the country was, as before, very open and covered with inferior jungle, evidently burnt over nearly every year. It was all the same type of *jhummed* land, without a vestige of *sal*. As in the case of Jirang, the siem of Khyrim had forbidden *jhumming* where there were *sal* trees, hence in a fairly large area here, there was no *jhumming* at all. The existing villages practised wet rice cultivation in the valleys.[64]

A Bengali firm had been given a lease of this area during this period. This lease was to expire in October 1916, and after that period the area was proposed to be divided into three or more blocks, according to the natural features of the ground. Siem encouraged them to open only one at a time. But if any attempt at regularizing fellings and at preservation of the forests was to be made, it was essential that markings be done by a trained officer on silvicultural principles. The firm naturally wanted to cut every mature tree, regardless of the situation. There were still plenty of young *sal* and seedlings on the ground, but in this way, they stood a good chance of being altogether choked out by the jungle which was bound to spring up consequent to the opening up of the forest in this haphazard way. The firm, now working, was not doing well as it cut only 316 and 471 trees in the first two seasons, and only supplied about 9,000 sleepers to the Assam-Bengal railway, instead of the 15,000 contracted for. Their expenses were heavy and they got profit of only about 5–6 paise out of a sale price of Rs.2–Rs.4 per sleeper (1½ c. ft. of timber) when delivered to the railway. They should however have been able to cut a good many more trees than this, but they seem to have had considerable trouble in inducing sawyers to stay and work. But if on the expiration of this lease, further work or leases were contemplated with preservation and improvement of the forest attempted at the same time, marking must have to be done by a competent officer and the lessee must be content only to take such trees, as it was permissible to fell from a silvicultural point of view.[65]

The country appeared to have all been *jhummed* at some period and had grown up in long grass, with bamboos and few inferior trees intermingled; higher up, there were not so many signs of *jhumming*, but the country was still bare. Bamboos were less frequent and trees isolated. The grass became shorter and finally the country resembled rather the 'down' lands of England. There were a number of Nepali graziers (*bathan*) at Mawlalut, and these men burnt off the grass annually for grazing purposes. As the country here was very thinly populated and there was very little cultivation, they did not appear to do any harm by grazing their buffalo over the hills.[66]

At Umsuai, the actual forests consisted of bamboos, grass and undergrowth of shrubs with a certain amount of scattered standard trees. Among these were some rubber trees—*koroi* (Albizia Procera), *lampalia* (Duabhanga Sonneratoides), *simul* (Bombax Malabaricum), *mallala* (Macaranga), *makria* (Schima Wallichii), etc., while there were reported to be a few trees of *poma* (Oedrela Toona), *awari* (Amoora Spectabilis), *gomari* (Gmelina Arborea), *cham* (Artocarpus Chaplasha) and on the banks of the Umiam, very few *nahor* (Mesua Forrca).[67]

These trees were, however, few and far between and the Umiam was not floatable up here, so that extraction was impossible. Moreover, there was no *jhumming* going on now, except just close to villages where they cultivated a little land for production of chillies and things which would not grow on flooded land. Otherwise, they all went for wet rice cultivation. Part of the forest too was a sacred grove, so there seemed no need to insist on further measures for its protection as it could not be called a valuable forest.[68]

Between Phlang and Barrato, evergreen forest was still more in evidence and gradually filled most of the *nullahs* and ridges on the hills, though the hill sides were often grasslands, the result apparently of *jhumming* in former times. There was, however, very little *jhumming* now and it was because the villages of Barrato had excellent lowland cultivation.[69] At Barrato itself, the higher hills were mostly open grass with very few trees. This was burnt off annually for the supply of young grass on which the villagers pastured their cattle.[70]

Conclusion

After necessary investigation and perusal of the reports that were filed by the local officers, the provincial administration temporarily shelved the project. As a result of the investigations and the reports thereof, the administration came to the conclusion that:

> with regard to terrace cultivation I am to say that the Chief Commissioner agrees with you in thinking that the time has not come either for the general prohibition of *jhumming* or for the imposition of an enhanced rate of house tax on those villagers which follow this method of cultivation. But he approves of your proposal that every effort should be made to induce the villagers to adopt the system of permanent cultivation by means of grant of revenue free term of three years succeeded by a term during which special concession rates of revenue will be charged. Agricultural loans should also be granted on a liberal scale for the purchase of seedlings, agricultural implements and plough cattle in places where these can be used.[71]

The instruction, however, still maintained that in the Jowai division enquiries should be continued to be made to find out if there was any suitable land for conversion to terrace cultivation and what forms of encouragement should be given for such conversions.[72] The government had already taken measures against the deforestations resulted from *jhumming*. The agreement of 1859 armed the government with the power to take over a forest also for its protection.[73] In the year of the agreement, the government persuaded the sirdar of Jeerung, whose petty state contained one of the richest *sal* forests in Khasi hills, to hand it over to the government.[74] In the Jaintia hills, forests were declared forest reserves.[75] A programme of restoration was also taken up in 1871 by planting pine trees around the civil station. There was a magnificent grove of Pine jungle surrounded south and north of Shillong, which was a consequence of these efforts of the government.[76] Although the state administration gave up the project of shifting to terrace cultivation temporarily, the investigation reports revealed interesting details about the existing state of peoples' subsistence endeavours, the arguments in favour of and against shifting cultivation by the people themselves, as well the administration, the state of forests and the transforming landscapes as well as property relations of the Khasi and Jaintia people. It depicted the stress the tribal agriculture was facing from the colonial commercial interest and how it was transforming the economy and society of the people.

It is interesting to note that the colonial administration had discovered the existence of subsistence agriculture in their colonies only. They were amazed at how minimalist Indian agriculturists were in the hills of north-east India. They not only produced little, vast lands were left uncultivated which, according to them, could have produced enough crops to not only sustain the local population but also export and earn profit. In characterizing such non-profitable agriculture, they fell back upon the dominant European discourse of slash and burn being the most primitive form of agriculture. It was first developed in England pioneered by Lewis Henry Morgan in his work *Ancient Societies or Researches in the Lines of Human Progress from Savagery through Barbarism to Civilization*.[77] However, it was printed in America and was very difficult to obtain in London.[78] The book associated slash and burn cultivation with primitive tribal cultures which was subsequently further legitimized and established by Friedrich Engels.[79] History was thus seen as a linear and progressive march of civilization and various stages were delineated in this universalist scheme of things. The earliest stage was seen where the tribal was associated with savagery, barbarism and low knowledge of science and technology. In this scheme, slash and burn cultivation was seen as the lowest form of agriculture pursued by savage and barbaric tribal communities. The colonialists appropriated this discourse. One way of introducing the tribals to civilization was to transform their agriculture. Hence, various constructed corollary devastations, like low productivity, destruction of forests, frequent migrations, lack of surplus for trade and commerce, low nutrition, etc., were invented to dissuade the tribes from practising such agriculture and go for settled rice or terrace cultivation. Often forced and coercive measures like imposition and enhancement of house taxes were levied to compel the tribal to opt for settled rice cultivation. There were not just proposals but actual initiatives of encouraging immigration of settled agriculturists into the tribal areas to train and inspire them to go for settled cultivation which was posited as a more civilized form of agriculture. Thus in the plain areas of Assam, among the Karbi (Mikir), Lalung and Mishings, attempts were not only made to encourage civilized agriculture, but such endeavours were made in the hills too. In the Lushai hills, the British introduced wet rice cultivation along with which came the system of ploughing, introduction of cows and bulls in the region and trained farmers were encouraged to migrate to the hills. However, due to inadequacy of lowland/plains area in the region, this system of cultivation could not take off.[80] Thomas Lewin tried every persuasion to convert the Mizo farmers into sedentary agriculturists. Like the administration in Khasi hills, he even offered loans to the Mizo farmers to settle down permanently in a place and become permanent owners of a piece of land where they could practise settled rice cultivation.[81] It did not produce much result though. Subsequently, Major John Shakespeare also tried to introduce settled rice cultivation in the Lushai hills in 1898, not with the Mizo farmers but imported Nepali and Santhal farmers. This too did not succeed, except for little flatlands available in the valleys of Champai and Vanlaiphai.[82] In north Vanlaiphai, the administration sold paddy plots to the farmers.[83] By 1913, the administration was leasing land for horticultural use for commercial purposes in Lushai hills.[84] The presence of settled and plough cultivation in south Mizoram are remnants of these endeavours of the British. However, the civilizational discourse failed to impress the tribal farmers as far as giving up slash and burn cultivation was concerned. Like all other tribals, the Khasis too showed little interest in terrace cultivation. However, the Jaintias did take it up more, but that was less due to the insistence of the colonial administration and more due to the availability of plain lands in their area, and also their proximity with Sylhet, from where they had already been familiarized with settled rice cultivation techniques.

Notes and References

1. Ajay Pratap, *An Ethnology of Shifting Cultivation in Eastern India*, Delhi, 2000, pp. 17–18.
2. Ibid, p. 19.
3. Ibid.
4. Ibid., p. 20.
5. It has recently been coined by Dutch scholar Willem Van Schendel who has extensively worked on the shifting cultivation tribes of South Asia. It was also used earlier by James C. Scott for the same people.
6. Most popular being the one given by colonial administrators like T.H. Lewin who felt it was a word given by the neighbouring Bengalees to those hill people who practiced shifting cultivation. T.H. Lewin, *A Fly on the Wheel or How I helped to Govern India*, London, 1912; repr., Calcutta, 1977.
7. C. Geertz, *Agricultural Involution*, Berkeley and Los Angeles, California, 1963, pp. 15–16.
8. K.J. Pelzer, *Pioneer Settlements in Asiatic Tropics*, New York, 1945, p. 16.
9. E.H.G. Dobby, *South East Asia*, London, 1954, pp. 347–9.
10. O.H.K. Spate, 'The Burmese Village', in *The Geographical Review*, no. 25, 1945, p. 527.
11. Geertz, *Agricultural Involution*, pp. 15–16.
12. Pratap, *An Ethnology of Shifting Cultivation in Eastern India*, pp. 17–18.
13. A. White, *Memoir of David Scott*, London, 1832, p. 21.
14. *Bengal Political Consultations*, 2 July 1830, no. 2, Scott to Swinton, 4 June 1830, Bengal Political Consultations, 13 August 1830, no. 65; Scott to Swinton, 17 July 1830. Cited in P.N. Dutta, *Impact of the West on Khasi and Jaintias: A Survey of Political, Economic and Social Change*, New Delhi, 1982, p. 159.
15. W.J. Allen, *Report on the Administration of the Cossyah and Jynteah Hill Territory*, Calcutta, 1858, p. 49.
16. A.J.M. Mills, *Report on the Khasi and Jaintia Hills*, Calcutta, 1853, pp. 3–4.
17. *Bengal General Proceedings*, 27 August 1868, no. 60, Carnegy to Secretary, Government of Bengal, 27 June 1868. Cited in Dutta, *Impact of the West on Khasi and Jaintias*, p. 159.
18. *Bengal General Proceedings*, 8 September 1873, no. 83, Hopkinson to Secretary, Government of Bengal, 30 July 1873. Cited in Dutta, *Impact of the West on Khasi and Jaintias*, p. 159.
19. J.P. Mills, *Some Recent Contact Problems in the Khasi Hills: Essays in Anthropology, Presented to Rai Bahadur Sarat Chandra Roy*, Lucknow, n.d., pp. 1–10.
20. Allen, *Report on the Administration of the Cossyah and Jynteah Hill Territory*, p. 273.
21. Assam Secretariat Records, Letter issued to Board of Revenue, vol. 23, 1872, Hopkinson to Secretary, Government of Bengal, 4 September, no. 173, IT. Cited in Dutta, *Impact of the West on Khasi and Jaintias*, p. 159.
22. *Bengal General Proceedings*, 8 September 1873, no. 83, Hopkinson to Secretary,

Barbaric Hoe and Civilized Plough 165

Government of Bengal, 30 July 1873. Cited in Dutta, *Impact of the West on Khasi and Jaintias*, p. 159.
23. L. Lindsay, *Lives of Lindsay or Memoir of House of Crawford and Balcarres*, vol. III, London, 1849, p. 174.
24. Mills, *Report on the Khasi and Jaintia Hills*, pp. 3–4.
25. Dutta, *Impact of the West on Khasi and Jaintias*, p. 161.
26. *Bengal General Proceedings*, 27 August 1868, no. 60, Carnegy to Secretary, Government of Bengal, 27 June 1868.
27. Ibid.
28. *Bengal General Proceedings*, 27 August 1868, no. 60, Carnegy to Secretary, Government of Bengal, 27 June 1868; Assam Secretariat Records, Letter issued to Board of Revenue, vol. 23, 1872, Hopkinson to Secretary, Government of Bengal, 4 September, no. 173, IT.
29. *Bengal General Proceedings*, 27 August 1868, no. 60, Carnegy to Secretary, Government of Bengal, 27 June 1868.
30. Ibid.
31. Secretary, Chief Commissioner of Assam to the Director, Land Records and Agriculture, Assam, Shillong, 28 May 1897, Revenue, A. Progs., March 1898, nos. 5–8, Assam Secretariat.
32. Ibid.
33. F.J. Monahan, Director, Land Records and Agriculture to the Secretary, Chief Commissioner of Assam, no. 406, 22 January 1898, Revenue, A. Progs., March 1898, nos. 5–8, Assam Secretariat.
34. Ibid.
35. Ibid.
36. Revd Stephens quoted in F.J. Monahan, Director, Land Records and Agriculture to the Secretary, Chief Commissioner of Assam, no. 406, 22 January 1898, Revenue, A. Progs., March 1898, nos. 5–8, Assam Secretariat.
37. Ibid.
38. Ibid.
39. Deputy Commissioner, Khasi-Jaintia Hills to the Director, Land Records and Agriculture, no. 3195, Shillong, 7 December 1897, Revenue, A. Progs., March 1898, nos. 5–8, Assam Secretariat.
40. Commissioner, Surma Valley and Hill Districts to A. Phillipson, no. 2998, 19 April 1915, Revenue Agriculture, A Progs., nos. 50–61, Assam Secretariat.
41. Ibid.
42. Ibid.
43. B.C. Allen to Chief Commissioner, 16 May 1915, Revenue Agriculture, A Progs., August 1915, no. 50, Assam Secretariat.
44. W.J. Reid to Chief Commissioner, Assam, no. 2998, 19 April 1915, Revenue Agriculture, A Progs., August 1915, no. 50, Assam Secretariat.
45. Ibid.
46. A.W. Dentith, Deputy Commissioner-in-Charge of Forests, Khasi and Jaintia Hills to the Commissioner Surma Valley and Hill Districts, Silchar, no. B 537,

Shillong, 17 March 1915, Revenue Agriculture, A Progs., August 1915, no. 50, Assam Secretariat.
47. Ibid.
48. Ibid.
49. Ibid.
50. Ibid.
51. Ibid.
52. The Subdivisional Officer, Jowai to the Deputy Commissioner-in-Charge of Forests, Khasi and Jaintia Hills, no. 125, 11 March 1915, Revenue Agriculture, A Progs., August 1915, no. 50, Assam Secretariat.
53. Ibid.
54. Ibid.
55. Ibid.
56. Ibid.
57. Ibid.
58. Ibid.
59. Ibid.
60. G.N. Simeon, Assistant Conservator of Forests, Khasi-Jaintia Hills, 4 March 1915, no. 53, Revenue Agriculture, A Progs., August 1915, no. 50, Assam Secretariat.
61. Ibid.
62. Ibid.
63. Ibid.
64. Ibid.
65. Ibid.
66. Ibid.
66. Ibid.
67. Ibid.
68. Ibid.
69. Ibid.
70. Ibid.
71. A. Phillipson, Under Secretary to the Chief Commissioner of Assam, Revenue Department to the Commissioner of Surma Valley and Hill Districts, no. 2589R, Shillong, 22 June 1915, Revenue Agriculture, A Progs., August 1915, no. 60, Assam Secretariat.
72. Ibid.
73. Assam Secretariat Records, Letter issued to the Government, vol. 34, 1867–8, Hopkinson to Secretary, Government of Bengal, 3 November, no. 160.
74. B.C. Allen, *Assam District Gazetteer*, vol. X, p. 81.
75. W.W. Hunter, *A Statistical Account of Assam*, vol. II, Calcutta, 1880, p. 21.
76. Ibid., p. 226.
77. Lewis Henry Morgan, *Ancient Societies or Researches in the Lines of Human Progress from Savagery through Barbarism to Civilization*, London, 1877.
78. Friedrich Engels, *The Origin of the Family, Private Property and the State*, Moscow, 1948; repr. 1985, p. 5.

79. Ibid.
80. Audrey Laldinpui and Laithangpuii, 'Rice Economy and Gender Concerns in Mizoram', in *North East India Studies*, vol. 1, no. 2, January 2006, pp. 107–27.
81. Lewin, *A Fly on the Wheel*, p. 294.
82. Major John Shakespeare, Superintendent, Lushai Hills to the Secretary to the Chief Commissioner of Assam, Aijal Political Department, CB 5, pol. 48, Aijal, 1904, Mizoram State Archives.
83. Aijal Political Department, CB 8, pol. 75, Aijal, 22 May 1911.
84. Under Secretary to the Chief Commissioner of Assam, Aijal Political Department, CB 19, Shillong, 4 March 1913.

8

Agricultural Knowledge and Practices in a Bengal District: Burdwan under Colonial Rule

Achintya Kumar Dutta

From time immemorial, agriculture has been the mainstay of livelihood of Indians and the 'vital industry' of India. Even on the eve of the coming of the Europeans, the Indian farmers knew various techniques related to agricultural practices, including selection and preparation of soil and seeds according to planned techniques, and also methods of sowing. They were familiar with rotation of crops and importance of using manures, thus making the agricultural production impressive.[1] Though in the eighteenth century, Indian agriculture was criticized as stagnant, the peasants of the subcontinent carried on these practices with their traditional knowledge even in the colonial times. These agricultural practices were more or less adopted in most provinces including Bengal. Burdwan, a district of south-west Bengal, also followed these practices and became one of the most productive regions of the country. Rice received considerable significance in the agriculture of Rarh Bengal districts—Burdwan, Bankura, Birbhum and certain parts of Midnapore. Burdwan had the advantage of huge fertile lands conducive to rice production. An appraisal of the agrarian history of Burdwan would give us an opportunity to recognize the diversities and distinct features of the agricultural practices and knowledge at a local level. A scrutiny of the agrarian changes in Burdwan, viz., the means and method of cultivation and production, use of implements, seeds and manures, irrigation system and so on, helps one to explain the nuances of agricultural practices and knowledge at the regional and nation level in a better way. It supplements our findings in macro-history and adds to historiography of agriculture in Bengal. Thus, a history of agricultural practices and knowledge of Burdwan would throw some light on how the peasants of a fertile region maintained the age-old practices in cultivation and production, and responded to the changes introduced in this field and to the extension of knowledge.

Debate on Agricultural Growth

Agricultural knowledge and practices of a country can best be understood in the light of the debate on the growth rate of food crop output and efficiency of the method and means of production. The classic work on Indian agricultural output and productivity in the colonial period is that of George Blyn, who argues, on the basis of the acreage and yield estimates collected by the colonial revenue administration, that average growth rate of food crop output was small in India during 1891–1947. Bengal suffered a much greater decline than any other region in India. In rice, the acreage declined and yield per acre fell sharply.[2] Observing that static overall yield figures do not mean that output everywhere was stagnant, B.R. Tomlinson has commented that Blyn's account of Indian agriculture is pessimistic. Blyn's estimates have been subjected to minute scrutiny, and the fragility of their empirical base expounded at length. Estimates of agricultural output, based on direct measurements derived from rigorous and wide-ranging crop-cutting experiments, were not widely available until the 1940s.[3] Tomlinson further argues that much of the data for crop output and yields was gathered very casually, as part of the fiscal system, and the linkage between land tax and output estimates may have encouraged under-reporting, especially as the British bureaucracy progressively gave up day-to-day supervision of rural administration, after the political reforms of 1919.[4] There is no denying the fact that, in the aggregate, agricultural output was largely stagnant in colonial India. But that does not mean that there was no dynamism in rural India. Punjab and parts of western United Provinces witnessed some agricultural growth (wheat in the Punjab, and cotton and groundnuts in western United Provinces), underpinned by technological change and capital investment, during the late nineteenth and early twentieth centuries. Even some parts of Gujarat and Bengal experienced, in the same period, an expanding acreage and growth of yield.[5] Obviously, Bengal was not in a better position than Punjab in the field of agriculture. Punjab had been spending Rs.40 lakhs a year on agriculture up to 1939, while Bengal was spending only Rs.9 lakhs. Consequently, the latter could not make much headway in agricultural research or in terms of producing agricultural graduates until 1940, or in establishing wider contact with the cultivators by the Agricultural Department, as was found in Punjab.[6]

Examining the situation in Burdwan, it is observed that with regard to acreage and output of rice in the district, from the beginning of the twentieth century till the end of World War II, there was no sign of progress. There was rather a declining trend with some sort of stagnation.[7] The area under winter rice was somewhat stagnant on an average till 1909. From 1910 to 1943, it could not reach the level of the first decade of the twentieth century. Nevertheless, it can be said that sometimes during this period, a little upward trend was found in the acreage of winter rice. A period of ten years from 1917 witnessed such a picture, though the move was not uninterrupted.[8] However, it was only from 1944 that a steady vertical move in respect of area under winter rice was noted in the district.[9]

However, the agricultural condition of Burdwan district was not as disappointing as was thought by Blyn in the case of Bengal. Surprisingly, the production of rice was more than enough for its people and there was huge surplus of rice (as shown in Table 8.1). It indicates that agricultural knowledge and practices in Burdwan were not inferior and inefficient. Though there was, as shown in Table 8.1, a significant decline in surplus of rice between 1911 and 1941, caused by a downward trend in rice acreage and output due to the occurrence of malaria, notorious floods, and insufficient rainfall, Burdwan did not face any scarcity of paddy or rice. Burdwan had a surplus of rice until World War II. Like manufactured goods, agricultural products, viz., rice and pulses, were also more than sufficient for local consumption and were largely exported to some districts of Bihar and Bengal, including Calcutta.[10] An official enquiry of the 1930s also revealed that Burdwan was a surplus district in Bengal.[11] Its people did not suffer from shortage of food crop except under abnormal circumstances caused by natural calamity or events of politico-economic degradation. Rice constituted the main item of trade of the district and there was considerable export of rice to the west of the district.[12]

Burdwan's rice was exported to the new sources of demand, viz., mining and industrial area and plantation sector in eastern India, including Raiganj-Asansol industrial area, Serampur industrial area, north Bengal and Calcutta, where population increased, and also to the deficit regions of south Bihar and

TABLE 8.1: Output, Consumption and Surplus of Rice (in tons) in Burdwan District (Decennial Statistics)

Year	Population as per census	Total output of rice	Consumption	Surplus
1901	15,28,290	2,97,522	2,53,314	44,208
1911	15,33,874	3,71,858	2,54,239	1,17,619
1921	14,34,771	2,46,148	2,37,813	8,335
1931	15,75,699	3,04,325	2,61,172	43,153
1941	18,90,732	3,09,186	3,13,388	–4,202

Source: *Census 1951, West Bengal, Burdwan*, p. XIII, from Manoj Kumar Sanyal, 'Peasant Paddy Production, Indebtedness and Dispossession: A Case Study of the Bengal Districts 1901–1941', Ph.D. thesis, University of Calcutta.

Assam.[13] Rice of Burdwan was good in quality and more reasonable in rate than the rice of most other districts in south-west Bengal. Bulk of the marketable surplus of paddy and rice could then easily find way from the villages to the newly arising trade centres and ultimately get exported to the bigger markets. Although no statistics regarding the quantity of rice exported from Burdwan district for the period 1901–47 are available, one can get an idea about it by looking at the data available in an unofficial source for the period from 1917–18 to 1921–2, when government restriction decelerated the movement of foodgrains from one place to another. It reveals that 37,575t. of rice and 1,26,300t. of paddy were supplied from Burdwan to Calcutta during this period.[14] Lack of data inhibits explanation of the rank occupied by Burdwan (in the western Bengal block districts, viz., Burdwan, Birbhum, Hooghly, Bankura, Midnapore, Murshidabad and part of Nadia) in terms of export of rice before and after this period. But it can be ascertained that Burdwan district contributed substantially in supplying paddy/rice to Calcutta. There also took place the diversion of rice to other places, during the control system prevalent at that time. For instance, during the seven months from December 1918 to June 1919, a total of 47,662t. of rice was exported from Burdwan to various provinces—Assam, Bihar, United Provinces, Central Provinces, Bombay, Madras and Punjab.[15]

Trade in paddy and rice became one of the major areas of the economic activities in this district. It was primarily inter-district and secondarily, inter-provincial. As a surplus district, Burdwan had helped Bengal over decades to keep her flow of rice supply unabated to the areas where it was in great demand. As to the movements of foodgrains, provincial statistics are kept by trade blocks and movement between the districts cannot be estimated accurately from them. No one had any idea what normal stocks of paddy/rice were in each district. Yet, the data available from various official and non-official sources throw some light on the volume of rice trade and movement of rice from the districts, and they reveal that Burdwan occupied an important place in the rice trade of Bengal. Thousands of tons of rice were exported from this district during colonial rule. The colonial government had to depend largely on the grains produced in south-west Bengal districts, in order to cater to the needs of areas of scarcity. The expanding rice trade of this region, directly or indirectly, helped the colonial rulers to take the village surplus away from the hinterland.

The ideas generally entertained in England at that time about Indian agriculture were unimpressive. Many even held the opinion in India that Indian agriculture was, as a whole, primitive and backward, and that little had been done to remedy it. But John Augustus Voelcker, consulting chemist to the Royal Agricultural Society of England, who was sent to India in 1889 for better working of the agricultural department, did not agree with this view.[16] He held that the conditions under which Indian crops were grown were good. The Indian raiyat or cultivator was quite as good as the average British farmer, and in some respects, even superior. H.T. Colebrooke talked about similarities of agricultural methods in India and Europe, particularly countries of south Europe, illustrating that the plough and spade of Bengal, and the coarse substitute for the harrow, would remind one of similar implements in Spain.[17] It is true that Indian cultivators could not grow larger crops, but neither the agricultural knowledge and practices, nor the cultivators were responsible for it. There were a lot of constraints on agricultural development. Facilities for improvement were limited, such as supply of water and manure, and the cultivators could have hardly had any access to them. The district of Burdwan could not escape these constraints, and yet it strove hard to maintain its reputation as the rice 'bowl' of Bengal.

Resource Endowments for Agricultural Production

The district of Burdwan[18] was not only famous for Burdwan fever but was also well known for its agricultural richness, for which it was called the 'granary' of Bengal. The agricultural efflorescence of Burdwan has been referred to in Kabikankan Mukundaram's *Chanimangal*, which is said to have been written in the sixteenth century. It is known that in the late sixteenth century, Sangam Rai, a Khetri of Kotli *mahalla* in Lahore, on his way back from pilgrimage to Jagannath temple at Puri, was so attracted by the great fertility of the soil and the luxuriance of the crops in the area of the province, now known as Burdwan district, that he settled and commenced business at a place called Baikunthapur, a village on the outskirts of Burdwan town. He succeeded so well that he amassed a large fortune.[19] Subsequently, the family acquired prominence under able successors and emerged as Burdwan Raj family. The productive agricultural value of Burdwan has also been recognized by the European scholars. It has been pointed out by a celebrated scholar in the early nineteenth century that Burdwan was one of the most productive and dearest districts in Bengal.[20] Burdwan in those days was prosperous, considered a healthy place and people from Calcutta needing a change would come here.[21] Burdwan's productive character has also been illustrated in the following words:

That this district continues in a progressive state of improvement is evident from the number of new villages erected and the increasing number of brick-buildings, both for religious and domestic purposes, nor is there any other portion of territory in Hindoostan that can compare with it for productive agricultural value in proportion to its size. In this respect Burdwan may claim the first rank; the second may be assigned to the province of Tanjore in the southern Carnatic.[22]

The eastern part of the district under the subdivisions of Burdwan, Kalna and Katwa, covering three-fourths of its total area, encompassed by three rivers—the Ajay, the Bhagirathi or the Hooghly and the Damodar—contains fertile alluvial plains and agricultural riches; while the western part under Asansol subdivision, covering the rest of the area, consisting of rocky laterite soil, bears mineral resources, dotted with coal pits and factories. In other words, the large area of rocks forming Chotanagpur hills constitutes the western part, while its eastern portion gradually merges into the alluvial belt of central Bengal, the great rice plain of Bengal. Agricultural lands in the eastern part were well irrigated, extensively cultivated and densely populated. This area was the production zone of the district. Though the onslaught of Burdwan fever devastated the district's agricultural prosperity in the 1860s and 1870s by decimating the rural population, especially the sizeable labour force, the district survived and its agricultural potential revived after the scourge of the epidemic fever ended. The process of recovery from the ravages is said to have started from the 1880s, when the inroad of the disease reduced and controlling measures, such as opening of dispensaries, filling up of the breeding places of mosquitoes, and destruction of jungles by the settlement of displaced persons, combated the fever.[23]

Two main resources of the district for agricultural production were land and water. Lands for agricultural purposes were situated around the villages. Such agricultural lands of Burdwan district were divided into two main classes—*sona* (highland) and *sali* (lowland). The former stood on the vicinity of the homestead site, while the latter was remotely situated.[24] The *sona* land was generally of loamy soil and lay above the ordinary inundation level. This was generally two-cropped land (called *aus* land) which grew a variety of crops, such as *aus* or autumn rice, pulses, oilseeds and potatoes. The *sali* land was a formation of rich clayey soil known as *metel* or *entel*, which became fertile by the diluvial detritus washed down from the higher level. It was water retentive and remained submerged under water during the rain. It was usually one-cropped land, commonly called *aman* paddy land, which produced mainly *aman* or winter rice for which the land was best suitable. Most of the cultivated areas belonged to this category of land, where paddy was planted considerably. Besides *sona* and *sali*, there were *diara* lands or alluvial river lands, formed by the deposition of river silt on the beds of the rivers. Such lands became fertile every year during the rain by a deposition of silt and required no artificial manures. These were suitable for growing winter and spring crops, such as wheat, barley and oilseeds. So far as soil is concerned as distinct from land, seven to eight types of soil are recognized in the district. Most common of them are *entel* (clay soil), *doansh* (loamy soil), *bele* (sandy soil) and *pali* (riverside soil). These are of better quality and are well suited for producing all types of crops, such as paddy, wheat, gram, potato, pulse and linseed. The cultivable lands of the district belonged to these types of soil.

In most areas of India, the availability of water determines the nature of agricultural production. Cultivators in regions and localities without artificial sources of water supply are constrained in their choice of crop cultivation. In most areas, the state of agriculture depends vitally upon the irrigational facilities, which in many parts of India, such as coastal Andhra, Tamil Nadu, Ganges-Jamuna doab and Punjab, made appreciable agrarian changes. Bengal also witnessed the availability of irrigational facilities. Easy supply of water facilitated expansion of cultivation and production here. The fertile paddy fields of the district of Burdwan could easily receive water for cultivation, as Burdwan had ample sources of water, like rivers, streams, tanks, pools, wells, and above all, rainfall. Water and silt for the riverside lands could easily be available from the Bhagirathi, Ajay, Damodar, Khari, Banka and Brahmani and consequently, cultivation could be carried on there. Rain water, undoubtedly, added to the cultivation in the district, but the distribution of rainfall over the season was often unsatisfactory. Artificial irrigation was, therefore, required for important crops and about half the cultivated land received some forms of irrigation.[25] Tanks were the chief source of irrigation for cultivable lands in the villages. Plenty of tanks were available in the district. In no other parts of Bengal were so many tanks found. Some of them were finest in Bengal and most of them well situated for irrigational purpose.[26] This popular mode of watering land covered 90 per cent of the total irrigation of Burdwan.[27] Almost every village of the district had a special tank for irrigating land. These were situated in the fields distant from village sites, acting as reservoirs of water for the cropped fields. Traces of such tanks are found even today in some of the well-known villages of Burdwan district.

Of the other sources of water, government canals, wells and creeks were available in the district. But they covered a small percentage of the total irrigated area, as their number was small. However, canal irrigation could not play any remarkable role in this regard till 1933, as the Eden canal, opened in 1881, could irrigate only a small area. There were no regular distributaries from this canal, and therefore, the economical distribution of water could not be properly maintained. Tanks and other sources, thus, played an important role. It was only after the opening of the Damodar canal in 1933, that the area irrigated by government canals showed an unprecedented increasing trend. It made headway from the mid-1930s, when out of the total irrigated area of 2,79,700 acres, canal irrigation covered 1,52,665 acres.[28] The Damodar canal supplied water for irrigating lands, so far not covered by canal water, and provided security in times of drought in the canal area of Burdwan. The silt-laden canal water was sure to give an impetus to the fertility of the land.

Crop Pattern

Unlike Punjab, Bengal did not see major changes in the crop pattern. It remained paddy-based. A variety of crops were planted and produced in the cultivated lands of Burdwan district. Besides paddy, some other crops of the district were wheat, barley, gram, maize, rapeseed and mustard, potato, jute and sugar cane. Potato became commercially a valuable crop in the district and its cultivation gradually increased from the 1870s in the eastern part of the district.[29] Sugar cane was produced in many parts but to a limited extent. Jute was also grown, though in small areas, in the Kalna subdivision. But rice was by far the dominant crop of the district, as it is found today. It is revealed by the *Settlement Report of the Burdwan Raj (1891–96)* that the net area cropped in the *raj khas mahals* and in four coparcenary villages in the district of Burdwan was 27,304.45 acres and out of it, a total of 24,492.35 acres were under rice cultivation.[30] The preponderance of rice cultivation was also the scene in this district in the first half of the twentieth century, as is evident from Table 8.2, occupying 80 per cent of the total cropped area. It was not simply a subsistence crop but the single-most predominant commercialized product, increasingly grown for the market. Rice cultivation was so popular that thirty-four varieties of rice were cultivated in the district.[31] Burdwan also produced a variety of autumn rice or *aus* which was consumed by the poor local people. But summer rice or *boro*, which was coarse and less nutritious, was sown in the marginal lands of the district.[32]

The question may here be raised as to why the cropped area of Burdwan district was dominated by rice cultivation. Rice was the main food crop of the people of Burdwan and, therefore, for subsistence the cultivators used to plant it much more than any other crop. This preponderance of rice cultivation was determined mainly by the provision of irrigation over large areas and the availability of suitable land as well. It has already been mentioned that irrigation was done mainly from tanks and *beels*. The soil which could retain rain water for a long time was the low-lying lands in the eastern part of the district, where paddy was planted considerably. Interestingly, the soil of the large area of Burdwan, as well as most of the cultivable land, was fit for growing paddy. The crops other than paddy, such as sugar cane, wheat, etc., grew mainly on double-cropped land or *aus* land. These crops required irrigation at a time when it was difficult to provide water for them. Hence, paddy was cultivated in most cases. Besides, the out-turn would be reduced if same crop was planted repeatedly. But the exception was in case of paddy. There was a general belief that paddy, instead of exhausting the land, improved it. This belief was founded on the fact that for bringing the newly broken soil to condition, paddy was one of the best crops.[33]

TABLE 8.2: Area (in acres) Cropped in Burdwan District (Decennial Statistics)

Crops	1901–2	1909–10	1919–20	1929–30	1939–40	1949–50
Rice: Winter or *aman*	7,19,000	8,40,000	7,36,000	6,09,700	5,72,900	10,31,600
Rice: Autumn or *aus*	1,62,000	1,32,200	1,21,300	59,400	31,800	54,900
Rice: Summer or *boro*	500	300	400	300	700	2,300
Total rice	8,81,500	9,72,500	8,57,700	6,69,400	6,05,400	10,88,800
Wheat	1,500	1,800	2,000	1,100	1,500	5,600
Gram	11,000	8,400	4,000	2,700	1,700	17,000
Maize	NA	3,000	1,200	1,900	3,300	200
Linseed	26,000	17,500	6,300	4,000	2,800	600
Rapeseed and mustard	23,800	20,200	8,500	7,200	7,700	2,300
Sugar cane	27,700	24,000	18,100	14,000	10,400	12,000
Jute	11,000	16,300	6,600	3,000	1,400	11,600
Total cropped area	11,28,300	11,78,300	9,88,700	7,59,900	6,69,800	11,80,500

Source: *Season and Crop Reports of Bengal from 1901 to 1950*, relevant years (appendix ii), appendices.
Note: NA: Not Available.

Another important point to be noted in this context is that of the main three species of paddy—*aman*, *aus* and *boro*—*aman* or winter rice was largely cultivated in the paddy producing area. In the densely populated eastern portion of Burdwan district, the low-lying tracts separating the village sites, which constituted the larger part of the cultivated area, were devoted almost entirely to winter rice.[34] The peasants and the farmers were aware of the fact that the output of *aus* or autumn rice, which was of inferior quality and coarse, was smaller and it brought a low price, whereas *aman* was of superior quality and easily accepted and consumed by the people. As the latter was very popular, they used to grow different varieties of the same in Burdwan. Some of them were of finest quality—*jhingesaal*, *badsabhog*, etc., which had a good demand in many places of Bengal. The determining factor was again the irrigation facility. *Aus* was generally produced on highlands, where provision for water supply was very difficult. More importantly, the out-turn of *aman* per bigha was higher than that of *aus*. In the late nineteenth century, the cost of cultivation of *aus* was Rs.11-3-0 (Rs.11 – Ana 3 – Pie 0) per bigha and the out-turn was 8 maunds of grain valued at Rs.7-0-0, and 8 *pans* of straw valued at Rs.1-8-0.[35] It thus gave a total of Rs.8-8-0 after harvest. On the other hand, the cost of cultivation of *aman* was Rs.8-12-0 per bigha at the same time and it yielded 10 maunds of paddy valued at Rs.10-0-0 and 10 *pans* of straw valued at Rs.2-0-0, i.e. a total of Rs.12-0-0.[36] An official report reveals that the yield per acre of unhusked winter paddy from tank irrigated land was 22.50 maunds, with 34 *pans* of straw which valued at Rs.95-12-0 and provided a net profit of Rs.54-0-0 before the Depression, and even a net profit of Rs.13-5-0 after the slump.[37] The yield of autumn paddy per acre, on the other hand, was 20 maunds and that of straw was 20 *pans*, which in total valued at Rs.67-8-0 and left a profit of Rs.30-2-0 before the slump and only Rs.1-4-0 after it.[38] *Boro* was a coarse rice and less nutritious, and therefore, had less demand and consequently the peasants did not grow it. All this explains why *aman* rice was largely cultivated in the cropped area of the district. Needless to say, with all this knowledge, the peasant inclined to produce rice, and specially winter rice, for sustenance and living in a better way.

Method of Cultivation and Agricultural Practices

The peasants of Burdwan were familiar with varieties of paddy or rice. The several varieties, adapted to every circumstance of soil, climate and season, might exercise the judgement of sagacious cultivators. The selection of the most suitable kinds was not neglected by the husbandman. Interestingly, the peasants of Burdwan mainly followed the traditional means and methods of cultivation. A brief account of them seems to be relevant in this context. Both the main crops, *aman* and *aus* paddy, were grown by transplanting. Both required a preliminary breaking of the soil known as *dhular chash* and puddling known as *kadar chash*, which were done by repeated and frequent ploughing. For both, the land was lightly manured with cow dung, six cartloads per acre appearing the most usual amount, though where tank mud or river silt was available, this might be omitted.[39] After breaking of the soil when rain set in, the clay would be softened. It was then worked into mud in which rice seedlings were transplanted. Both *aman* and *aus*, after transplanting, were weeded according to local need, twice being most usual. For *aman*, the land was in most seasons irrigated thrice if water was available; for *aus*, irrigation was not usually given unless the season was found to be unfavourable. It is through this method that the seedlings grew and crop was reaped as and when matured. To remove the grain from the straw, three different methods were used by the peasant in accordance with their convenience—the crops were trodden by bullocks, beaten by a stick, and struck against a plank. The grain was winnowed in the wind, and was stored either in the jars of unbaked earth or in baskets made of twigs or of grass.

Manures used by the cultivators in this district were cow dung, cow dung ashes, oilcakes (both castor cake and mustard cake), pond mud and hide salt. In most of the cultivating fields of the villages, manures were applied and these were inevitable for growing rice, sugar cane, potatoes and onions. Almost every raiyat had his dung heap, and cow dung was used on all crops, except pulses. But there was no uniformity of its application to the land. It was used more for producing sugar cane and potatoes than *aman* or *aus*.[40] Oilcakes were applied for growing potatoes, sugar cane, etc. Pond mud, which was then regarded as the most important manure, was used for producing all crops. But it required huge quantity per bigha, viz., 40 cartloads per bigha.[41] It also had evil effects on paddy sometimes, causing a disease of paddy, known as *kadamara*, and in such cases, hide salt was applied. Green manuring was also practised by a limited number of raiyats in Burdwan district. Horn shavings were used only in a few villages of the district.[42]

Rotation of crops was practised but no regular system was followed in the district.[43] The condition of the field at the time, the state of the weather, the demand in the market and the individual means of the particular farmer were the considerations that determined the cultivation of a particular crop. The general practice in Burdwan district was to grow *aman* paddy repeatedly on the low-lying lands, and *aus* paddy on the higher grounds surrounding the village sites in the rainy season and one of the pulses as winter crop. Potatoes, onions, etc., sometimes took the place of the pulses. Sugar cane was a special crop requiring a full year to ripen, and was grown at intervals of three or four years.[44]

The majority of the raiyats cultivated with their own hands, with the help of indigenous agricultural implements which were few in number and very simple in construction, such as ordinary country plough, ladder (used for breaking the clods, pressing the soil, gathering the weeds, levelling the ground, and burying the seeds when sown), *phor* (a weeding hook used in weeding and stirring the soil), *pashuni* (very much resembles an English hand hoe), *kodali* or spade and sickle (which were used in harvesting). Their prosperity depended upon their doing so. An official report reveals that indigenous implements of India were ingenious and effective and there was hardly any scope for improvement.[45] The implements used in irrigation in Burdwan and other districts of Burdwan division were very simple and inexpensive, but fairly efficient.[46] Water was raised from wells by means of buckets or earthen pots with a rope which was occasionally put round the pulley on a wooden bar fixed on supports.[47] For irrigating from fields, tanks or shallow depressions, a *donga* made of wood and subsequently made of iron was used to raise the water.

Such tenants had their own plough and oxen. If a raiyat had to borrow one bullock, he did so with the understanding that, for each day that he assists the bullock owner in ploughing his land, he would have to do two days of ploughing for himself. Thus, these raiyats often cultivated on the cooperative principle called *ganta*, under which system each member of the association helped the other in the same manner and to the same extent during ploughing, transplanting, weeding, reaping and gathering.[48] The higher castes whose social position forbade them to handle the plough cultivated lands through servants or hired agricultural labourers. Another practice related to cultivation and production that was extremely common in Burdwan district was bhagjote or sharecropping system. The usual practice was that the sharecroppers had to bear all the expenses of cultivation, supplying plough, cattle, seed and manure, and the crop was shared equally between the sharecropper and the original holder of the land. There were also other methods of division, but those were far less common in this district. This system was freely used by the better classes of tenants, the big and rich peasants, *bhadraloks* or higher castes, and by the widows, absentees and others, who were unable to personally manage their lands.[49] Under this system, the sharecroppers or bhagjotedars had to reap and bind into sheaves after due notice to, and in the presence of, the owner. While crops were growing, the owner usually visited the field to form an estimate of the out-turn. Interestingly, not less than a quarter of the paddy lands, and in places, over one-third, were cultivated by sharecroppers.[50] They were deprived of the status of tenant or a raiyat, having no rights on land, neither security of tenure, nor fixity of rent. The period for which the sharecropper held the same land varied. Cases were not uncommon where they had held the same land even for generations. On the other hand, some landlords changed them every year or after a few years, though the same sharecropper continued under the same landowner. In fact, they were at the mercy of the landowner, as the tenancy legislations, introduced by the colonial administration, did not give them tenancy right. As the practice was highly profitable and hassle free to the superior or rich peasants or the landlords, the latter strove hard to retain it and neither the colonial rulers nor the Burdwan Raj dared to defy them, because they were the real controller of agricultural land and production and hence, highly powerful in the villages.

The *saja* or *dhanthika* system was adopted frequently when lands could not be cultivated under personal supervision. By this arrangement, the cultivator had to pay a certain fixed quantity of actual produce of particular seasons, anything above and over that fixed quantity being appropriated by himself. These cultivators had a more permanent interest on land than the bhagjotedars.[51]

Irrigational Knowledge and Practice

The Damodar and its tributaries bestowed agricultural prosperity on this district. The supply of silt-laden water to the land by the inundation of rivers, which was facilitated by the low banks of the rivers, known as 'overflow irrigation', fertilized the cultivable land. This natural and comparatively easier irrigation was retained in the eighteenth century, providing benefits to the cultivators.[52] This traditional system of irrigation declined during the colonial rule when the north bank of the Damodar was heavily embanked in the mid-nineteenth century for protecting railways and the Grand Trunk Road from the floods of the Damodar. Moreover, the neglect of *pulbandi* works (maintenance of embankment works of the rivers and other channels of irrigation) caused the accumulation of silt, blocking the mouths of the old channels of irrigation, including that of the Damodar. The collection of revenue became the main consideration of the colonial rulers, and steps were hardly taken to restore the decaying channels of irrigation.

Consequently, the cultivators had to depend mainly on rain water and the tanks which served as reservoirs of water in the villages and the main sources of irrigation. Irrigation from the tanks was done by two methods known as *melan* and *sech*.[53] *Melan* was done by cutting the bank of the tank and allowing the water to flow by gravity from field to field. *Sech* was done by raising water usually by means of a *duni*, a boat-shaped implement made of iron (formerly made of a narrow single piece of wood) and operated on the lever principle by one man. This was raised and lowered by means of a bamboo working on a pivot. The water was then sent as rule by well-defined channels. It was more under control and therefore, more economically and widely used. Of the other sources of irrigation, government canals, wells and creeks were prevalent in Burdwan. But they covered a small percentage of the total irrigated area till 1940 (as is evident from Table 8.3). Wells were not numerous and the cultivators had a superstitious dread of irrigating lands with water raised from them.[54] In low-lying areas of the district, where the banks of the creeks were low, the common practice was to temporarily dam such creeks for the purpose of watering the neighbouring fields.[55] Water was taken out from the creeks with a *duni* or with a *chheni*, a sort of scoop of matting with strings attached and worked by two men. In the hill tracts of the western part of Burdwan, the practice of storing up rain water through *bandhs* or terraced embankments was well understood, and the whole system of cultivation there seemed to have been dependent upon it.[56] The hillsides were converted into rice fields, which were embanked along their lower edges. The rain water in its downward course was arrested and, instead of being allowed to pass down the hillside in a torrent, was made to irrigate the field one after another.[57] Water could also be derived from dry beds along the banks of the smaller streams by making holes in the coarse red sands of which riverbeds were made.[58]

Modern scientific irrigation system or canal irrigation was unknown in Burdwan until the opening of the Eden canal in 1881. Originally, its

TABLE 8.3: Area (in acres) Irrigated in Burdwan District
(Decennial Statistics)

Area irrigated from	Years					
	1900–1	1910–11	1920–1	1930–1	1940–1	1950–1
Government Canal	19,358	16,711	40,832	17,312	1,63,952	1,95,720
Private Canal	NA	20,000	NA	20,000	200	19,791
Tanks	NA	3,47,844	4,19,696	2,71,995	1,10,000	79,640
Wells	NA	4,155	400	350	1,750	4,000
Other Sources	NA	1,54,517	1,94,946	12,964	22,100	45,017
Total Area Irrigated	NA	5,43,227	6,55,874	3,22,621	2,98,002	3,44,168

Source: *Agricultural Statistics, Bengal for the Year 1900–1941*; *An Account of Land Management in West Bengal for the Year 1945–46*; *Season and Crop Report of West Bengal for the Year 1950-51*, relevant tables.
Note: NA: Not Available.

construction was not intended for irrigation; its purpose was to throw a supply of fresh water for sanitary purposes from the Damodar into the natural channel and old riverbeds of Burdwan and Hooghly districts that were invaded in the 1860s and 1870s by a fatal epidemic fever, which was attributed to the stagnant and insanitary condition of the water courses. However, the practice of irrigating from this canal sprang up after its opening, when a demand for irrigation arose by the peasants, and the government hesitatingly agreed to it.[59] But it proved to be unsatisfactory because of certain inherent defects. There were no regular distributaries from this canal and, therefore, the economical distribution of water could not be properly maintained. The old riverbeds serving as its distributing channels were more or less silted up or choked up with reeds and hence could not carry adequate water to meet the requirement of the cultivators. The canal only possessed two distributaries constructed in 1896–8.[60] Moreover, at the critical time, usually in the month of October, when more water was required for paddy crop, the absence of a weir at the intake prevented any supply reaching the canal. The Indian Irrigation Commission was of the opinion that irrigation system of the Eden canal could never be satisfactory without a weir across the river at its head sluices, for without a weir there could never be any certainty of securing the supply required for the canal.[61] Irrigation from it was restricted to the early and later parts of the season. Insecure supply of water was the main complaint. Besides, it did not serve the large unprotected area between the Damodar and the Ajay rivers. As it had no weir at its head, its supply depended on the level of the river and was, therefore, uncertain. To overcome these problems, the government adopted a project of constructing the Damodar canal, which was opened in 1933. This was really an improvement on the existing pattern of agriculture in Burdwan, contributing to an increase in fertility of the soil and bringing a large cultivable area under modern irrigation.

In fact, a large part of the area irrigated in Burdwan district, as shown in Table 8.3, was covered by water from the tanks and other sources till 1932. The role of canal irrigation was not found to be impressive. For instance, in the areas under fourteen police stations of the three subdivisions of Sadar, Kalna and Katwa, out of 39.50 per cent of the cultivated area irrigated, the government canal had a share of 1.20 per cent only, whereas 34.30 per cent was occupied by tanks.[62] It was only after the opening of the Damodar canal that the area irrigated by government canals showed an uninterrupted and unprecedented increasing trend. Yet, the significance of tank irrigation cannot be underestimated. It occupied an important place till 1939–40, when out of the total irrigated area of 3,39,509 acres, 1,54,079 acres were covered by government canals and 1,60,327 by tanks.[63] It was from 1940 that influence of tanks in the field of irrigation gradually reduced and canal irrigation made headway. But the irrigation system by the government canals in Burdwan was still incomplete because no water could be diverted into the canal during low flow seasons.[64]

Critique of Government's Role

Interestingly, irrigation in Burdwan during the period under review cannot be said to have been satisfactory. There was further scope for its improvement which could have made remarkable agricultural developments. Many of the existing tanks required renovation, but were neglected. Neither the government nor the landlords spent anything for their maintenance. Excavation of new tanks was also required for the district.[65] But that was lacking. In fact, scientific and technological development of irrigation in Burdwan was not found until 1933. Though a large area was brought under irrigation by the Damodar canal, it also had certain limitations. The existing irrigation works were neglected and this affected production as a whole. The exorbitant revenue demand by the British often compelled the landlords to reduce their investments for the maintenance of proper irrigation.[66]

In fact, the government did not demonstrate any commendable gesture for irrigational works in Bengal and this is illustrated by the fact that government canals covered only 0.40 per cent of Bengal's net sown area, as contrasted with 39.20 per cent in the Punjab in 1928.[67] The government seemed to have believed that the province's relatively plentiful rainfall rendered any need for irrigation superfluous. The irrigation levels in the province of Bengal were quite low indeed. The colonial administration had little interest in public investment in irrigation.[68] New scientific irrigation system combined with modern technological know-how started in Burdwan district and other areas of Bengal from 1957 to 1958, when Damodar Valley Corporation started supplying water for irrigation through the Durgapur barrage.[69]

Little had been done to transform the system and method of cultivation in Burdwan in terms of introducing new appliances, new crops or manures to improve the productivity of the land. Interestingly, an attempt to bring about a change in this field was made in 1885 when Burdwan agricultural farm was established at Pallah, about two miles to the south-east of Burdwan town. This farm was set up by the Government of Bengal on the recommendation of the Burdwan Raj and the Court of Wards.[70] The cost of running the firm was incurred by the Burdwan Raj, while its management was entrusted to the agriculture department. The object of this farm was primarily to carry out field investigation into crops and manures; to propagate pure seeds of recommended varieties for distribution; and to give trials with implements and to cultivate a certain area of the farm as a demonstration to cultivators. Successful experiments were carried on in the farm, from 1891 to 1892, with paddy, sugar cane, jute, cotton, potato and other crops in which bonemeal and saltpetre were used and Shibpur-wrought iron plough was tested.[71] The result of the experiments in the case of paddy was impressive. Seed was accordingly distributed from the farm to the cultivators and private persons for experimental purposes.[72] The manurial experiments with *aus* and *aman* paddy indicated the superiority of bonemeal and saltpetre to cow dung. During 1894–5, the experiment demonstrated that an admixture of bonemeal and saltpetre enhanced the yield from 17 maunds 29 seers of paddy to 56 maunds 32 seers per acre and gave a net increase of profit estimated at Rs.95 per acre.[73]

This farm was one of the oldest private experimental farms in India. In October 1921, the Maharajadhiraj of Burdwan handed over the farm to the government and undertook to pay Rs.2,800 per annum towards the expenses for running it.[74] The farm continued its demonstration work and distribution of paddy seeds among the cultivators. But it could not produce the desired result. The farm could not spread and popularize the newly adopted experiments among the raiyats. The officials of the farm failed to encourage the cultivators for using new manures and iron plough in their lands. The farm was situated at such a lonely place that the public could not take full advantage of the valuable experiments and demonstrations in the farm.[75] In fact, a few experimental farms, which were essential for agricultural improvement, were established in Bengal, but their position was not favourable. The condition of Seebpore (Shibpur) farm (established in 1887) was so deplorable that it was contemplated to give it up in the 1890s.[76] Experimental works in these farms were actually done on a very limited scale and in a crude way, and that could do but little real good. The peasants of the district naturally resorted to the old practices using the traditional seeds and manures at a moderate price. Moreover, the raiyats were said to have been prejudiced against rejecting the age-old agricultural practices that they learnt from their forefathers and had been carrying out for years. They feared that ignoring traditional practices and using a new method, which was unknown to them, might affect the cultivation and production and therefore hardly felt any urge to adopt new methods of cultivation and ingredients of agriculture.[77] The prejudices should have been broken for agricultural improvement, and the zamindars or landlords or superior peasants could take initiative in doing so. But that was not in the agenda of the landlords.

There was also the District Agricultural Association in Burdwan with the district magistrate as its president, which distributed seeds and manures among the poor cultivators for demonstration works.[78] But they could not lead to any change in this field. Moreover, financial stringencies crippled the Association in the 1930s to do any work.[79] Besides, teaching of agriculture in the school was attempted in Burdwan district. In a secondary school of Amarpur, teaching of agriculture was initiated as an experimental measure in the 1910s, which was financed by local effort.[80] One gentleman, Babu Manmatha Nath Pal provided money for the maintenance of the school and a suitable area for land as a demonstration farm. In this case, the government only agreed to appoint a district agriculture officer to Amarpur for agricultural teaching to students.[81] But, it is not known whether the government made any serious effort in this direction to influence the local people with new teaching or technique. It appears that as there was no organized system of agricultural enquiry, the spread of agricultural education was hardly possible.

However, these experimental works in Burdwan district proved to be inadequate and could not bring about any change in the system of cultivation and production. Active prosecution and encouragement of positive measures could not be undertaken. An official report reveals that the number of improved agricultural implements that had been tried in Burdwan division was very limited and of these very few had met with any degree of success.[82] The American plough had been tried with success by some of the well-to-do farmers and planters. As its price was high, the ordinary farmers or raiyats were not in a position to buy it.

The raiyats were not very responsive to the new experiments including new manures. They allowed the agricultural department to apply the manures only to the most backward and sickly fields, and would not tolerate any interference with fields promising to give a fair average out-turn.[83] The raiyats of Burdwan were willing to use dung or oilcake on their fields. It was only from the end of the 1940s that the raiyats of the district became more open to the idea of using manure. Chemical manures like ammonium sulphate, phosphates and standard mixtures were then becoming increasingly popular.[84] On the other hand, it should be noted that there was a lack of artificial manures and as manufacturing them was expensive, the government declined to induce the peasants for their use. Consequently, organic manures, such as cattle dung, were used for fertility of soil. In fact, better varieties of seeds, better use of manures and fertilizers, better control of water supply for crops, better implements, and better agricultural practices could have improved agriculture, but the government was hardly enthusiastic in this regard. It is revealed by an estimate made by an English scholar that the yield of rice could be increased by 20 to 30 per cent by manures, 5 per cent by improved varieties and 5 per cent by protecting from pests and diseases.[85]

In fact, the colonial state's outlook towards agriculture was limited until World War II, which necessitated a new agricultural policy and consequently, efforts were made to change the existing situation. It was much more concerned with law and order and left the people to shape their agricultural destinies. The colonial state's interest was to collect land revenue and to spend as little as possible on improving agriculture. Agriculture could have been improved further by extending modern scientific knowledge to each district of Bengal and by making all possible attempts to use it successfully. But that was rarely seen in Burdwan. Though modern scientific methods to improve agriculture were welcomed by Burdwan Raj, who encouraged scientific experimental works of Burdwan Agricultural Farm, the agricultural department could not manoeuvre it to give agriculture a new dimension. Although scientific research in the agricultural sector was being carried out in certain institutes like the agricultural research institute at Pusa, agricultural departments in the provinces were not well equipped with men of scientific attainments to bring science close to agricultural development. The agricultural department of Bengal endeavoured to carry on experiments in a few experimental farms, but it had certain limitations and therefore could not make much headway in the agricultural development. The close connection between science and agricultural improvement, as pointed out by Voelcker, should have been emphasized,[86] but that was rarely found. Agriculture was not helped by the land reforms including the Permanent Settlement which had rather promoted absentee landlordism and rack-renting. More importantly, the zamindars extracted as much rent as possible, but spent almost nothing for the development of agriculture. Though few attempts were made for irrigational improvement, no design for agricultural development was introduced in a consistent manner in India.[87] Though considerable efforts were made in the 1930s to improve the implements, the new implements were not always more effective, compared with the old ones.[88] The peasants were not properly informed about what retarded agriculture and prevented progress in agriculture. In Burdwan, manure heaps were not well kept, and sometimes the heap was very carelessly managed, and allowed to get too dry.[89] More importantly, demonstration at experimental farms, of the value of manure and ways of preservation, seems to have been useless since most cultivators burnt cattle manure as fuel. New appliances had scarcely been introduced in Burdwan. The only exception was iron pans for boiling sugar cane juice and irrigation vessels made of iron. Attempts have from time to time been made to introduce iron ploughs, but without much success.[90] The common raiyats were not even made self-sufficient in the existing agricultural implements. The agricultural department could not introduce new implements or modify the existing plough. It was in the post-1947 years that modern implements like American hand hoe, tractor and other power tools became common with big farmers.[91]

Conclusion

It is true that there were constraints of irrigation in Burdwan and the condition necessary for further growth of production was absent. But it is hard to deny scientificity in indigenous systems of irrigation. Traditional indigenous knowledge about irrigation helped produce crops, though it lacked potential for higher production. Before the facilities of canal irrigation were available, agriculture had been extensive in Burdwan district, and market centres and towns had developed long ago. Arguably, if indigenous knowledge and technique of agriculture had lacked potential for progressive growth of output, it may be observed that the application of modern technology could not really yield better results during colonial rule. Rather, unwise and uneconomic public works were found to have disrupted the traditional potential of indigenous technology. The embankment on the Damodar had stopped overflow irrigation, caused water stagnation, and provided favourable habitats for anopheles mosquitoes and thereby helped diffusion of epidemic fever which had adversely affected the agriculture of the district. The age-old practices that held the ruler and the ruled in a common interest towards the management and improvement of land and crop, the two basic things in rural husbandry, were never scientifically tested by the foreign masters, to whom the prosperity of Bengal husbandry was entirely a gift of nature.[92] During British rule, the zamindars gradually neglected the practice of management of agricultural resources, land and water. Unfortunately, the traditional system was not given due care and patronage for further improvement. Nevertheless, it can be ascertained that indigenous agricultural knowledge and practices of the people of Burdwan had potential, though condemned as backward by the colonizers, which was proved by its ability to produce more than was required for consumption and having accumulation of surplus of food crop which was supplied regularly to the deficit region. However, any knowledge and system in any branch of science needs periodic experiment and observation for its better performance. Surprisingly, that was rarely done in this district in the colonial period. Therefore, the husbandmen carried on the age-old system and practices in the field of agriculture.

Notes and References

1. Satpal Sangwan, *Science, Technology and Colonization: An Indian Experience, 1757-1857*, Delhi, 1991, p. 4.
2. George Blyn, *Agricultural Trends in India, 1891-1947: Output, Availability and Productivity*, Philadelphia, Pennsylvania, 1966, Chaps. V–VII.

3. B.R. Tomlinson, *The Economy of Modern India 1860–1970*, New Delhi, 1998, p. 31.
4. Ibid.
5. Ibid., p. 32.
6. N.M. Khan, *Post-war Agricultural Development in Bengal*, Calcutta, 1945, p. 2. This book contains the speech of the author at the meeting of the Royal Asiatic Society of Bengal on 1 November 1945.
7. Achintya Kumar Dutta, *Economy and Ecology in a Bengal District: Burdwan, 1880–1947*, Kolkata, 2002, pp. 93–5.
8. Ibid., p. 94.
9. *Season and Crop Reports of Bengal for the Year 1944–45* and *1949–51*, Calcutta, 1949 and 1953, pp. 6–7, Appendix II.
10. *Settlement Report of Burdwan Raj and Certain Other Estates in Districts Burdwan, Hooghly and Bankura in 1891–1896*, Calcutta, 1897, pp. 23–4.
11. *Report of the Bengal Paddy and Rice Enquiry Committee*, vol. I, Calcutta, 1940, p. 23.
12. J.C.K. Peterson, *Bengal District Gazetteers: Burdwan*, 1910; repr., New Delhi, 1985, p. 126; Achintya Kumar Dutta, 'Rice Trade in the "Rice Bowl of Bengal": Burdwan, 1880-1947', *Indian Economic and Social History Review*, vol. 49, no. 1, 2012, pp. 73–104.
13. Dutta, 'Rice Trade: Burdwan', pp. 85–90.
14. Santosh Nath Seth, *Bange Chaltattwa* (All about Rice in Bengal), Chandennagar, 1332 BS, 1925, pp. 403–6.
15. *Report on the Operation of Director of Civil Supplies, Bengal from 1918 to 1920*, Calcutta, 1922, p. 21.
16. J.A. Voelcker, *Report on the Improvement of Indian Agriculture*, 2nd edn, Calcutta, 1897, pp. 10–11.
17. Henry Thomas Colebrooke, *Remarks on the Husbandry and Internal Commerce of Bengal*, Calcutta, 1804; repr., Calcutta, 1884, p. 23.
18. The district, as was finally formed in 1881, was bound on the north by the Santal parganas, Birbhum and Murshidabad; on the east, by Nadia; on the south, by Bankura and Midnapore and on the west by Manbhum.
19. Somerset Playne, *Bengal and Assam, Bihar and Orissa: Their History, People, Commerce and Industrial Resources*, London, 1917, p. 524.
20. Colebrooke, *Remarks on the Husbandry and Internal Commerce of Bengal*, p. 20.
21. *Settlement Report of Burdwan Raj*, pp. 20–1; Government of Bengal, Municipal Department, Progs. 22–5, June 1899.
22. Walter Hamilton, *A Geographical, Statistical and Historical Description of Hindoostan and the Adjacent Countries*, 1820; first repr., 1971, vol. I, Delhi, p. 157.
23. *Census, 1951, West Bengal: Burdwan*, p. xv; *Burdwan Division Municipal Report 1918-19 and 1919-20*, pp. 8–9.
24. *Settlement Report of Burdwan Raj*, p. 29.
25. K.A. Hill and P.K. Banerjee, *Final Report on the Survey and Settlement Operations in the District of Burdwan 1927-34*, Calcutta, 1940, p. 12.

26. A.C. Sen, *Report on the Agricultural Experiments and Enquiries in Burdwan Division*, Calcutta, 1897, p. 5.
27. Hill and Banerjee, *Final Report on the Survey and Settlement Operations in the District of Burdwan*, p. 12.
28. For details, see Dutta, *Economy and Ecology*, pp. 119–20.
29. *Settlement Report of Burdwan Raj*, p. 29.
30. Ibid., Appendix III.
31. For details, see Dutta, *Economy and Ecology*, pp. 88–93.
32. Sen, *Report on the Agricultural Experiments and Enquiries in Burdwan Division*, p. 41; *Settlement Report of Burdwan Raj*, pp. 16–17.
33. Sen, *Report on the Agricultural Experiments and Enquiries in Burdwan Division*, p. 14; for details about paddy cultivation, see Raj Narayan Biswas, *'Bardhamaner Dhaner Chas'* (Paddy Cultivation in Burdwan), *Krisak*, no. 11, Falgun 1320 BS, p. 325.
34. Birendra Nath Ganguli, *Trends of Agriculture and Population in the Ganges Valley: A Study in Agricultural Economics*, London, 1938, p. 248.
35. An anna was a currency unit formerly used in India, equal to 1/16 rupee. It was subdivided into 4 pice or 12 pies (thus there were 64 pice or 192 pies in a rupee). In other words, 1 rupee = 16 annas; 1 anna = 4 pice; 1 pice = 3 pies. *Pan* is a rough measure of volume of straw. 1 *pan* weighed on an average about 18 seers. Sen, *Report on the Agricultural Experiments and Enquiries in Burdwan Division*, p. 18.
36. Ibid.
37. Hill and Banerjee, *Final Report on the Survey and Settlement Operations in the District of Burdwan*, p. 10.
38. Ibid.
39. Ibid., p. 9.
40. Sen, *Report on the Agricultural Experiments and Enquiries in Burdwan Division*, pp. 11–12.
41. Ibid.
42. Ibid., p. 13.
43. *Census, 1951, West Bengal: Burdwan*, p. xxi.
44. Ibid., p. xxii.
45. Voelcker, *Report on the Improvement of Indian Agriculture*, p. 228.
46. Sen, *Report on the Agricultural Experiments and Enquiries in Burdwan Division*, p. 5.
47. *Census, 1951, West Bengal: Burdwan*, p. xviii.
48. *Settlement Report of Burdwan Raj*, p. 28.
49. Hill and Banerjee, *Final Report on the Survey and Settlement Operations in the District of Burdwan*, p. 32.
50. Ibid.
51. *Settlement Report of Burdwan Raj*, p. 28.
52. William Willcocks, *Lectures on the Ancient System of Irrigation in Bengal and its Application to Modern Problems*, Calcutta, 1930, pp. 20, 60.

53. Hill and Banerjee, *Final Report on the Survey and Settlement Operations in the District of Burdwan*, p. 12.
54. Peterson, *Bengal District Gazetteers: Burdwan*, p. 88.
55. *Settlement Report of Burdwan Raj*, pp. 19–20.
56. *Census, 1951, West Bengal: Burdwan*, p. xviii.
57. Peterson, *Bengal District Gazetteers: Burdwan*, p. 88.
58. Sen, *Report on the Agricultural Experiments and Enquiries in Burdwan Division*, p. 62; Peterson, *Bengal District Gazetteers: Burdwan*, p. 88.
59. G.C. Maconchy, *Report on the Protective Irrigation Works in Bengal*, Calcutta, 1902, p. 224; *Selections from the Records of the Bengal Government relating to the Damodar Canal Project from May 1867 to January 1870 and July 1903 to July 1925*, Calcutta, 1927, p. 263.
60. *Selections from the Records relating to Damodar Canal*, p. 263.
61. *Report of the Indian Irrigation Commission 1901–03*, Part II, Calcutta, 1903, pp. 155–6.
62. Hill and Banerjee, *Final Report on the Survey and Settlement Operations in the District of Burdwan*, p. 12.
63. Census 1951: A. Mitra, *An Account of Land Management in West Bengal 1870-1950*, p. 243; *Season and Crop Report for the Year 1939-40*, relevant tables.
64. W.L. Voorduin, *Preliminary Memorandum on the Unified Development of the Damodar River*, Calcutta, 1945, p. 8.
65. A.C. Sen, *Report on the Agricultural Experiments and Enquiries in Burdwan Division*, p. 6; *Settlement Report of Burdwan Raj*, p. 20.
66. B.B. Chaudhuri, 'Rural Power Structure and Agricultural Productivity', in *Agrarian Power and Agricultural Productivity in South Asia*, ed. Meghnad Desai, Susane Howber Rudolph and Ashok Rudra, Delhi, 1984, p. 167.
67. James K. Boyce, *Agrarian Impasse in Bengal: Institutional Constraints to Technological Change*, New York, 1987, pp. 164–5.
68. Ibid., p. 196.
69. P.S. Rau, *All about D.V.C.*, Calcutta, 1955, p. 14; S.K. Basu and S.B. Banerjee, *Evaluation of Damodar Canal (1959–60): A Study in the Benefits of Irrigation in the Damodar Region*, Bombay, 1965, p. 11.
70. *Annual Report of the Burdwan Experimental Farm for the Year 1900–1901 to 1904–05*, p. 1.
71. *Annual Agricultural Report of the Department of Land Records and Agriculture, Bengal*, Revenue Department, Government of Bengal, Progs., November 1895, p. 110.
72. *Settlement Report of Burdwan Raj*, p. 30; for details of experiments, see *Annual Report of Burdwan Experimental Farm*.
73. *Annual Agricultural Report of Land Records*, p. 110.
74. Agriculture and Industrial Department, Government of Bengal, Agricultural Branch, Progs. 1, December 1928.
75. Agriculture and Industrial Department, Government of Bengal, Agricultural Branch Progs. 1–2, June 1933.
76. Voelcker, *Report on the Improvement of Indian Agriculture*, p. 373.

77. *Settlement Report of Burdwan Raj*, p. 30.
78. Agriculture and Industrial Department, Government of Bengal, Agricultural Branch, Progs. 9–10, January 1917; Progs. 5–6, September 1919.
79. Agriculture and Industrial Department, Government of Bengal, Agricultural Branch, Progs. 11–12, August 1935.
80. Agriculture and Industrial Department, Government of Bengal, Agricultural Branch, Progs. 3–4, June 1921.
81. Ibid.
82. Sen, *Report on the Agricultural Experiments and Enquiries in Burdwan Division*, p. 9.
83. Ibid., p. 67.
84. *Census, 1951, West Bengal: Burdwan*, p. xxi.
85. S.B. Singh, *Second World War as Catalyst for Social Changes in India*, Delhi, 1998, p. 37.
86. Voelcker, *Report on the Improvement of Indian Agriculture*, pp. 334–5.
87. Jayashree Sengupta, *A Nation in Transition: Understanding the Indian Economy*, New Delhi, 2007, p. 49.
88. Sir John Russell, *Report on the Work of the Imperial Council of Agricultural Research in Applying Science to Crop Production in India*, Delhi, 1937, p. 59.
89. Voelcker, *Report on the Improvement of Indian Agriculture*, p. 126.
90. Peterson, *Bengal District Gazetteers: Burdwan*, p. 96.
91. *Census, 1951, West Bengal: Burdwan*, p. xxi.
92. Arabinda Biswas and Swapan Bardhan, 'Agrarian Crisis in Damodar-Bhagirathi Region 1850-1925', *Geographical Review in India*, vol. 37, no. 2, 1975, pp. 132–50.

9

Strains of Settlement: Reclamation and Cultivation in the Sundarbans—Myth and History

Sutapa Chatterjee

Settlements in the Sundarbans have a long antiquity. Marshy tracts, saline creeks, occasional cyclones, large trees and dense undergrowth have characterized the Sundarbans. Here, the jungle and wildlife reign supreme. This vast mangrove forest is the haunt of the Royal Bengal Tiger; crocodiles and other animals abound; and in these environs live a community of brave humans whose courage is manifest in their daily battle against nature, seeking as they do to eke out a living by cultivation, some by venturing deep into the forest to collect honey and at times ending up as victims to the lord of the jungle. Faced by such challenges of nature, collective life in the region is marked by an unusual degree of fluidity, and often ephemerality. Those who made this region their home, cleared forest and brought land under the plough, hailed almost entirely from the margins in more senses than one— they endowed the social fabric of a hostile geographical category with a historical unity and an element of popular spontaneity. This essay is about the formation of settlements and the spread of habitation; it is also crucially about the natural and human strains to which they are subject. In other words, it is both about the imprint of recurrent hazards on settled human existence and the chain of implosions endemic in a frontier agrarian society.

Life in the Mangroves

Bangāla zamīn kharāb va dil-tang;
Rū fātha khwān va manumā; mzī dirang.
Andar bar-o bahr jā-yi āsāyish nīst;
Yā hast dahan-i shīr, yā kām-i nahang.[1]
Bengal is a ruined and doleful land;
Go offer the prayers to the dead, do not delay.

Neither on land nor water is there rest;
It is either the tiger's jaws, or the crocodile's gullet.

Jale Kumir Dangai Bagh ('crocodile in the water and tiger on land') is a popular expression in Bengali about the marshy lowlands of deltaic Bengal. The Sundarbans, a cluster of over a hundred islands, form the largest mangrove delta in the world and is home to a wide variety of flora and fauna. It is widely known for its intriguing beauty; it is also a region of considerable natural, anthropological and archaeological interest. 'Sundarbans' literally means 'beautiful jungle or forest', but the name may also have been derived from the Sundari trees that are characteristic of the region. Other explanations suggest a derivation from 'Samudra Ban' (Sea Forest), or from 'Chandrabhanda' (name of a primitive tribe). However, the generally accepted view is attributed to Sundari trees. The brackish Sundarbans freshwater swamps are a tropical, moist, broadleaf forest region. Further down, especially to the west, lie the Sundarban mangroves, where salinity is more pronounced. In the former, the water turns fresh during the rainy season, when the thrust of the Ganges and Brahmaputra push the intruding salt water out, and bring on a deposit of silt.[2] The Sundarbans cover an area of 14,600 sq. km. (5,600 sq. m.) spread out across the vast Ganges-Brahmaputra delta, and extends from the southern part of Khulna and Bakharganj districts (now in Bangladesh) to the mouth of the Bay of Bengal; scattered portions extend into the state of West Bengal in India. The Bengal delta is a unique land water scape, elucidating binaries of adversity and opportunity, wilderness and habitation, vulnerability and resilience. Settlers in the Sundarbans had to brave lot of unknown situations.[3] The jungle and the rivers fail to provide comfort. Apart from storms, which occur all through the year, the fishermen are always wary of crocodiles. In short, the Sundarbans have been and shall always remain synonymous with fear and anxiety. This essay deals with the relations between man and nature, the reclamation and settlement debate in this area, the history of which is the subject of much controversy. The struggle between man and nature took the form of reclamation, and despite the slow pace of development, changes in settlement patterns and the stratification of society informed the structuring of social relations. The region suffered denudation and deforestation, leading to a continuing expansion of the agrarian frontier.

The diverse and unique nature of this land had struck outsiders long ago. Travelling through the Sundarbans, Francois Bernier (1665–6)[4] wrote:

Among these islands it is in many places dangerous to land and great care must be had that the boat, which during the night is fastened to a tree, be kept at some distance from the shore, for it constantly happens that some person or another falls prey to tigers. These ferocious animals are very apt, it is said, to enter into the boat itself while the people are asleep and to carry away some victim who, if we are to believe the boatmen of the country, generally happens to be the stoutest and fattest of the party.[5]

Nearly 350 years have passed since Bernier made his observation. Today, while tigers have been given protection under wildlife preservation laws, not just in the Sundarbans but elsewhere too, the story of the boatmen is still eerily similar to the story of those who had rowed Bernier across the rivers.[6] In the eighteenth century, the entire region was referred to as *bhati*, or lowlands, subject to the influx of tides.[7] There is, however, unanimity over the fact that the Sundarbans is no ordinary forest.

The Lowlands

The Sundarbans stretch from the Hooghly on the west to the Meghna, the estuary of the Ganga and Brahmaputra, on the east. It covers the southern portions of the district South 24 Parganas in West Bengal, Khulna and Bakharganj in Bangladesh. There is evidence which suggests that due to geomorphological processes, the Bengal basin is tilting, diverting fresh water through the Ganga-Brahmaputra river system to the east. It was thus only natural that east Bengal was marked by its greater agricultural produce and greater population growth compared to the western part. This was made possible by the rich silt deposits because of the shift in the river system.[8] The Indian part of the delta is thus being denied its fresh water from upstream, resulting in increased salinity.[9] The entire region is subject to violent storms, particularly during the monsoon months, as the Bay of Bengal frequently sees the development of cyclonic depressions.[10] This frequent battle with the elements makes life in the Sundarbans much more difficult. The northern part of the Sundarbans is dotted with agrarian settlements, the fertile land allowing the growth of rice.[11] The northern most point is the Hooghly where the river widens; its navigability led to the development of many port towns.[12] The southern part is the unfinished seaboard where the deltaic formation is yet to be completed. Thus, the Sundarbans stretch out as one vast alluvial plane abounding towards the southern fringes in morasses and swamps, which are now filling up.[13] The rivers in the north are stagnant, generally beyond the reach of tides, almost frozen into channels by high banks of silt.[14] The principal rivers here are Hooghly, Sattarmukhi, Jamira, Matla, Bangaduni, Gosaba, Raimangal, Malancha, Bara Panga, Marjata, Bangara, Haringhata, Rabnabad and Meghna.[15] A study of maps and charts makes it evident that the rivers have been constantly changing their courses in this region. A comparison of Rennell's Atlas of 1779 with the Morrison map of 1811 or of later periods makes this clear. The north has always been more suited for human inhabitation than the south, which posed great threats and challenges to settlers.[16]

The topography of the Sundarbans changes as one travels east from the Hooghly, towards the Meghna.[17] Between the Hooghly and the Jamuna, or Kalindi, the water is mostly saline, while east of the Jamuna, the water is generally sweet during the rains.[18] The Bakharganj portion of the Sundarbans was more fertile and considering the elevation of the land, it was not necessary to have embankments, which were otherwise essential to protect the crops.[19] Here, agriculture was predominant and there was large-scale human settlement. As geographer R.K. Mukherjee observes, 'From the fifteenth century, man has carried out the work of reclamation here fighting with the jungle, the tiger, the wild buffalo, the pig and the crocodile, until at the present day, nearly half of what was formerly an impenetrable forest has been converted into gardens of graceful palm and fields of waving rice.'[20]

Initially, the population of the Sundarbans consisted entirely of marginalized people. 'Nearly all the inhabitants' of the Sundarbans were 'either Hindus or Muslims', according to Hunter. 'There were, of course, a few Magh Buddhists and native Christians who also came up with missionary penetration of Portuguese and later British power. The rest belonged to the Hindu and Muslim communities, mostly of low status.' The Hindus were mainly of the shudra castes: Napit, Kaibarta, Kapali, Pod, Chandal, Jalia, Bagdi, Tior, Dhoba, Jogi, Suri and Kaora. Of these twelve castes, the Pods and the Chandals were the 'most numerous'. Next came the Bagdis whom Hunter described as 'rather numerous'. Thereafter came the Kapalis who were neither 'numerous nor few'. The others were either 'few or very few' in number. All of these castes pursued a mix of occupations, such as 'cultivation, woodcutting and fishing' as means of subsistence.

The largest group among the Muslims of the Sundarbans, according to Hunter, was the Shaikhs (cultivators and woodcutters). The Sayyids and Pathans were higher in status than the Shaikhs. They were cultivators, and were said to be 'very few in number'. Besides these, Hunter noted, the Mirshikaris (hunters and fishermen), the Sapurias (snake catchers and snake charmers) and the Bediyas, all outcastes or gypsy tribes, had 'professed Muhammadanism'. The native Christians, as recorded by Hunter, were all cultivators.[21] All these people depended on the forest for their livelihood, though they had to perpetually struggle against the hazards of nature. The geographical and ecological evidence would indicate that most of the area was covered by dense and impenetrable forest but patches of cultivation sprang up from time to time.[22] Struggle between man and nature was continuous in the region. Recent archaeological findings have proved beyond doubt that ports and settlements had sprung up and disappeared recurrently in deltaic

Bengal in the last two thousand years.[23] Clearings would spring up from time to time, only to be overtaken by some natural or political calamity.[24]

The Old Settlements and the Debate

Romantic speculation has identified this terrain as the site of a lost civilization.[25] Different authors have tried to draw different conclusive ideas about the nature of the settlements in the Sundarbans from studies of various sites of excavated settlements. Authors like Satish Mitra and Kalidas Dutta are of the opinion that settlement and civilization in the Sundarbans were of the highest order.[26] European scholars, however, differ from such a conclusion.[27] Beveridge refuses to believe that the Sundarbans was ever highly populated and that its inhabitants enjoyed any kind of an urban existence.[28] The available evidence indicated the presence of sparsely populated pockets, their inhabitants struggling against the elements and hazards posed by wild animals. The earliest reference of this land can be traced back to the epic *Mahabharata*, where the eldest and second Pandava brother Yudhistir and Bhima visited Gangasagar during their pilgrimage.[29] *Vayu Puran* refers to the flourishing trade and commerce in the region[30] and old Bengali literature speaks of settlements among swamps visited by ships from abroad. The poem of Bipradas dated 1495 describes the voyage of a merchant called Chand Saudagar from Burdwan to the sea, mentioning several riverside villages from Bhatpara to Baruipur.[31] It tells us that he passed by Ariadaha on the east and Ghusuri on the west, then kept along the eastern bank of the river and passed Calcutta. He then proceeded down the Adi Ganga, which was formerly the channel by which the water of the Ganga found its way to the sea. Chand Saudagar, however, could only have had a glimpse of the mangrove forest as it stood shrouded in mystery. Even the settlement efforts of Todar Mal in 1582 did not take him beyond Hathiagarh, Medninal (near Canning), Maihati and Dhuliapur, the four southernmost mahals (subdivisions of districts) recorded in Abul Fazl's *Ain-i-Akbari*.[32] Even during the rule of the East India Company, these places constituted the northernmost boundary of the Sundarbans, as is evident from Hunter's *Statistical Account of Bengal*.[33] I would like to put forward the arguments of the British administrators in connection with the settlement debates in order to understand their view of the past human inhabitations in this area.

In 1857, Ralph Smyth conceded that there was an old settlement there, judging from the ruins of several temples and tanks, but, it was impossible to ascertain their exact age.[34] Rennell's map shows that between 1764 and 1772, very little land was under cultivation.[35] The villagers complained to the survey team of constant attacks on them and their cattle by tigers.[36] Gastrell also saw no reason why the legends of past glory should be believed, as there were only the ruins at Kabadak to support any claim of antiquity.[37] Further south, towards the sea, however, there were many clearings, which were made chiefly, if not wholly, by the Maghs of Arakan.[38]

These people crossed over from the Arakan coast and being fearless in nature penetrated deep into the islands,[39] cleared the forests and began cultivating the soil to grow rice. In the process, they also depopulated the land and together with the Portuguese pirates became an additional scourge which the original inhabitants had to suffer. Reference to Khadi-Visheya or Mandal as a flourishing district in the Sena period (AD twelfth century) which in later periods became a dense forest, and the region between Bishkhali and Rabanabad, which was depopulated by the Maghs, may be recalled in this connection.[40] In Bakharganj in 1862, there were many small clearings.[41] Despite such evidence, Gastrell stuck to his theory that the Sundarbans never enjoyed any civilized population and that the settlements were too small and scattered to justify any such tales from the past. The impermanence of some of the settlements, as mentioned earlier, was noticed by Westland, who had also made this point while writing about Jessore.[42] The history of Khanja Ali illustrates this. The local tradition at Bagahat and Masjidpur shows that he came to reclaim the lands in the Sundarbans,[43] a large part of which again became a jungle after his death in 1458. The land was again reclaimed at a later date.[44] While reclaiming such land a second or even a third time, it was natural to come across evidence of earlier habitation, like the brick remains of a bathing ghat or traces of a tank. Oral tradition has it that these were places which had once come under the farmer's plough but got abandoned, one of the reasons for this being the marauding raids of the Marathas who forced the locals to run away and seek shelter elsewhere. Many of such settlements were to return to the fold of the jungle.[45] Beveridge was always very doubtful whether the Sundarbans was ever heavily populated and well cultivated. He had his reasons for such an opinion. Evidences suggest that large quantities of salt were manufactured in Sandwip. This militates against the view of extensive cultivation, for the salt could not have been made without great expenditure of fuel and this, of course, implied the existence of large tracts of jungle.[46] Du Jarric speaks of Sandwip as being able to supply the whole of Bengal with salt. It seems that in olden times salt was reckoned as the most valuable production from this part of the country. In fact, salt manufacture was always a great obstacle to the clearing and colonizing of the *chars* (sandy loams/sand bars/riverine islands) and islands in Bakharganj and Noakhali.

According to Beveridge, 'it is true that Sandwip was cultivated in Caesar Frederick's time (1569), so it is now and there is no reason to suppose that its civilization was greater than it is at present.'[47] It may have had at that time, as it certainly had some thirty or forty years later, one or more forts, but those were marks of insecurity rather than of prosperity, and they did not exist simply because there was peace in the land (during Beveridge's time) and the Arakanese pirates were no longer formidable.[48] Beveridge next relates an incident of the terrible hardship which the crew and passengers of the ship *Ter Schelling* suffered on the coast in 1661. *Ter Schelling* was a Dutch ship which sailed from Batavia for Angula in Bangla (on the Medinipur coast) on 3 September 1661, and was wrecked off the coast of Bengal on the 8th of the following month. The narrative of the sufferings of the crew and passengers was written by one named Glanius. The ship seemed to have landed on an island near Sandwip, and the sufferings of the passengers were most terrible. Their only drink was salt water. They saw very few inhabitants and those whom they did come across were almost as wretched as they themselves were.

The copperplate inscription found at Idilpur in Bakharganj and described by Babu Pratap Chandra Ghosh in the *Journal of the Asiatic Society* (1838) indicates that the inhabitants of that part of the country belonged to a degraded tribe called the Chandabhandas, a fact which is not favourable to the supposition that the Sundarbans was inhabited by a high-caste population in the past. Beveridge has quoted from a letter from Fonseca Jesuit, who was a Jesuit priest and who visited Bakharganj and Jessore between 1598 and 1600: 'The king (of Bakla), after compliments asked where I was bound for, and I replied, I am going to the king of Ciendecan who is to be your Highness' father in law.'[49]

Beveridge here concludes that Ciendecan and Chandecan are identical to Dhumghat and Jessore, and the boy king of Bakla can be no other than Ramchandra Rai, who married the daughter of Pratapaditya. Fonseca had also given a description of the route from Bakla to Chandecan, which would be used to describe the route from Barisal to Kaliganj, near which Pratapaditya's[50] capital was situated. According to Beveridge, Fonseca's description shows that the Sundarbans was much in the same condition in 1599 as it was during his time.[51] He points out the fact that Bikramaditya's choice of Jessore as a safe retreat is the strongest possible evidence of the forested nature of the surrounding country. In all probability, it was cultivated in the previous century by Khanja Ali; but, the experiment had proved a failure and the land had again relapsed into jungle during the time of his successor Chand Khan (sixteenth–seventeenth centuries).[52] Thus, the main contention of Beveridge and other Western historians was that no civilization worth naming had existed in the Sundarbans. Contrary to the opinions expressed by Gastrell, Westland and Beveridge, nationalist historians like Satish Mitra and Kalidas Dutta firmly believed that the Sundarbans was a well civilized and densely populated region. The largely unsettled physical condition of this area with subsidence, cyclone, and above all, ever changing river courses, made many localities vanish periodically, along with their population. Satish Mitra differs sharply from Beveridge and argues that the Sundarbans was inhabited by fairly civilized people. He puts forward many arguments to prove his point. According to him, the island Sandwip at that time produced salt in big quantities and it could annually load as many as 200 ships. Moreover, the Turkish Sultan of Alexandria thought it prudent to get his ships built cheaper from Sandwip. Such a place, Satish Mitra argues, which could build excellent ships and manufacture enormous amounts of salt, must have been inhabited by civilized people. He is of the opinion that people of both higher and lower orders inhabited the place and the manual labour was done by the lower orders who were called Chandabhandas. According to Mitra, the debris of civilization can still prove the glorious days of the Sundarbans with a population of civilized people.[53]

In addition to this, Mitra has given another example to prove that the Sundarbans had innumerable sites of cities. In December 1868, in a meeting of the Asiatic Society, Revd Long stated that in 1848 he saw a Portuguese map of India in Paris that was drawn in the year 1648. This map pointed out five famous ports in Sundarban area named Pacaculy, Cuipitavaz, Noldy, Dapara and Tipara. According to Mitra, the pronunciation of the Portuguese and the Dutch seriously distorted the name of some places that still existed in the Sundarbans. Pacaculy was a pargana in the 24 Parganas. He identified Cuipitavaz with Khilafatabad, which was associated with Khan Jahan Ali Khan. Dapara was identified with Daspara, a place to the north of Rabanabad. As for the other two ports, Mitra is of the opinion that Noldy could be identified with Nalua near Mathurapur in South 24 Parganas and Tipera was identified with Tripura, thereby proving to Mitra's satisfaction that the Sundarbans at that time extended up to Chittagong.[54]

Loins or scanty clothing—which was interpreted by Beveridge as indicative of uncivilized people—was defended by Satish Mitra on the ground that this was usual and normal in a tropical country. In the village Bichhat in Bhuluka pargana, ruins of a huge dock were found where big ships were built. Near it, in Basudevpur, coins of Danujamardanadeva were available. These evidences, as described previously, are sufficient to prove the existence of a civilized people in the Sundarbans and Mitra firmly sticks to this opinion. The study of the ruins of ports and temples suggests that 200 years ago, the riverbanks were teeming with activity. The relics discovered belonged to the Maurya, Sunga, Gupta, Pala and Sena periods.[55] The most important places from where these archaeological findings have been excavated in and around lower Bengal are Chandraketugarh, Harinarayanpur, Deoulpota, Atghetha, Sita Kundu, Karanjali, Roydighi, Tota, Krishnachandrapur, Chandkhal, etc., of which, Chandraketugarh deserves special mention.[56]

Repeated excavations in this area have unearthed a treasure trove of archaeological findings from the pre-Maurya to the Gupta period. The terracotta seals and coins with ship motifs bore a striking resemblance to the coins of Imperial Rome in the early Christian era. Archaeologists have concluded that Chandraketugarh was a flourishing coastal town from the fourth century BC to the post-Gupta era (AD fourth to sixth centuries), having trade contacts with foreign countries. Temples belonging to different cults and types were excavated in the area known at present as the 24 Parganas. A considerable number of them still survive. The *Revenue Survey Report* of 1857 mentions the existence of several ruined temples and two huge tanks in this area, which Hunter also refers to in his *Statistical Account*. In the lot no. 22 in Bakultala, a copperplate grant of Lakshman Sen was found. Two Jain images, one of a Svetambara and the other of a Digambara of Adinath have been found some miles north-west of Raidighi. In Mograhat, some relics of a church were discovered. The folk architecture, as can be seen here, possesses a freshness and charm of its own. The discovery of more archaeological evidences in the course of scientific excavations and accidental findings during the reclamation of the Sundarban delta give us a much more detailed information about the cities, ports, rulers, kings and inhabitants.[57] The deul or monument of lot no. 116,[58] dilapidated old buildings in 'G' plot almost near the Bay, the mosque in lot nos. 129 and 58, the fort of Pratapaditya,[59] scattered temples and ancient settlements in the Chandraketugarh epoch, lead one to believe that the vast tract designated as the Sundarbans was once thriving villages during the early period. It seems that while we do not need to be as sceptical as the I.C.S. Magistrates, we do not have to accept the local chauvinism of Sundarban nationalists (patriots) either.

Life in the Region

Whatever the nature of its population, civilized or tribal, sparse or thick, the Sundarbans with its unique physical features offered a tough proposition to human habitation throughout the ages. In fact, battling with hostilities of nature was so overwhelming an aspect of the settlers' lives in the Sundarbans, that it led to the evolution of deities to whom they could seek refuge psychologically. These deities were not the regular godheads of the Indian pantheon. On the other hand, they were fashioned from day-to-day experiences and realities that revolved around conflict and strife. The attributes of the deities reflected the angst of the settlers who battled against nature and whose survival was by no means easy.[60] The deities, thus, became essentially patron saints whose benevolence could effectively ward off the dangers of the jungle and its dreaded fauna. What, therefore, evolved here was not so much a civilization as a way of living but a culture emerging from responses to the challenges of nature in various forms. Between the seventeenth and nineteenth centuries, in the lower deltaic Bengal, there thrived a *punthi* literature in Bengali verse devoted to the gods and goddesses of the Sundarbans. This literature reflected elements of the surroundings in which it emerged. Its theme was the struggle between humans and nature.[61] *Punthi* literature arose to cater to the most marginal sections of the population.[62] Their beliefs stood apart from mainstream Hinduism and Islam. Folk religion here, as represented by local syncretic cults, had a distinctive aura of its own. The deities worshipped in the Sundarbans had a standing below that of the Bengali pantheon. They were the gods and goddesses of woodcutters, honey gatherers, beeswax gatherers, boat builders, and the most desperate cultivators. They were deities with whom the man in the forest could identify himself. This literature came into existence much before the British became fully hegemonic in Bengal, and was, therefore, not surprisingly free from colonial influence. In fact, *punthi*, in particular, is a literature of transition prevalent during late pre-colonial and early colonial Bengal. The life and condition of the habitants of the Sundarbans are explained here as they were before the advent of British rule in Bengal. The various sources of danger in this area formed the underlying theme of the *punthi* literature of the Sundarbans. The people here had not been equipped with firearms to any appreciable extent, as late as the end of the nineteenth century. They were always at the mercy of wild animals who ventured into the inhabited villages in search of prey. The Royal Bengal Tiger of the Sundarbans was said to be a habitual man-eater, unlike other tigers. One of the central motifs of the *punthis* was man's struggle against wild animals, especially the tiger, which was idealized as a monstrous foe or, at times, even a subordinate deity.

The texts were the *Ghazi-Kalu-Champavati-Kanyar-Punthi* composed by Abdur Rahim. The date of composition is unknown. But reference to the subdivision in which the author was born makes it clear that the text could not have been written before the Mutiny. It was presumably a late nineteenth-century *punthi*. The other text is the *Banabibi Jahuranama* which was composed by Banayuddin in 1877.

There is another version of the tale by Marhum Munshi Muhammad Khater entitled *Banabibi Jahuranama* written in Kartik 1287 bs, i.e. 1880. These texts are all written in simple verse. It is well known that the people chanted some verses before they entered the forest so that no danger would befall them. At first glance, these texts appear to be fantastic narratives about gods, goddesses and their interactions with humble men and women.

The world of the folk deities offers important insight into the character of past human settlements in the Sundarbans. It tells us that, in all probability, the land was sparsely populated.[63] Prosperous cities, as Satish Chandra Mitra and Kalidas Dutta would have us believe, imply social relations and religious beliefs of a higher order. But in the folk tales of the Sundarbans, the prevalence of tiger gods and goddesses point to a more primordial context.[64] More research is needed before we arrive at conclusive statements. The lost civilization of Chandraketugarh cannot be ignored, thus, the debate continues.

The New Settlers

Down the ages there have been numerous attempts at reclaiming forests and setting up human habitation in the region.[65] Before the advent of British rule, such efforts were largely individual and sporadic. The advance of wet rice agriculture into formerly forested regions is one of the oldest themes of Bengali history. In Mukundaram's *Chandi-Mangala*, composed around 1590, the goddess Chandi gives the poem's hero, Kalaketu, a valuable ring and tells him to exchange it for cash.[66] With the money thus obtained—70 million *tankas*—Kalaketu is to clear the forest and establish a city and temple in honour of the goddess.[67] Once the land is ready for agricultural operations, Kalaketu promises to advance Kayastha landlords as much cash as they need for their own thousands of labourers (*prajā*; lit., 'subjects') to come and settle on the newly claimed lands.[68] Such contemporary literary evidence points not only to the high level of monetization in the late sixteenth century, but to the role that cash played in transforming virgin jungle into settled agrarian communities. A number of factors—natural, political and economic—combined to create the booming rice frontier in the east in the seventeenth century, the eastward movement of Bengal's rivers and hence of the active delta, the region's political and commercial integration with Mughal India, and the growth in the money supply with the influx of outside silver in payment for locally manufactured textiles. We shall see that the high volume of cash circulating in Bengal during the Mughal period not only contributed to the movement of men and resources to and within the frontier. Wang Ta-yüan, the Chinese merchant, who visited the delta in 1349–50, observed that the Bengalis 'owe all their tranquility and prosperity to themselves, for its source lies in their devotion to agriculture, whereby a land originally covered with jungle has been reclaimed by their unremitting toil in tilling and planting. . . . The riches and integrity of its people surpass, perhaps, those of Ch'iu-chiang (Palembang) and equal those of Chao-wa (Java)'.[69] Although peoples of the delta had been transforming forested lands to rice fields long before the coming of Muslims, what was new from at least the sixteenth century onward was the association of Muslim holy men (*pīr*), or charismatic persons popularly identified as such, with forest clearing and land reclamation.

In popular memory, some of these men swelled into vivid mythico-historical figures, saints whose lives served as metaphors for the expansion of both religion and agriculture. They have endured precisely because in the collective folk memory, their careers captured and telescoped a complex historical socio-religious process, whereby a land originally forested and non-arable, became arable.

After the Mughal period, land reclamation was taken up by the British. They were largely successful in bringing in people of various origins into the region. Despite the slow pace of development compared to the rest of Bengal, the face of the Sundarbans changed. There was in place a process of transformation of settlement patterns and a continuing stratification of agrarian society. As a fait accompli, the region witnessed denudation and deforestation. As social stratification increased and exploitation intensified, tensions finally exploded in the form of the Tebhaga movement of 1946. The Sundarbans, acquired in 1757 as part of the 24 Parganas, drew the immediate attention of the English and schemes for reclamation came under consideration. A backdrop to this is provided by a recent work by Robert Travers, *Ideology and Empire in Eighteenth-Century India: The British in Bengal* (2007), which suggests that the Company's chequered attempts at establishing legitimate forms of governance, after it began to collect revenue from the province at large, were 'rooted in a series of fiscal crises' (1765–72). Company officials in Bengal were under immense pressure from London for ever greater quantities of revenue. It is worth noting, however, that there are numerous instances of leases being granted to reclaimers in Mughal and nawabi Bengal. The British began the process of pushing the forest back through clearance and settlement operations. The receding of the jungle followed the logic of revenue maximization and the rhetoric of improvement. The rationale behind this was that a land 'covered with impenetrable forest, a hideous den of beasts and reptiles', could only be improved by deforestation.[70] The question of reclaiming the Sundarban jungles was taken up by the Collector General Claude Russell and by Tilmann Henckell, Magistrate of Jessore, in 1783. Their work in the Sundarbans forms a part of a larger story of early British administration in India, the antecedents of which, in terms of ideology and practice, have been analysed by historians like Eric Stokes and Ranajit Guha.[71]

The uncertainties of empire building have been commented upon more recently by Jon Wilson who has argued in favour of the similarities between British and Mughal forms of government. His work draws on and builds upon the work of Robert Travers, who depicts a central engagement in early colonialism with the ways and means of translating Indian law and custom into policy, of securing political legitimacy through a qualified appeal to 'ancient constitutionalism'. Men like Russell and Henckell have escaped the scrutiny of scholars working on the making of the Company Raj and its ideological apparatus, although they were considered to be pioneers in many ways and played a major role in devising land settlements, engaging directly with the cultivators of the soil, and articulating a paternalist position for their own administration. In the Sundarbans, the colonial ideas about rural settlements were often dictated by prevailing conditions of life which, to a large extent, were guided by the influence of a sense of wilderness and frontier. Initially, the government in the frontier part of Bengal found it difficult to identify the social categories which actually formed the intermediary landholding class and this obviously led to a construction of the dominant and the privileged in rural society. This construction of a local society went through a process of rigid classification of agrarian classes, leading to great divisions between the privileged and underprivileged, which later found expression in the radical Tebhaga movement. Changes in settlement patterns in the frontier society brought in stratification, an inevitable outcome of colonial land policy. The social class formation and the gradual sub-infeudation of rights relating to land led to the emergence of a sharecropping stratum. Ultimately, towards the end of our period, the region witnessed socio-economic conflict leading to peasant upheaval. The conflict between man and man overshadowed that between man and jungle.

The Sundarbans as a tract was ever moving southward and had, in its northern limits, lands whose assessment required special knowledge and effort to ascertain the revenue potential of the areas. The zamindars of the north of the Sundarbans always claimed the forests as their land. Both Henckell and Russell came in conflict with the claims of these zamindars, who were eager to add all such wastelands, as they could bring under cultivation, to their estates. The Sundarbans, as such, did not form a separate district, with revenue, magisterial and civil jurisdiction of its own. The Collectors of the 24 Parganas, Jessore and Bakharganj exercised concurrent jurisdiction with the local commissioner in the revenue matters of the Sundarbans.[72] The revenue of all Sundarban estates was paid into the collectorates mentioned earlier. The magisterial and civil jurisdictions of these districts included the Sundarbans.[73] Henckell in his attempts at dealing with the problem of revenue collection essentially encountered a frontier society, but one that had already developed a measure of stratification. The complexity of the situation meant that Henckell had to balance paternalist authoritarian strategy with a more formal structure of revenue and judicial management. In retrospect, one may conclude that Henckell's administrative tenure (1783–89) in the region had to accommodate various factors at work, like the Magh piratical incursions and the hostility of the neighbouring zamindars. The boundary dispute between the neighbouring zamindars and the grantees in this region was not easy to solve. The boundary line was in a state of flux as it was subject to periodic redefinition under the supervision of the various Collector Generals in this area. Henckell wanted to draw the attention of the Board regarding this problem but did not find adequate help forthcoming—the reason being attributed to the heavy expenses the project would entail. Henckell thus failed in his efforts. However, Henckell succeeded in making an important and positive impression in the minds of the local people—the ryots who had received lands on lease from the British government and who were convinced that the Englishmen had their welfare in mind. Thus, not for nothing did the people of this part of the Sundarbans consider Henckell to be their saviour and named the place Hingulgunge after him. Actually, the significance of Henckell's settlement lies in the fact that he was the pioneer in colonizing a part of the Sundarbans and set a precedent for later colonialists to follow. The task of reclaiming the jungle continued after Henckell, and by the twentieth century, there was much visible transformation in the region.

The colonial policies with regard to the Sundarbans's reclamation moved through several phases. At each phase, the Bengal government became more aggressive and interventionist. In the course of this period, the official policy gradually turned away from bestowing huge land grants on speculative entrepreneurs to favouring smaller grants to less wealthy applicants who would actually manage or even cultivate the lands themselves.[74] These changes partly reflected wider currents in official policy and ideology and partly the increasing cost and difficulty of reclaiming remote Sundarban tracts. In 1829, the Bengal government set out terms for grants to be opened in the newly demarcated Sundarban forest. The land grants, free of any revenue demands for twenty years, were made available to individuals who had the means to carry out reclamation.[75] One-fourth of the land granted had to be cleared and put under cultivation within five years. At the end of the twenty-year period, three-quarters of the land became liable for land revenue at moderate fixed rates (the remainder being settled and built up or considered to be so). In 1844, an interim report showed slow progress: 95 persons had taken up 138 surveyed tracts totalling 4,91,000 bigha in 24 Parganas and Bakharganj districts. Most of these speculators hired woodcutters and labourers from Hazaribagh or Magh immigrants from the eastern coasts to develop their grants.[76] But after five years, only one-tenth (49,000 bigha) of the granted land had been placed under active cultivation (usually wet rice) by the new tenants. The results did not improve and eight years later, in 1852, only a few additional allotments had been taken up in the Sundarban forests. The Santhal, Mundas, Oraons and Mahatos were the other settlers who came and settled in this region, presumably in the hope of a better life which promised arable land and livelihood.[77] Oppression of the landlords and moneylenders prompted the tribals (Oraons, Mundas, Bhumijes, etc.) to seek new fortune in the Sundarbans where the terms of settlement were easy. According to the traditions, the Santals and their ancestors have been wanderers from one land to another, at times, staying for generations in one place.

In 1915, C.J. Stevenson-Moore, after an extensive tour through the Sundarbans, recorded a note on the development of the area.[78] He objected to a revival of the system of lease to *lotdars* and propounded a scheme for ryotwari colonization. In forwarding the draft rules to the government, it was proposed that while ryotwari settlements should be the ordinary method of reclamation, the Board of Revenue should be authorized to make settlement with capitalists in special cases. The Government of India accepted the proposal and the new rules were published. These rules were enforced soon afterwards and ryotwari settlements ordinarily became the means of reclamation in the whole of the Sundarbans.[79]

A seventy-year history of land use for the three districts in which the Sundarbans are situated (24 Parganas, including the city of Calcutta, in West Bengal, India and Bakharganj and Khulna in Bangladesh) reveals massive transformation of the land; but its recent history is one which largely turns around individual islands, usually large ones. The islands of Canning, Gosaba, Kakdwip and Marichjhapi saw massive changes during nineteenth and twentieth centuries. It is in Canning that we see the establishment of port, laying of railway tracks, formation of cooperative at Gosaba, and the emergence of socio-economic conflict leading to the peasant movement of Tebhaga at Kakdwip. Between 1880 and 1950, cultivated land in these districts expanded by 45 per cent. Natural processes added to the alluvial wetland formations along the water channels and at the sea face of the Bay of Bengal. During this period, the human population of these districts increased from 5.60 million to 12.90 million persons.[80] Large-scale land clearance occurred between 1940 and 1950, resulting in the increase of cropland by 23 per cent. This reflects a response to two large-scale traumatic events—the Bengal famine in 1943 and the massive movement of refugees in both directions across the newly created India-Pakistan border following the 1947 Partition.

During the nineteenth century, as sub-infeudation expanded, the numbers of sharecroppers on the khas lands of tenure holders increased.[81] In 24 Parganas, the men of substance who had leased government lots in the Sundarbans for reclamation were somewhat similar to the big jotedars (rich tenants) of the north. The lots were large blocks of land held at easy rents, which progressively enhanced as reclamation proceeded, and were subject to forfeiture if clearance conditions were not fulfilled. Often speculators and landholders obtained such lease and sublet them to smaller leases on cash payments with the result that the work of reclamation was actually carried out by small peasant cultivators paying rack-rents.[82] The process of dispossession of what was initially a relatively free peasantry, and their conversion into near-serfs paid with a share of the crop, began in these areas and gathered pace during the first half of the twentieth century. The cultivators in the Sundarbans were generally an oppressed lot. What worsened things for them was that the leases or tenure holders stayed away from the estates which were run by unscrupulous and ruthless agents (*naibs*) who subjected cultivators to oppressive demands. Besides rent, these hapless people were made to pay a number of illegal cess. Whatever be the changes that the Sundarbans underwent during the late eighteenth to early twentieth centuries, it hardly needs to be stressed that the welfare measures for the peasantry were minimal. Their erosion in all respects continued till the widely fermenting discontent broke into the final outburst of the Tebhaga Movement, which found its greatest intensity in the large island of Kakdwip, immediately before Indian Independence. Although the 1946–7 sharecroppers' movement was limited in its spread and intensity, it had the most powerful lines of continuity with radical agrarian campaigns in post-independence West Bengal. In spite of this, the forest became an inseparable part of the life of the inhabitants. Nature shaped the life cycle of the people and their human bondage; it also explained the social dependence on the jungle for livelihood.[83]

The last important memory from the region is the story of refugees in one of the islands, Marichjhapi, where there was a forcible eviction of Bengali refugees between January and June in 1979, bringing starvation, exhaustion and abject misery in its wake.[84] The refugees who had fled East Pakistan in the 1960s had settled in Dandakaranya. In the 1970s, the leaders of the newly elected Left Front launched a campaign for the return of the Bengali refugees to their native land. By April 1978, around 30,000 Bengali refugees had settled in the island of Marichjhapi in the Sundarbans, moving into the reserves.[85] The then government, having failed to reason out their relocation, took drastic measures and eventually imposed economic sanctions and cordoned off the island with the police. When the inhabitants tried to swim across to other islands, a bloody skirmish ensued resulting in the loss of lives. This clearly exemplifies the saga of the land—twin and conflicting narratives about the human struggle for material survival, and the imperatives of preserving the rich flora and fauna of the world's largest mangrove biosphere.

Conclusion

Settlements in the Sundarbans thus have a long history, subject to continual strain and susceptible to reverses. This is an ever changing tract, usually moving southward; what is now the Sundarbans was probably inside the Bay of Bengal hundreds of years ago, and what is now old land north of Calcutta, was once the heart of the Sundarbans. Efforts at reclaiming forest have been recurrent and patches of habitation have appeared in this region time and again. The principal strands in the story of the region centre around the contest between the tiger and the forest dweller, for whom reclamation meant means of survival. With the advent of new settlers, there was a bloody contest between man and man in a rapidly receding jungle. The story was in more ways than one a transition from an inaccessible jungle to an intensively cleared and cultivated area, with its attendant social changes—in the form of stratification, exploitation, the workings of greed and commercialization, inevitable harbingers of modern development, so to say. But there has also been an articulated concern in the more recent past that human settlements should also incorporate measures and practices to secure the protection of the ecosystems out of which they are born, and that limits should be placed on the exploitation of natural resources. A victim of large-scale clearing and settlement to support one of the densest human populations in Asia, this ecoregion is posed with the threat of extinction; hundreds of years of habitation and exploitation have exacted a heavy toll on its habitat and biodiversity. Historically, it would appear that whenever large-scale economic activity has been thrust upon its ecological fragility, the outcome has either been abject surrender to the greed of an exploitative economy or the eruption of embattled political conflict. Both have ultimately caused immense human suffering, and have been the cause for fear about the future. Thus, never in its history has the Sundarbans mangrove provided a setting for man and the jungle to coexist at peace and in harmony. The way out, thus, inevitably lies in a shift from a man against nature context to one where man lives in harmony with nature. In 2009, Aila, a relatively strong tropical cyclone caused extensive damage in India and Bangladesh—a disaster that once again underscored the fact that struggle here is a way of life. Parallels to the life of the swampy marshy inhospitable terrain of the south would never be found in the more settled lands of the north.

Notes and References

1. Richard M. Eaton, *The Rise of Islam and the Bengal Frontier, 1204–1760*, New Delhi, 1994, p. 169; Imam al-Din Rajgiri, *Manāhij al-shattār*, Persian MSS. nos. 1848, 1848-A, 2: fol. 383b, Khuda Bakhsh Library, Patna.
2. Kumudranjan Naskar and Dwijendra Narayan Guha Bakshi, *Mangrove Swamps of the Sundarbans: An Ecological Perspective*, Calcutta, 1987, pp. 3–4. This

evergreen forest is the abode of many animals like crocodiles, buffalo, wild hogs, cats, deer, porcupines, sharks, Gangetic dolphins, water monitor, snakes, the king cobra, common cobra, banded krait, red crabs, oysters, etc. Numerous birds are also found in the region. The open bill stork, little and large egret, grey heron, purple heron, cormorant are local birds and the most common migratory bird is the pelican. The most important animal in this region is the Royal Bengal Tiger. They are the biggest predators, occupying the pinnacle of the food web of both the aquatic as well as terrestrial ecosystems.

3. Shyamali Gupta, '*Aranyer Thaba*', in *Eksathe* (cultural monthly for women, Paschim Banga Ganatantrick Mahila Samiti), 1 August 2007, Kolkata, p. 40; Deborah Pasmantier, 'Man-eating Tigers Wreak Havoc on India's Island of Widows', *Daily Times—Leading News Resource of Pakistan*, 22 December 2004; Sutapa Chatterjee Sarkar, *The Sundarbans: Folk Deities, Monsters and Mortals*, New Delhi, 2010, p. 9; W.W. Hunter, *A Statistical Account of Bengal, Districts of the 24 Parganas and Sundarbans*, vol. I, 1st edn, London, 1875; repr., Delhi, 1973, pp. 286–7; L.S.S. O'Malley, *Bengal District Gazetteer, 24 Parganas*, Calcutta, 1914, p. 2.
4. François Bernier (1625–88) was a French physician and traveller, born at Joué-Etiau/Anjou. For twelve years, he was the personal physician of the Mughal Emperor Aurangzeb. He wrote *Travels in the Mughal Empire* which is mainly about the reigns of Dara Shikoh and Aurangzeb.
5. Rathindranath De, *The Sundarbans*, Calcutta, 1990, p. 1.
6. Shib Shankar Mitra, *Sundarban Samagra*, Calcutta, 1988, p. 7. Portions of this article are available in Sarkar, *The Sundarbans*, pp. 9–14.
7. L.S.S. O'Malley, *Bengal District Gazetteer, Khulna*, Calcutta, 1908, p. 198.
8. Eaton, *The Rise of Islam and the Bengal Frontier*, pp. 194–5.
9. Hunter, *A Statistical Account of Bengal*, vol. I, pp. 287–8.
10. Ibid., p. 335.
11. O'Malley, *Bengal District Gazetteer, 24 Parganas*, pp. 2–13.
12. Ibid., pp. 2–7.
13. Kanangopal Bagchi, *The Ganges Delta*, Calcutta, 1944, p. 67.
14. O'Malley, *Bengal District Gazetteer, 24 Parganas*, pp. 2–5.
15. Hunter, *A Statistical Account of Bengal*, pp. 293–9. Other large rivers connected were: Passar, Bishkhali, Thakuran, Kabadak, Hariabhanga, Kholpetua, Ichhamati, Sibsa, Bhadra, Bhola, Buriswar, Andharmanik and Bahadur.
16. O'Malley, *Bengal District Gazetteer, 24 Parganas*, p. 2; Hunter, *A Statistical Account of Bengal*, p. 287.
17. Hunter, *A Statistical Account of Bengal*, p. 287.
18. Ibid.
19. Ibid., pp. 288–9. Every well-to-do peasant had his thatched hut and granaries, surrounded with a garden of palms, coconut, betel nut and other trees.
20. R.K. Mukherjee, *The Changing Face of Bengal: A Study of Riverine Economy*, Calcutta, 1938, p. 137.
21. Hunter, *A Statistical Account of Bengal*, pp. 317–18.

210 *Sutapa Chatterjee*

22. *Report on the Survey and Settlement Operations in the District of 24 Parganas,* June 1938, File no. 11R-12(1) of 1937, West Bengal State Archives (WBSA), Revenue Department (RD), Land Revenue Branch (LR).
23. Hunter, *A Statistical Account of Bengal*, pp. 382–3; H. Beveridge, *The District of Bakarganj: Its History and Statistics,* London, 1876, pp. 178–80.
24. A.F.M. Abdul Jail, *Sundarbaner Itihas,* 2nd edn, Dacca, 1986, pp. 3–4; Walter Hamilton, *A Geographical, Statistical and Historical Description of Hindoostan and the Adjacent Countries,* London, 1820; repr., Delhi, 1971, pp. 125–6; C.J.C. Davidson, *Diary of Travels And Adventures in Upper India,* vol. 2, London, 1843, p. 199.
25. Satish Mitra, *Jassahaur Khulnar Itihas,* vol. I, Calcutta, 1914, pp. 68–73.
26. Kalidas Dutta, *Dakhin Chabbis Parganar Ateet,* vol. I, Baruipur, South 24 Parganas, 1989, pp. 19–22; Mitra, *Jassahaur Khulnar Itihas,* p. 73.
27. Beveridge, *The District of Bakarganj,* pp. 169–80; J. Westland, *A Report on the District of Jessore: Its Antiquities, Its History, and Its Commerce,* Calcutta, 1871, p. 21.
28. Beveridge, *The District of Bakarganj,* p. 169. According to Beveridge, the 'Bengali mind, as being prone to the past at the expense of the present, has answered the question in the affirmative and maintains that there were formerly large cities in the Sundarbans'.
29. Dutta, *Dakhin Chabbis Parganar Ateet,* p. 27; Bagchi, *The Ganges Delta,* p. 42.
30. Dutta, *Dakhin Chabbis Parganar Ateet,* p. 23.
31. Ashutosh Bhattacharya, ed., *Manasa Mangal,* Calcutta, 1954, p. 149.
32. O'Malley, *Bengal District Gazetteer, 24 Parganas,* p. 26.
33. Hunter, *A Statistical Account of Bengal,* p. 381.
34. Ralph Smyth, *Statistical and Geographical Report of the 24-Pergunnahs District,* Calcutta, 1857, pp. 69–81.
35. Col. J.E. Gastrell, *Geographical and Statistical Report of the Districts of Jessore, Fureedpore and Backergunge,* Calcutta, 1868, p. 23.
36. Ibid., p. 24.
37. Ibid., p. 25. According to Gastrell, traces of former inhabitants in the Sundarbans were also found when tanks were being dug on the Morellgunge estate.
38. Hamilton, *A Geographical, Statistical and Historical Description of Hindoostan and the Adjacent Countries,* pp. 84–5.
39. Ibid., p. 85. Sundarbans have always been regarded as peculiarly adapted for the reception and concealment of river pirates.
40. Jamini Mohan Ghosh, *Magh Raiders in Bengal,* Calcutta, 1960, pp. 50–1.
41. Gastrell, *Geographical and Statistical Report of the Districts of Jessore, Fureedpore and Backergunge,* pp. 24–5.
42. Westland, *A Report on the District of Jessore,* pp. 20–21.
43. Ibid., pp. 11–21. Some of the earliest traditions and some of the oldest ruins in the district of Jessore connect themselves with the name of Khanja Ali. Ruins of Khanja Ali were found near Baghahat on the outskirts of the Sundarbans, the place which was declared by tradition to have been his residence. The largest of

Khanja Ali's buildings was the Shat Gumbaj or sixty domes. Another beautiful mosque was found at Masjidpur. This shows that parts of the Sundarbans were prosperous during the time of Khanja Ali.

44. Westland, *A Report on the District of Jessore*, p. 21. Westland had doubts about the Maratha invaders who, after all, as we know, never crossed the Bhagirathi during the *bargi* raids of the 1740s against *Rarhdesa* in West Bengal. But he was certain that Maghs once settled in these regions and that they had reclaimed southern portions of Bakharganj area.
45. Westland, *A Report on the District of Jessore*, p. 21.
46. Ibid., p. 169.
47. Ibid., p. 170.
48. Beveridge, *The District of Bakarganj*, pp. 169–80; Ralph Fitch visited Bakla in 1586, and described the country as being prosperous and fruitful. He did not mention that Bakla was a city. Beveridge dismissed Fitch as being not very observant. His descriptions were often incomprehensive and his itinerary not clearly known. Fitch mentioned nothing about the storm that devastated Bakla only twelve months or so before his arrival. However, the question of Fitch's credibility and intelligence is not very relevant as there is no evidence to suggest that Bakla was part of the Sundarbans.
49. Beveridge, *The District of Bakarganj*, p. 175.
50. Towards the close of the sixteenth century, Pratapaditya of Jessore established his kingdom in the Sundarbans. Nihar Ranjan Ray, *Bangalir Itihas*, Calcutta, 1939, pp. 46–50.
51. Ibid., pp. 175–6.
52. In *History of Bakarganj*, Beveridge states, 'Chandecan is evidently the same as Chand Khan, which, as we know from the life of Raja Pratapaditya by Ram Ram Basu (modernized by Haris Chandra Tarkalankar) was the name of the former proprietor of the estate in the Sundarbans, which Pratapadiyta's father Vikramaditya got from King Daud.'
53. Mitra, *Jassahaur Khulnar Itihas*, p. 87. In lot no. 26, two big tanks named Raidighi and Kankandighi were found. There was a temple of Biranchi in lot no. 127. In lot no. 128, a fort was found which was called Bharat Garh. Near this fort, there is a temple of Raja Bharat. The grove of Pir Gorachand can still be seen in Haroa whose annual fair is famous. In between the rivers Jamuna and Ichamati in plot no. 165, evidence was found of the existence of the city of Jessore, fort of Jessore and the temple of Jashohareshwari north of Dhumghat. The Dhumghat fort was situated at the south.
54. Mitra, *Jassahaur Khulnar Itihas*, pp. 87–90.
55. Dutta, *Dakhin Chabbis Parganar Ateet*, pp. 28–35. Among relics of earlier times, there were hundreds of silver punch marked and rectangular cast copper coins, clay models inscribed with Brahmi scripts, a fragment of a small stone pillar with Asokan inscription, Mauryan and Kushana terracotta figurines, and toy carts of animals depicting Jataka stories. These discoveries reveal that this part of Bengal witnessed the development of civilization from the New Stone Age.

56. Bratindranath Mukherjee, 'Decipherment of The Kharoshti-Brahmi Script', *Monthly Bulletin,* Asiatic Society, August 1989, pp. 1–6; Gourishankar De and Shubhradip De, *A Lost Civilization,* Kolkata, 2004, p. 109.
57. Aparna Mandal, 'Sundarban, A Socio-Eco-Cultural Study, 1757–1947', unpublished M.Phil. dissertation, Jadavpur University, 1990, p. 3; A.K. Mandal and R.K. Ghosh, *Sundarban, A Socio Bio-Ecological Study,* Calcutta, 1989, pp. 94–100.
58. Dutta, *Dakhin Chabbis Parganar Ateet,* pp. 19–28.
59. Aniruddha Roy, 'Case Study of a Revolt in Medieval Bengal: Raja Pratapditya Guha Roy of Jessore', in *Essays in Honour of Professor S.C. Sarkar,* ed. Barun De et al., New Delhi, 1976, p. 146.
60. Dulal Chaudhuri, 'Folk Religion', in *A Focus on Sundarban,* ed. Amal Kumar Das, Sankaranda Mukherji and Manas Kamal Chowdhuri, Calcutta, 1981, p. 74.
61. Abdur Rahim, *Ghazi-Kalu-Champavati-Kanyar-Punthi,* repr., Gaudia Library, Calcutta, 1987, p. 14.
62. Khater, Marhum Munshi, *Banabibi Jahuranama,* repr., Gaudia Library, Calcutta, 1987, pp. 39–40.
63. Sarat Chandra Mitra, *On Some Curious Cults of Southern and Western Bengal,* Calcutta, 1918, p. 441; Ashutosh Bhattacharya, 'The Tiger Cult and its Literature in Lower Bengal', *Man in India,* no. 1, March 1947, pp. 49–50; Sarkar, *The Sundarbans,* pp. 32–9; Annu Jalais, 'Bonbibi: Bridging Worlds', *Indian Folk Life,* no. 28, January 2008, pp. 7–9.
64. Chaudhuri, 'Folk Religion', pp. 74–8.
65. Richard M. Eaton, 'Human Settlement and Colonization in the Sundarbans, 1200-1750', *Agriculture and Human Values,* Spring 1990, pp. 6–7.
66. Ibid.
67. Ibid.
68. Ibid.
69. Ibid.
70. Goutam Chattopadhyay, ed., *Bengal: Early Nineteenth Century, Selected Documents,* Calcutta, 1978, pp. 95–9.
71. Sarkar, *The Sundarbans,* pp. 55–6.
72. Hunter, *A Statistical Account of Bengal,* p. 286.
73. Ibid.
74. John F. Richards and Elizabeth P. Flint, 'Long-term Transformations in the Sundarbans Wetlands Forests of Bengal', *Agriculture and Human Values,* vol. VII, no. 2, Spring 1990, pp. 17–20.
75. F.E. Pargiter, *A Revenue History of Sundarbans, 1765-1870,* Calcutta, 1934, p. 23. Pargiter writes, 'In the bestowal of the grants, however, the European residents of Calcutta largely predominated, and few, if any, were given to the border Zamindars.'
76. Hunter, *A Statistical Account of Bengal, Districts of the 24 Parganas and Sundarbans,* pp. 318–19.

77. Pargiter, *A Revenue History of Sundarbans*, p. 90; Richards and Flint, 'Long-term Transformations in the Sundarbans Wetlands Forests of Bengal', p. 22. According to the traditions, the Santals and their ancestors have been wanderers from one land to another, at times, staying for generations in one place, but while migrating always more or less towards the east, not to the west. Almost all their traditions maintain the same direction and the Santals believe that they have come from the west. The earliest settlements of which Santal tradition speaks of, especially those in Hihiri Pipiri and Cai Campa, lie on the north-western frontier of the tableland of Hazaribagh and in the direct line of advance of the numerous Hindu immigrants from Bihar, who, according to the report of the Revd J. Philip, undoubtedly drove the Santals eastward. Whatever the Santal's original habitat might have been, they were found in large numbers in Chotanagpur, especially in the districts of Hazaribagh, Palamau and Singhbhum and in the neighbouring districts of Midnapur and Birbhum around the middle of the eighteenth century. Towards the end of the eighteenth century, they began to migrate to the Rajmahal hills, situated on the north-eastern side of the Chotanagpur plateau. In 1833, the British government demarcated the area of Rajmahal hills, comprising 1,366 sq. mi. which came to be known as the Damin-i-Koh.
78. Letter from Secretary, Board of Revenue, Bengal to the Secretary, Government of Bengal, 6 April 1915, File no. 2–1/10(1), Progs. (Proceedings) nos. 21–22, W.B.S.A., R.D., L.R. Branch.
79. Letter from Secretary, Government of Bengal to the Secretary, Board of Revenue, Bengal, June 1916, File no. 14S-2, Progs. no. 97–100.
80. Richards and Flint, 'Long-term Transformations in the Sundarbans Wetlands Forests of Bengal', p. 22.
81. Radhakamal Mukherjee, *Land Problems of India*, London, 1933, p. 99.
82. Sugata Bose, *Agrarian Bengal*, Cambridge, 1984, p. 16.
83. Sarkar, *The Sundarbans*, pp. 163–5.
84. Annu Jalais, 'Dwelling on Morichjhanpi: When Tigers Became "Citizens", Refugees "Tiger-Food"', *Economic and Political Weekly*, vol. 40, no. 17, April 2005, pp. 57–62.
85. Ross Mallick, 'Marichjhapi Massacre', *The Journal of Asian Studies*, vol. 58, no. 1, February 1999, pp. 105–8.

10

Cattle, Cruelty, Cow Doctors: Examining Animal Health in Rural Bengal, 1850–1920

Samiparna Samanta

The material culture of nineteenth- and twentieth-century Bengal revolved around domesticated animals. Agricultural cultivation, its mainstay of economy, depended primarily on cattle. The urban city life relied on draft animals which also featured in food and rituals of the indigenous Bengali subjects. Yet, in spite of the socio-economic implication of animals in Indian culture, contemporary South Asian scholarship is somewhat silent on the importance of studying non-human animals as historical actors. This study is a modest intervention that might arguably fill in this gap in South Asian historiography by co-opting non-human players into the study of empire building. Domesticated animals reflected both nature and culture, and hence an analysis of animal treatment across indigenous traditions and British practices reveals compelling insights into a colonial society. To that end, this essay elucidates how in the late nineteenth and early twentieth centuries, domesticated animals were at the centre of a heated debate in Bengal as farmers, veterinarians, humane societies, and the Hindu middle class viewed cattle health and animal cruelty with varying degrees of keenness and intensity.

We notice three major cross-currents on the question of cattle health that indicate an increasing preoccupation with domesticated animals in late nineteenth- and early twentieth-century Bengal: first, an intervention from colonial authorities that primarily informed veterinary initiatives in combating epizootics; second, a renewed protectionist stance that manifested itself in the form of colonial legislations and a surge of anti-animal cruelty literature among Bengalis; and third, a collective paranoia among the Bengali Hindu middle class regarding a crisis of cattle health that needed to be addressed immediately. Closely examining these trends, this essay shows how the concern over cattle health can be studied as a 'window' to a colonial society ridden with hidden tensions of class, race and science. The outbreak of animal diseases, like cattle plague, in mid-nineteenth-century India triggered a series of responses and initiatives towards improving agriculture and the condition of livestock in general. However, as this essay argues, the path of improvement was fraught with contradictions that brought out the tussle between 'Western' values/technology and Hindu knowledge systems. By examining those fault lines, this essay attempts to explicate the multiple meanings of animal health and material culture in a colonial society.

Cattle Ecology in Rural Bengal: Murrains and Mortality

The rural economy of Bengal suffered a series of setbacks from the mid-nineteenth century onward owing to severe bout of cattle plague or 'rinderpest' that broke out in different districts.[1] This story of cattle plague has unfortunately not received adequate attention in histories of disease and medicine in South Asian scholarship. Clive Spinage was one the earliest scholars to provide an overall sketch of the trajectory of rinderpest in India.[2] In recent years, Saurabh Mishra in his article, 'Beasts, Murrains and the British Raj' has demonstrated that like human disease, animal disease that threatened the 'health of the prized British horses received greater attention'. He has argued that cattle diseases that were quite widespread did not receive adequate attention in spite of the disastrous impact that they had on the rural economy.[3] This chapter questions this particular assumption to show how the outbreak of cattle murrain in parts of Bengal did impact colonial policy making and stimulated a number of interventions that reflect the immediate urgency to combat epizootics.[4] In particular, by focusing on Bengal, I study the colonial debates surrounding cattle murrains to investigate how an empire grappled with the dilemma of tackling animal mortality and preventing an agricultural economy from jeopardy.

Animals died in huge numbers in mid-nineteenth-century Bengal. It is indeed true that many of these figures might have been under-reported, as the officials themselves claimed. Colonial authorities, often sceptical of the high mortality rates, were eager to play down the statistics. They were equally concerned to make the rinderpest appear 'manageable'. But devastation was not only massive, the heavy loss of cattle often occurred within a brief span of only fifteen days (see Table 10.1). Epizootics often led to 'cattle famine', which, in turn, in many cases, led to a food famine due to the intimate link between cattle and cultivation.[5] In certain exceptionally bad years, when epizootics spread over larger areas, agricultural production over large tracts of the country could suffer. This was certainly the case in 1870, when Clive Spinage estimates that the total number of dead cattle and buffalo within India reached the figure of nearly one million.[6]

TABLE 10.1: Animal Mortality Rates in Assam from 16 to 31 May 1869

Type of animals affected	Number of deaths reported
Buffaloes	737
Cows and Bullocks	6,536
Goats	10
Horses	45
TOTAL	7,328

Source: Statement showing the progress of cattle disease from 16 to 31 May 1869, as in the Commissioner's letter no. 699, 19 May 1869, Political (Medical Branch), 27 July 1869, no. 66–7, West Bengal State Archives.

If the dread of rinderpest was a constant presence, as also the perpetual concern for the British in India throughout the late nineteenth and mid-twentieth centuries, their response in checking the spread of rinderpest was uncertain.[7] The symptoms of epizootics were not unknown to the Victorian medical practitioners.[8] Yet, despite this familiarity, there were elements of uncertainty about the nature and cause of cattle plague in India. British veterinarians in India often sought information from England about the nature of rinderpest and ways of dealing with it. However, while there was sufficient official confusion over the first occurrence of the disease in India, the British officials consistently claimed to have first 'discovered' the disease.[9] Judging from the official reports, it seems a somewhat partial consensus was reached by 1865 that 'rinderpest' was indeed a new disease that was 'unrecognized in the neighbourhood by either cattle proprietors or by civil surgeons'.[10]

Animal Disease, Human Health

Once the prognosis of rinderpest in Bengal was established, all focus was diverted towards its control. The last decade of the nineteenth century saw frantic colonial attempts to produce systematic knowledge about cattle diseases through surveys and reports, and the ultimate need to create a network of veterinary boards, dispensaries and qualified surgeons who could substitute the village 'quacks' by their professional doctors and, thus, save cattle lives. Ensuring that 'English remedies' reached the indigenous agrarian societies, British veterinarians pushed to circulate adequate Bengali translations of J.H.B. Hallen's (Inspecting Veterinary Surgeon of Bombay Army) *Manual of the More Deadly Forms of Cattle Disease in India* to different officers under the government.[11] The government even aspired to introduce veterinary education as curriculum in colleges and create a class of professional veterinarians who could treat animal diseases—the Belgachia Veterinary College being one such institution established in 1896. While local knowledge of the environment, animals and the indigenous communities was highly sought after, colonial authorities strove to apply metropolitan disease control measures to the Indian colony.

Saurabh Mishra in his study of cattle murrains and veterinary science has argued that the colonial veterinary policy, at least till the end of nineteenth century, was geared solely towards protecting military livestock (horses) owned by the state. According to him, cattle upkeep was so removed from colonial concerns that it hardly ever received a passing mention. Even by the end of the nineteenth century, the Civil Veterinary Department paid very little attention to the question of epizootics among 'public cattle'. He contends that in budgetary terms, the department spent very little toward disease prevention and invested most of its resources on horse breeding measures.[12] In that respect, he borrows Radhika Ramasubban's phrase to argue that the colonial veterinary policy was somewhat 'enclavist', i.e. meant to preserve the health of the white civilians and military quarters.[13] Some questions are however left unanswered in Mishra's analysis. Mishra himself acknowledges the adverse impact that cattle murrain often had on a predominantly rural economy. However, he downplays the severity of cattle mortality in British minds by arguing that it 'appeared to have little direct impact on the state of colonial coffers as long as the mortality did not spread to military animals'.[14] But if the veterinary policy of the Raj was indeed exclusively 'enclavist' as Mishra claims, then how can we explain the widespread colonial paranoia surrounding enormous mortality of cows, bullocks and goats in parts of Bengal, that we find evidence of in official correspondence on rinderpest? Additionally, military horses were perhaps no longer deemed to be as useful once the British conquest of India was almost complete by the year 1818. Barring the Anglo-Afghan and Anglo-Burmese wars, which were fought outside India, we hardly notice any major military confrontations in India involving horses in huge numbers. The empire was firmly in place by 1858, thus removing any need for immediate military combat. Hence, the argument that the importance of safeguarding health of the horse stemmed largely from the colonial desire to go to wars holds good only for the early colonial phase.

Mishra argues that while the military motivation of the Raj might not be an entirely novel idea, it was the close nexus between military and veterinary that appeared striking. The relationship between military and veterinary hardly seems surprising, given the fact that the British Empire in India, at least till the late nineteenth century, was a military state not built upon the notion of promising 'public good'.[15] If early colonial veterinarians were turned into horse breeders, even in the Indian Medical Service the doctors were all army doctors. However, acknowledging that the colonial veterinarian policy was definitely narrow, I argue that it was not confined solely to military circles. Inadvertently, colonial and indigenous concerns surrounding animal disease affected the realm of public health as well. The colonial authorities were worried not only about combating glanders and surra affecting horses, but also controlling rinderpest and animal-borne diseases. This anxiety became increasingly visible towards the beginning of the twentieth century as medical doctors (British and indigenous) began to speculate upon an ominous possibility—the transmission of disease from animals to humans through meat. The fear of diseased meat created intense panic in the colonial establishment, leading to a more stringent inspection of Calcutta's slaughterhouses. In fact, by the mid-nineteenth century, most of the slaughterhouses had been driven out from Calcutta to the outskirts of the town. Slaughterhouse regulations undoubtedly privileged the cantonments and 'white towns' as it reinforced the colonial logic of a discriminatory sanitary order.[16] While such hygienic enclaves were expected to reduce the threat of disease among Europeans, the British, however, soon realized that their sanitary protection could not be fully secured unless the 'black towns' and Indian habitations were also regulated and insulated. Such fears explained why the municipal authorities attempted to extend sanitary regulations and regulate slaughterhouses in the 'black towns' which had acquired a menacing meaning since the Revolt of 1857. The colonial anxiety concerning diseased meat, thus, profoundly affected the spatial configuration of Calcutta, as it attempted to redefine the contours of the city. Diseased meat also entered into the *bhadralok* mental worlds as Hindu-Bengalis began to give out a sudden, collective call for vegetarianism.[17] New Western notions of 'nutrition science' began to influence these upper-class Hindu men in their understanding of diet and germs.

Death and Distress

A cure for rinderpest was hard to find. The initial remedies thought upon were quarantines, slaughter and culling which hit the farmers hardest. From the 1860s onward, unlike their medical counterparts, veterinarians accepted that most epizootics were contagious and spread by the transmission of some disease matter from beast to beast.[18] However, the boundary between animal and human health was not so secure in the Victorian era. The problem was aggravated by the fact that there was no medical consensus on rinderpest until the 1890s; indeed, there were many different groups producing understandings from different starting assumptions, by different means and to different ends.[19] By the end of the nineteenth century, in a disease-ravaged Bengal, a general sense of helplessness prevailed. The unpopularity of inspection and slaughtering, along with the economic losses borne by farmers, were powerful reminders of the failure to check the outbreak. The *Evening Mail* captured the despair and anguish in 1885 in a very dramatic way.

> What strikes us most is the facility with which farmers and veterinary surgeons abandon themselves to despair, so far as regards all hope of successful treatment, and resort to the extreme measure of prohibition and destruction. We could do this, of course, if we were merest savages, ready to believe a demon had passed over our cattle and glad to fall back on our yards and plantains; but we are rather better than savages, and we profess to have some power over the diseases of man and beast. It is a humbling confession that is made by our medical authorities when they tell us to kill at once, for there is nothing at all to be done.[20]

David Arnold has examined how the early plague (human) years in Bombay were a major crisis point in the history of state medicine in nineteenth-century India, occasioned by people's hostility against anti-plague measures.[21] Even in Calcutta, prophylactic intervention by the government in terms of disinfection, serum therapy, rat killing, segregation, plague hospital and ambulance van raised storms of popular protest which resulted in frequent street riots.[22] Likewise, in the case of rinderpest, helplessness of the colonial government is evident as it faced a crisis of control. Despite the quarantine measures in the villages and the city of Calcutta, the epidemic went on unabated, affecting Bengal, Punjab and Madras. From 1865 onward, textual evidence indicates the official anxieties concerning rinderpest, which might eat into the profitability of the empire. By 1869, it was evident in the official circles that there was 'a considerable decrease in cultivation on account of the want of plough bullocks'.[23]

Epizootics, Empire, Agriculture

What then lay behind the urgency of the state's rinderpest operation? Was it prompted by a colonial imperative to showcase the superiority of 'Western science' or was the government genuinely interested in improving the health of cattle to put its agricultural returns in place? Given the context of cattle murrains, this seemingly simple question defies an easy answer. Historians of empire over the past few decades have been engaged in studying Indian and African experiences with Western science and medicine.[24] In the South Asian context, historians have argued that more than an ideological 'tool' of the empire, science was also used for sustaining that very empire through a hegemonic project.[25] Scholars like Ian Inskter, Deepak Kumar and Dhruv Raina have demonstrated how the empire in India literally used science to legitimize power relations in a hierarchical colonial world and thus establish their hegemony over the subjects. Inskter has emphasized that the British interest in India was largely restricted to commerce and the natural history project.[26] Others like Zaheer Baber and Roy Macleod have argued that in the eighteenth and nineteenth centuries, 'the loss of political and economic sovereignty produced a situation wherein the diffusion of modern science was more or less decided by imperial colonial policy'.[27] Kumar has testified that geology, mineralogy, botany and the material sciences were honeypots for a colonial regime that was built upon the principle of maximization of revenue.[28] To echo Kumar's argument, our discussion of the hysteria surrounding rinderpest attacks has demonstrated that the Raj did have a major economic stake in protecting livestock (draft animals and agricultural cattle) that were fast diminishing. Cattle and agricultural science were too profitable to be ignored. In fact, the colonial interest in livestock is evident from the numerous tracts on cattle rearing which churned out from the nineteenth century onward.[29] Therefore, a more crucial question might be in order: did the Raj make a concerted effort to improve the breed of cattle in Bengal in view of the recurring rinderpest attacks?

In the context of central India, Laxman Satya has argued that an 'anti-cattle ideology' of the British had led to a general deterioration in the condition and population of cattle in the nineteenth century.[30] Berar, for instance, that was known for its distinctive breeds of cattle and cattle fairs, by the beginning of the twentieth century, faced a progressive decline in the available pasturage for cattle. According to Satya, even as signs of cattle mortality increased, the British did little to counter it. When the British did act in the name of controlling cattle disease, they did so using methods of quarantine and segregation that only reflected, in Satya's words, the 'authoritarian, patriarchal, and patronizing nature of the officials and the state'.[31] In Bengal, on the other hand, rinderpest outbreak galvanized the creation of a Cattle Plague Commission in 1870, vernacularization of 'Western' remedies, and the Contagious Diseases (Animals) Act in 1869. The Act of 1869, however, seemed limiting in its operation, as the Assistant Commissioner of Golaghat reported that, 'from all that I can learn I finally believe that the disease has spread entirely by infection, and had I possessed some such summary powers as those conferred by Act I of 1869, I have little doubt that I might have lessened the severity of the infliction'.[32] Veterinary institutional infrastructure developed gradually in the late nineteenth century. In 1871, a hospital was opened at Culna (Kalna) in Burdwan for treatment of rinderpest-affected cattle.[33] The Lahore Veterinary College (with a dispensary and hospital) was established in 1882, and by 1897, it boasted of 90 students. At Poona College, a veterinary course was introduced, and students who passed it were qualified to take charge of the local dispensaries at Ahmedabad, Nadiad and other towns in the Bombay Presidency.[34] These dispensaries, Voelcker reported, were used to some extent by the different municipalities for the treatment of their working cattle. According to Voelcker, however, the most important step taken towards improving cattle health was the appointment of Dr. Lingard, 'a man of established scientific reputation' as Imperial Bacteriologist to the Government of India.[35] Voelcker considered this appointment to be 'one of great importance' because it was 'almost the first in which a man trained in scientific investigation has been brought to India and enabled to follow original research'.[36]

Where Compassion Meets Science

If rinderpest wreaked havoc in the countryside and Calcutta, in the midnineteenth century, had to be treated with the aid of science, then compassion towards domesticated animals—diseased, worked and tortured—found a new favour during the same time period. The sentiment of compassion towards animals came to be institutionalized in India in 1861 with the Calcutta Society for Prevention of Cruelty to Animals (CSPCA)—the earliest humane society in Asia founded by Colesworthy Grant. Concern for animals was, however, not a European import into India. Rather, animal shelters and regular caring for unproductive cattle can be traced back to as early as the fourth century BC, and animal homes for aging and enfeebled cattle, *goshalas* (cow homes) are attested in India by the sixteenth century.[37] In fact, Wendy Doniger has argued that the term 'ahimsa' or 'non-violence' in Vedic ontology originally applied not to the relationship between humans alone, but to the relationship between humans and animals.[38] Scholars examining Sanskrit texts of classical Hinduism, like the Dharmashastras, epics, Puranas and the Vedanta have noted that Hindu attitudes towards non-human animals were neither constant nor monolithic; rather they were frequently shifting, immensely complex and often antithetical.[39] While the sentiment of compassion towards non-human animals was not a novelty in India, its contact with the Raj lent a different hue to it. Compassion was no longer a commitment to the virtue of 'ahimsa', but implied a loyalty to 'science'. The best example of the mingling of ahimsa and *bigyan* (science) is the Belgachia Veterinary Infirmary. The Belgachia Hospital was instrumental in dealing with epizootics and animal cruelty alike. Animals with 'contagious cases' were often removed from Calcutta *gowkhana* (cow sheds) to the Belgachia Hospital for treatment.[40] A veterinary assistant picked out all infectious cases, and an ambulance from the College was permanently stationed in the southern *gowkhana* for the transport of the worst cases. All newly purchased ponies received special attention. They were malleinized and kept separate, so that they, at any rate, they should be free from glanders.[41]

The cattle, thus, removed from the Calcutta *gowkhana* were all inoculated with anti-rinderpest serum to prevent further spread of the disease in the city. In 1900, a humane dimension was added to the veterinary initiatives as the veterinary hospital attached to the Bengal Veterinary College came to be registered as an infirmary under the Prevention of Cruelty to Animals Act XI of 1890.[42] This was a significant step towards science and compassion as the twin ideas of veterinary science and protection of animals merged together in the Belgachia infirmary, now treating both diseased and abused animals. Science and compassion, thus, came to share a common cause in the newly established veterinary hospital. Nevertheless, the quality of colonial mercy was not unqualified. Animals receiving treatment at the Belgachia infirmary were categorized according to their 'utility'. Draft animals topped the priority list, followed by cattle, cats and dogs. 'Class III' animals comprised of sheep and goats, and the bottom of the list featured poultry, camels, elephants and ostriches.[43] Horse was evidently the prized animal of the Raj, as were cattle (though for a different reason) that were diminishing fast in huge numbers. There were, however, strict instructions concerning 'wild animals'. The manual made it clear that wild animals were not to be admitted to the infirmary unless 'properly qualified attendants' accompanied them. Additionally, the infirmary reserved the right to 'destroy at once any animal' that became 'dangerous or unmanageable'.[44] The question that bothers us here is why the sudden spurt of enthusiasm for transforming the Belgachia Veterinary College into an infirmary for animal protection. The answer perhaps calls into question the role of the Raj as also of the Bengali literati.

Torture or Treatment?

If the Raj had been consistent in its attempts to protect the much valued horse, middle-class Bengalis had their own favourites to shelter. As stated earlier, from the late nineteenth to early twentieth centuries, the sympathy for the domesticated animals, mainly cattle, manifested itself with renewed vigour in several Bengali tracts and farmers' manuals. An interesting case that reflected the growing Bengali unease surrounding cruelty to animals and an abiding faith in science was a pamphlet published by Hridoy Nath Ghosh, the secretary of the Jnandayini Sabha, in 1911. The author complained to the district magistrate of Howrah to take immediate action to rescue animals (especially cows) from the clutches of the illiterate *go-dagas* (cow doctors) who falsely professed to be 'veterinary surgeons' and, thus, posed a serious threat to cattle. According to the author, the *go-dagas* 'have no knowledge nor slightest insight whatever into the Science of medicine; but they have invented several processes of cruelty in the name of treatment. The public is quite ignorant of the tricks played upon them by these inhuman wretched'.[45] 'Proper' treatment of animals, therefore, necessitated adequate knowledge over 'medical science', and not stray experience in dealing with animals.

Deeply sceptical of the skills of such *go-dagas*, the Sabha meticulously detailed the way in which the animals were 'treated' by these cow doctors, who lacked any veterinary training. These doctors were believed to make their rounds once annually, from mid-August till the following summer, touring the villages of Bengal.[46] Rash Bihari Ghosh, member of the Sabha, composed a pamphlet titled, 'An Appeal for the Prevention of Cruelties to the Bovine Species', where he described the 'tricks' of the *go-dagas* as they travelled treating the sick animals from one village to another.[47] According to Ghosh, most of Bengal's cattle suffered from ill health and exhaustion owing to overwork. When their owner called for a doctor, the *go-dagas* arrived at the village, hollering loud. Inspecting the animal, the doctor would often proclaim the whole flock to be diseased or at times exclude two or three from a flock consisting of twenty, in order to avoid suspicion. Ghosh added that to avoid being caught in their fraud, 'every one of them has his own jurisdiction, upon which none of his party will encroach'.[48] The amount of fee determined the pain of the animals. The more an owner paid, the more was the pain the animal had to suffer. Many of the diseases, Ghosh believed, were simply the creation of the doctors. Ghosh also outlined the treatment for *sajondaga* offered by the *go-dagas*:

> The poor animal under so-called treatment is first made to fall on the ground, legs bound with a stout piece of rope and then branded mercilessly with two red-hot iron bars called 'Dagni' giving understanding to the owner that the animal would not be attacked with Rheumatism in future. They make use of certain iron instruments in the name of surgical instruments. No doctor has ever seen or heard or such weapons; but they can be found only in the bags of the so-called doctors. Some of these instruments are pointed while others are blunt. These iron bars, as they really are, are made re-hot and the poor animals' limbs are branded and pierced through with them.[49]

Ghosh submitted the pamphlet to the CSPCA hoping it will be circulated among the district magistrates of Bengal who would then identify the 'quack' animal doctors and take appropriate measures. While that did not yield immediate results, the Magistrate of Howrah, however, did take heed to the repeated plea from the Sabha and appointed a municipal veterinary surgeon,

K.M. Chatterjee to verify the cruelty charges against the *go-dagas*. Chatterjee enquired and filed a report of cruelty on two cart bullocks in the Haraganj bazaar, Howrah. The subdivisional officer and the Howrah SPCA took up the matter and prosecuted the three *go-dagas* under Section 34 of the Prevention of Cruelty to Animals Act of 1861. The cow doctors pleaded that the bullocks were ill and hence their tongues had to be pierced which was part of the 'treatment'. Their case was, however, overturned for lack of evidence to prove that the bullocks were really ill.

Go-Dagas: 'Thugs' or Veterinary Surgeons?

If the verdict favoured the tortured animals in the previously mentioned case, the colonial government, generally speaking, was not very keen to intrude into these 'malpractices' in the villages. On the other hand, the Jnandayini Sabha, loyal to the colonial government, spared no effort to attack the credibility of the 'heartless' *go-dagas*. Lauding the British government for its protection of animals through humane societies and laws, the Sabha pleaded for its intervention in suppressing the *go-dagas*, the 'thugs.' It implored:

> The kind and benign Govt. have suppressed the 'Thugs' and done away with many malpractices practiced upon both men and cattle. Now it may be inferred from the facts put together that these men are no better than the 'Thugs' who used to rob people of their money by force, but these men cheat the owners of cattle of their money and also put the dumb creatures into excruciating pain for the gratification of their excessive greed for money, as they have no other means of subsistence but this cruel trade.[50]

To exonerate themselves of the vices attached to them and prove that they were not thugs, the *go-dagas* needed to validate their skill through the stamp of science. They had to substantiate their methods of treatment by submitting their 'so-called surgical instruments' to the district magistrate for inspection. No longer was it sufficient to win the confidence of the ignorant villagers, 'proper' medical knowledge of animal ailments was required.[51] The 'quacks', according to the Sabha, 'could not be tolerated to do mischief both upon cattle and men in the British regime when the British Government has suppressed such barbarous customs as Sati Rite and Infanticide'.[52] The animals needed rescuing, just like the helpless satis. The benevolent Raj was their only hope.

Despite relentless requests to rescue the animals from the village cow doctors, the Sabha did not, however, receive necessary support from the government. In December 1911, the government, while acknowledging the presence of village 'quacks' that caused unnecessary pain to animals, announced that it was difficult for them to protect animals against the *go-dagas*. Commenting on the gullibility of uneducated villagers, the Director of Agriculture (Bengal) argued, 'It must be remembered that these quacks are believed in by the ignorant villagers who allow their animals to be treated. It is also notably the case that the quacks believe in their own treatment.'[53] The colonial policy of non-interference is hardly surprising. To frame legislations to persecute the cow doctors might also mean alienating the villagers. Such an eventuality was not desired, as we notice in case of colonial attitudes towards sati.[54] Hence, non-interference in 'native' practices had to be accommodated with the 'civilizing mission'. As the Director of Agriculture, W.B. Heycock expressed his pious hope that 'with the increase of the Veterinary Establishment and the spread of education in all probability these quacks will disappear'.[55] Thus, the colonial logic was simple. Rather than passing legislations to suppress the village cow doctors, institutional training and professional education could help discipline them.

It is evident that the colonial government cared deeply for its horses and rinderpest-affected cattle, but refused to meddle in 'native' practices concerning cows and bullocks. It even refused to prosecute the cow doctors under the Prevention of Cruelty to Animals Act of 1891, arguing that for the Act to be implemented, 'the crime must be committed in a street as defined in the Act, or in any other place to which the public have access, or within sight of any person in any street or any such other place'.[56] Additionally, the government claimed that since the cow doctors received full consent of the animal owners and also payment from them, the police could not arrest the offenders, unless the offence was committed in its view. Prosecution was, thus, rare because the cows were generally treated in closed places. Under such circumstances, the government reported that 'the proper remedy is to warn the literate villagers against the doings of the cow-doctors and this can best be done by the circulation of notices and pamphlets, but preferably in vernacular. In any case government interference does not seem called for'.[57]

While it is difficult to retrieve the voices of the farmers who resorted to multiple 'treatments' of the *go-dagas*, one way of doing so is to examine the agricultural manuals. The farmers' manuals like *Krishi Lakshmi* or *Krishi Gazette*, however, do not mention *go-dagas*, but do engage with the methods of treatment of what they call *go-baidyas*. These manuals seem divided in their attitude towards village cow doctors. On the one hand, they are sceptical of these doctors' knowledge. On the other hand, they often request the state agriculture department to seek help from indigenous cow doctors. For instance, the author of *Krishi Gazette* argued in 1885 that rather than introducing British ploughs, the agriculture department should focus its energies on saving the lives of cows and combating cow diseases.[58] The author further insisted that there was no need for highly salaried British doctors from veterinary colleges to treat animal diseases. Instead, the department should try to consult indigenous *go-baidyas* and use their knowledge to spread awareness among local farmers on cattle diseases.[59] These writers certainly privileged local knowledge against the use of 'Western' agricultural tools and equipment.

Seeking 'Science'?

The Sabha's continual insistence to prove that the *go-dagas* possessed insufficient knowledge of veterinary science to treat cattle reveals a compelling story. It reflects their confidence in colonial veterinary infrastructure and distrust in indigenous practices. As stated earlier, the Sabha repeatedly complained that the 'veterinary surgeon' had no practical skills, but only one remedy to cure sick animals—branding their bodies with red-hot iron.[60] But more importantly, it also depicts how the Bengalis had internalized the scientific language through their dialogues on animal cruelty. Nineteenth- and twentieth-century Hindu-Bengali debates concerning animal cruelty were not entirely new. Scholars have studied the inbuilt Vedic ambivalence on cruelty and compassion.[61] The tension between cruelty and empathy for animals also found expression in Bengali literature that tended to humanize animals, or use animals as metaphors.[62] The strange paradox of primate compassion and cultivated cruelty strikes us in several twentieth-century Bengali animal narratives like Saratchandra Chattopadhyay's 'Mahesh', Prabhat Kumar Mukhopadhyay's 'Adorini', Tarashankar Bandyopadhyay's 'Gobin Singher Ghonra' and Bibhutibhusan Bandyopadhyay's 'Budhir Bari Phera'. So, if the conflict between brutality and empathy towards animals had long been debated in Vedic-Hindu socio-philosophical discourses, what was novel during this period was the language of the debate.

Straddling two very different worlds—insistence on an innate Hindu/ Vedic compassion and the relevance of western science—the twentieth-century Bengali Hindu middle class eventually appropriated a scientific language in their understanding of human and non-human worlds. No longer did they take refuge in the Hindu doctrine of ahimsa, but spoke in a language nuanced in *bigyan* that stemmed from empirical tests and inspection. For instance, a passage in *Krishi Lakhsmi* in 1937 reads, '*Aaj bigyaner yug esheche* (The age of science has stepped in). We have to embrace science in this new age and introduce modern farming techniques and education. We have to educate the farmers.'[63] Discourses on animal cruelty, thus, reappeared with renewed vigour in the colonial period as a new paradigm entered the larger intellectual scheme—Vedic ahimsa came to be replaced with *bigyan*. The Hindu-Bengali middle classes mediated the language of science in their own mental worlds.

Cruelty and Crisis of Cattle Health

Scholars have shown how, by the end of the nineteenth century, the educated Bengalis started imagining the Indian village in a very different way.[64] Influential *daktari* author and the first Chemical Examiner to the government, Rai Chunilal Basu Bahadur, in his 'Village Sanitation and a Manual of Hygiene' appealed to the villagers to take up the project of restoring the health of their villages.[65] Central to this imagination of a 'national culture of hygiene' was a notion of decline. The idea of decline from a glorious past had permeated in a wide variety of discourses in the late colonial South Asia, and even formal political nationalism. It argued that the nation had, in the past, been much healthier and stronger and that the health problems faced in contemporary times was a consequence of modern changes rather than long-term factors.[66] Veterinary ideas too revolved around notions of decline of Indian cattle health.

As early as September 1820, William Carey founded the Indian Agricultural Society in Calcutta with the objectives inter alia of improving the material condition of Indian cattle.[67] In his presidential address, Carey deplored the archaic condition of Indian agriculture that he believed was conditioned by the neglect of cattle.[68] In 1863, the Lt. Governor of Bengal even proposed to organize an agricultural exhibition and expected the zamindars to participate and popularize it.[69] But, unfortunately, the zamindars were not adequately responsive. Even the agricultural and horticultural societies, it was argued, were more interested in exhibiting Western fruits and agricultural products, than improving indigenous ones.[70] The *Sambad Prabhakar* lamented that the government and the zamindars were more interested in exacting revenue from the raiyats, but none were keen on agricultural improvement.[71]

Concerns over cattle health and torture of animals came together in the analysis of Babu Joteendro Mohun Tagore, Honorary Secretary to the British Indian Association and Landholders and Commercial Association in 1864. In his letter to the Government of Bengal on the 'best means of improving the breed of Bengal Cattle' which was also forwarded to the Agricultural and Horticultural Society, Tagore pointed out the reasons for the 'visible deterioration in the breed' of Bengal cattle as want of good pasture, overwork, periodical murrains and defective breeding.[72] The enemy was, thus, the entire colonial state—British land revenue administration and Permanent Settlement were viewed as reasons behind shrinking pasture grounds. Romanticizing a pre-colonial idyllic Hindu past, Tagore pointed out that 'in the days of the Hindoo Kings of this country there were cattle ground, cattle roads, and cattle tanks', and 'pasture lands were kept aside by the Zamindars exclusively for the grazing of the cattle, on the charge of a moderate rent or quota of ghee.'[73] To Tagore, therefore, the 'indifference and apathy' towards cattle came with British rule. Furthermore, Tagore enquired before the government, how far the animal diseases had led to the physical deterioration of cattle in Bengal. Veterinary thinking among the Bengali landowning classes, thus, structured more around the health of the livestock economy, rather than diseases of organs and tissues. Tagore's letter, while attacking colonial land revenue policies, reinforced the perception of the vulnerability and fragility of Bengal livestock economy.

Joteendro Mohun Tagore's line of argument was, however, quickly brushed aside by J. Beckwith, Secretary to the Landholders and Commercial Association, and refuted the very idea of 'degeneration' among Bengal cattle, by pointing out that the 'increased difficulty in procuring draught bullocks' due to periodic murrains spread the idea that the breed of cattle had degenerated.[74] As stated earlier, Indian cattle health did attract the attention of the colonial authorities, as it was too valuable an economic resource to be ignored for agriculture or revenues. John A. Voelcker, thus, composed one of the earliest manuals on Indian cattle breed in 1897, where he enumerated the ways to improve Indian cattle health.[75] Colonial views on improvement of cattle health tended to revolve round the superiority of Western practices and the backwardness of native breeding techniques. Voelcker in his monumental volume, *Report on the Improvement of Indian Agriculture*, mirrored this sentiment as he commented at the very outset of his chapter on 'Livestock and Dairying', that on this subject 'there is not much to be learnt from the ordinary cultivator and his methods, and in attempting improvement. The experience of Western practice will have to be drawn upon largely.'[76] Voelcker was especially categorical in his recommendations to improve Indian cattle health when he suggested that 'in effecting any improvement in cattle the examples of native practice will not suffice, but the experience of Western practice must be applied also. The enforcement of regulations for affected animals will have to be firmly carried out, even if opposition be at first shown by the people'. In sharp contrast to Joteendro Mohun Tagore's arguments, Voelcker stated that the reason why better agricultural cattle were not to be found in parts of India was because of 'the inattention paid to the matter of breeding and selection'.[77]

Bengali periodicals like *Krishak* (Agriculturist), *Krishi Lakshmi* and *Krishi Gazette* that dealt with techniques of cultivation and gardening routinely published writings from the twentieth century onward that demanded greater government supervision into the health of Indian cows. For instance, in one such article entitled 'Gorur Durabostha' (The Miserable Plight of Cows) the author commended the ways in which the cows in the 'civilized' countries of Europe and the United States (US) were taken care of. Lamenting upon the deteriorating condition of cows in Bengal, the writer reminded his readers that the Aryan-Hindu lands had once boasted of *kamdhenu* and *kapila*.[78] Expressing serious concerns over the deteriorating health of cows in Bengal, another author in 1913 noted that a possible way to improve Indian cattle health was to introduce 'scientific education' in India.[79] Such an education meant reading works like *The Origin of Species* and Alfred Wallace's *Animals and Plants under Domestication*, which, in turn, would enable Indians to effectively learn techniques of cow rearing and protection. The author further demanded appointment of professional cow doctors in villages, who would tend to the diseased cows, given the huge cow mortality in several districts. Cow, the favoured animal of the Hindus, definitely found a privileged space in the Bengali periodicals over other diseased domesticated animals. Advertisements appeared in *Krishakin* in 1913 where farmers were offered to register for copies of the newly published pamphlet, *Go-Bandhab* (Friend of Cow), edited by Prakash Chandra Sarkar, that sought to detail scientific techniques for improvement of the breed of cows, cow protection and care.

Closely following this line of argument, many of these manuals were deeply sceptical of *krishi-mela* or cattle fairs. They believed that 'the farmers would not gain much from knowing who won what prizes for a certain agricultural innovation. What would really benefit the farmers is the outcome of that innovation'.[80] The writers, therefore, advocated that the agriculture department should shift the focus from implementing new ploughs to actually reducing the cost (*mojuri*) of agricultural equipment for the farmers. Only then could the Raj benefit the rural community. So what was the remedy? How could cattle health be improved? This question brings us to a strange story of a tussle between modernity and tradition as expressed in these writings. On the one hand, the Bengali authors of the farmers' manuals valorised the Hindu-Aryan virtues and lamented the fall from the pristine glorious Hindu past. On the other hand, they emphasized the need to introduce better *krishi-shikhya* (agricultural education) through the aid of science to improve the health of cows. This tension was pervasive throughout the early twentieth century, as noticed in several agricultural periodicals. To give an example, one writer in *Krishi Gazette* wrote:

With the progress of modern values, many of us today are distrustful of cow worship. We ought to take care of mother cow with true reverential love and protection as bhagawati, and not merely worship her by chanting mantras. Only then can she be appeased. Why cannot we Indians grasp this simple fact of cow worship?[81]

The author, thus, desired cow worship to assume a central role in the farmers' care for their animals. To support this claim, he alluded to the *Mahabharata*, arguing that in ancient times, Maharaja Dilip ardently used to worship saint Vaishishtha's cow Nandini, who was a *kamdhenu*. In a similar vein, Maharaja Virat believed in taking exceptional care of his cows; Yudhisthir nurtured around 800 cows. Having emphasized the need to continue with the tradition of cow worship, the author urged the colonial government to follow the example of the US in matters of farming and animal protection, as both the countries have 'striking similarities' in natural reserves. It is this dichotomy—the conflict between 'Western' values and tradition—that reveals itself in the early twentieth-century Bengali mind.

Conclusion: Education, Empathy, Science?

The attitude towards domesticated animals and the even larger battle to improve cattle health in colonial Bengal was not a linear narrative. It was heavily variegated and textured as different interests simultaneously vied for attention, further complicating the story of cattle health. The colonial government panicked about the loss of agricultural cattle and draft animals to a lethal disease like rinderpest. It initiated legislations and veterinary education and restructured the city and its slaughterhouses to tackle the disease. However, even with all its initiatives, the government did not directly want to intervene in indigenous 'malpractices' on animal treatment, as that would have allegedly antagonized the 'natives' against them. The Bengali *bhadralok* was, thus, caught between the reliance on science and the Hindu past and local knowledge in his understanding of cattle improvement. Examples abound of Bengali authors writing extensively in farmers' manuals that encouraged cow worship and used the Puranas and the *Mahabharata* as sources of legitimacy.[82] In matters of farming techniques too, these writers believed the government should prioritize local knowledge and collective wisdom, and not indiscriminately introduce British ploughs. They valorised the Vedic past and cattle breading of the 'Hindu kings'. However, when it came to knowledge of diseases and germs, they privileged 'Western' knowledge, science and the expertise of British veterinarians like Rutherford, than the indigenous *go-baidyas* who did not happen to possess adequate 'professional' veterinary training. But what is the 'science' that these men talked about? Was it a science that grew in a colonial milieu? Indeed this very question—what is colonial about colonial medicine—has engaged scholars for over a decade.[83] One can demonstrate how the Bengali *bhadralok* came to terms with modern science and imagined it in his own cultural context.[84] The Bengali *bhadralok* defended his indigenous knowledge system and, at the same time, appropriated new knowledge. It is this interesting moment of borrowing that I have attempted to highlight. As Projit Mukherji rightly points out, 'western medicine's links with repressive dimensions of (colonial and more recently postcolonial) power have been explored in depth, while its productive role in constituting new subjects and subject-positions have been relatively unexplored.'[85] Thus, it is demonstrated that there were also moments when the *bhadralok* translated Western notions of science and health into his own mental world. The *bhadralok* was thus not a passive recipient in the story.[86] By looking at late nineteenth and early twentieth-century notions of cattle health, through the lens of a colonial government and the voices of Bengali writers who campaigned for cattle improvement, we notice an uneasy and subliminal cross-fertilization of tropes between the ethical/religious and the scientific, between ahimsa and *bigyan*. Perhaps the only area where colonial authorities and the Bengalis seemed to converge was in their insistence on educating the lower class farmers. Class divisions ran deep and debates concerning cattle and agriculture often became the site for different parties to assert their authority.

Notes and References

1. European accounts often referred to rinderpest as cattle plague. The disease affected all cloven-hoofed animals, including domesticated cattle, African buffalo, and various species of antelope. It was the most lethal plague known in cattle. See Karen Brown and Daniel Gilfoyle, eds., *Healing the Herds: Disease, Livestock Economies and the Globalization of Veterinary Medicine*, Athens, 2010. The disease was known in Bengal by varying, locally circulated, indigenous names, like *mata, ghotee, puschima*, etc.
2. Clive A. Signage, *Cattle Plague: A History*, New York, 2003, pp. 447–96.
3. Saurabh Mishra, 'Beasts, Murrains and the British Raj: Reassessing Colonial Medicine in India from the Veterinary Perspective, 1860-1900', *Bulletin of History of Medicine*, vol. 85, no. 4, Winter 2011, pp. 587–619. Mishra argues that the colonial authorities did not intervene adequately in combating epizootics as the Raj was preoccupied with 'areas of military interest such as horse breeding'. Ibid., p. 589.
4. An epizootic is the equivalent in animal population of a human epidemic, i.e. high mortality due to a disease that is normally not present in that region.

232 *Samiparna Samanta*

5. Saurabh Mishra, 'Cattle, Dearth, and the Colonial State: Famines and Livestock in Colonial India, 1896-1900,' *Journal of Social History,* vol. 46, no. 4, Summer 2013, pp. 989–1012. In this article, Mishra looks at two massive famines that broke out in India at the end of nineteenth century, 1896–7 and in 1899. He argues that the huge cattle mortality 'not only created an ecological imbalance, but also had a massive impact on the livelihood of peasants, on agrarian structures, on agricultural productivity, and on cropping patterns.'
6. Signage, *Cattle Plague,* p. 471.
7. The numerous letters and official correspondence can be found in the West Bengal State Archives (hereafter, WBSA), Kolkata.
8. Keir Waddington, *The Bovine Scourge: Meat, Tuberculosis and Public Health 1850-1914,* Woodbridge, 2006, p. 31.
9. Samiparna Samanta, 'Dealing with Disease: Epizootics, Veterinarians, and Public Health in Colonial Bengal, 1850-1920', in *Medicine and Colonialism: Historical Perspectives in India and South Africa,* ed. Poonam Bala, London, 2014, pp. 61–74.
10. *Report on the Calcutta Epizootic or Cattle Disease of 1864,* Letter from Dr C. Palmer, Presidency Surgeon to S.C. Bayley, Junior Secretary to the Government of Bengal, 7 October 1865, General (Medical Branch), November 1865, no. 71, WBSA.
11. J.H.B. Hallen, *Manual of the More Deadly Forms of Cattle Disease in India,* Calcutta, 1885, pp. 1–57. The correspondences can be found in Municipal Department (Medical Branch), November 1884, File no. 2, Proceeding no. B31-62, WBSA.
12. Mishra, 'Beasts, Murrains and the British Raj', p. 617. Mishra argues that even the Imperial Bacterial Laboratory continued to conduct research mainly on diseases affecting horses.
13. Mishra, 'Beasts, Murrains and the British Raj', p. 595. The term 'enclavist' was first used by Radhika Ramasubban. See Radhika Ramasubban, *Public Health and Medical Research in India: Their Origins under the Impact of British Colonial Policy,* Stockholm, 1982, pp. 1–48.
14. Mishra, 'Beasts, Murrains and the British Raj', p. 602. Mishra believes that 'the expenses involved in controlling such outbreaks would perhaps have far outweighed the losses they caused to the treasury'.
15. The notion of 'public health' has been widely debated by scholars studying the relationship between colonial medicine, disease and power in the South Asian context. Some prominent works are Mark Harrison, *Public Health in British India: Anglo-Indian Preventive Medicine 1859–1914,* Cambridge, 1994; Poonam Bala, *Medicine and Medical Policies in India: Social and Historical Perspectives,* Lanham, 2007; Kabita Ray, *History of Public Health in Colonial Bengal 1921–1947,* Calcutta, 1998; Anil Kumar, *Medicine and the Raj: British Medical Policy 1835–1941,* New Delhi, 1998. For an overview, see Birdie Andrews and Andrew Cunningham, eds., *Western Medicine as Contested Knowledge,* Manchester, 1997.
16. The Slaughterhouse Act of 1865 categorically stated that no person should kill,

within the limits of the cantonment, any animal for public slaughter except at the public slaughterhouse (to be built at Tangra).
17. For an analysis of how rinderpest affected public health, and the culture of diet and sanitary science in colonial Bengal, see Samanta, 'Dealing with Disease'.
18. Michael Worboys, *Spreading Germs: Disease Theories and Medical Practice in Britain, 1865–1900*, Cambridge, 2000, p. 44.
19. The pioneering work after 1870s on how germs caused infection began with studies of diseases that crossed species barriers—anthrax, tuberculosis and rabies. See Neil Pemberton and Michael Worboys, *Mad Dogs and Englishmen: Rabies in Britain, 1830–2000*, Basingstoke, 2007.
20. *London Times/Evening Mail*, 21 August 1885, Calcutta, *The British Newspaper Archive*, http://www.britishnewspaperarchive.co.uk/ (accessed on 19 December 2015).
21. David Arnold, *Colonizing the Body: State Medicine and Epidemic Disease in Nineteenth-Century India*, London, 1993.
22. Arabinda Samanta, 'Plague and Prophylactics: Ecological Construction of an Epidemic in Colonial Eastern India', in *Situating Environmental History*, ed. Ranjan Chakrabarti, New Delhi, 2007, pp. 221–42.
23. Captain A.E. Campbell, Deputy Commissioner, Seebsaugor to the Personal Assistant to the Commissioner of Assam, Seebsaugor, 22 June 1869, Political Department (Medical Branch), 27 July 1869, Proceedings 66–7, no. 384, WBSA.
24. Megan Vaughan, *Curing their Ills: Colonial Power and African Illness*, Stanford, 1991; John Iliffe, *East African Doctors: A History of the Medical Profession*, Cambridge, 1998.
25. Daniel R. Headrick, *The Tools of Empire: Technology and European Imperialism in the Nineteenth Century*, New York, 1981.
26. Ian Inskter, 'Prometheus Bound: Technology and Industrialization in Japan, China and India Prior to 1914: A Political Economy Approach', *Annals of Science*, vol. 45, 1988, pp. 399–426.
27. Roy Macleod and Deepak Kumar, eds., *Technology and the Raj: Technical Transfers to India 1700-1947*, New Delhi, 1995; Zaheer Baber, *The Science of Empire: Scientific Knowledge, Civilization and Colonial Rule in India*, New York, 1996.
28. Deepak Kumar, *Science and the Raj: A Study of British India*, New Delhi, 1995.
29. Some examples are John A. Voelcker, *Report on the Improvement of Indian Agriculture*, Calcutta, 1897; Isa Tweed, *Cow-keeping in India, A Simple and Practical Book on their Care and Treatment, their Various Breeds, and the Means of Rendering them Profitable*, Calcutta, 1900.
30. Laxman D. Satya, *Ecology, Colonialism, and Cattle: Central India in the Nineteenth Century*, New Delhi, 2004.
31. While an exhaustive work on cattle ecology, Laxman Satya places too much emphasis on colonialism and almost treats it as a monolith. Barring a few evidence of resistance, the hapless farmers often appear as silent 'victims' of a hegemonic colonial power in his study.

234 *Samiparna Samanta*

32. Letter from Captain L. Blathwayt, Assistant Commissioner of Golaghat to the Deputy Commissioner of Seebsaugor, 18 June 1869, Political Department (Medical Branch), 27 July 1869, Proceedings 66–7, File no. 123, WBSA.
33. Letter from the President, Cattle Plague Commission, Culna to the Home Department, General Department (Medical Branch), December 1870, WBSA.
34. Voelcker, *Report on the Improvement of Indian Agriculture*, pp. 7, 301.
35. Ibid.
36. Ibid.
37. Deryck Lodrick, *Sacred Cows, Sacred Places: Origins and Survivals of Animal Homes in India*, Berkeley, California, 1981, p. 89.
38. Wendy Doniger, *The Hindus: An Alternative History*, New York, 2009, p. 9.
39. For an analysis of Hindu attitudes towards non-human animals, see Lance Nelson, 'Cows, Elephants, Dogs, and Other Lesser Embodiments of *Atman*: Reflections on Hindu Attitudes Toward Nonhuman Animals', in *A Communion of Subjects: Animals in Religion, Science, and Ethics*, ed. Paul Wadu and Kimberly Patton, New York, 2006, pp. 179–93.
40. Municipal Department (Medical Branch), October 1901, File no. 7R/10, Proceeding nos. 30–2, WBSA.
41. *Annual Report of the Civil Veterinary Department, Bengal, and of the Bengal Veterinary College, for the Year 1900-1901*, National Library of Scotland, digital collections http://digital.nls.uk/indiapapers/browse/pageturner.cfm?id=7634258 4&mode=transcription (accessed on 30 January 2012).
42. Judicial Department (Judicial Branch), November 1900, File no. 3 4C/2 1.3, Proceeding no. 1.4, WBSA.
43. 'Instructions to the Magistrates acting under Act I (B.C.) of 1869 (An Act for the Prevention of Cruelty to Animals) as amended by Act III (B.C.) of 1900', Letter from P.C. Lyon, Director of the Department of Land Records and Agriculture, Bengal to the Secretary to the Government, Revenue Department, Judicial Department, Judicial Branch, November 1900, File no. 239A, Proceeding no. 1–2.44, WBSA.
44. Ibid.
45. Letter from the Secretary of Jnandayini Sabha to the Secretary to the Government of Bengal, Judicial Department, Judicial Branch, 24 July 1911, Writers' Building, Proceeding no. B.304–12, WBSA.
46. Reports mainly came from the following districts of Bengal—Midnapore, Hooghly, Burdwan, 24 Parganas, Nadia, Murshidabad, Jessore, Khulna, Rajshahi, Pubna, Dacca, Rungpore, Dinajpore, Tippera, Barisal, Jalpaiguri, Cooch Behar, Purnea, Bhagalpore, Chapra, Durbhanga, Mozaffarpore, Munghyr. According to R.B. Ghosh, this fraud was not practised in Calcutta or in 'villages inhabited by educated gentlemen'.
47. R.B. Ghosh, 'An Appeal for the Prevention of Cruelty to the Bovine Species', Calcutta, July 1910.
48. Ibid.
49. Ibid.
50. Ibid.

Cattle, Cruelty, Cow Doctors 235

51. It is interesting to compare this gradual shift towards a 'scientific' enquiry among the Bengalis with similar concerns among social reformers surrounding foot binding in late nineteenth-century China. The 'godly' was being superseded by biomedical science. The Chinese reforming elite, very much like the Bengalis, was a hybrid of religion and science. See Angela Zito, 'Secularizing the Pain of Foot Binding in China: Missionary and Medical Stagings of the Universal Body', *Journal of the American Academy of Religion,* vol. 75, no. 1, 2007, pp. 1–24.
52. Letter from H.N. Ghosh, Secretary of the Jnandayini Sabha to W.B. Heycock, 5 March 1912, no. 126, WBSA.
53. Letter from W.B. Heycock, Director of Agriculture, Bengal to the Secretary to the 'Jnan Dayinee Sava', 1 December 1 1911, Agriculture Department, no. 12492, WBSA.
54. Lata Mani, *Contentious Traditions: The Debate on Sati in Colonial India,* Berkeley, California, 1998.
55. Ibid.
56. Judicial Department, Judicial Branch, File: I 4C/4 of 1912, WBSA.
57. Ibid.
58. *Krishi Gazette,* Calcutta, April 1885, National Library, Kolkata. CrossAsia-Repository, http://www.ub.uni-heidelberg.de/Englisch/fachinfo/suedasien/zeitschriften/bengali/overview.html.
59. *Go-baidyas* are described in these agricultural tracts as lower class, itinerant doctors who travelled during winter and spring months to treat cattle in the villages of Bengal. They were believed to be familiar with treatments of boils, eruptions and fever affecting cattle in Bengal. My understanding is that the *go-baidyas* were different from what the people termed *go-daga.*
60. Interestingly, this was also the first time that the category 'veterinary surgeon' was used to denote the village doctors.
61. Jan E.M. Houben, 'To Kill or Not to Kill the Sacrificial Animal (*Yajna-Pasu*): Arguments and Perspectives in Brahminical Ethical Philosophy', in *Violence Denied: Violence, Non-Violence and the Rationalization of Violence in South Asian Cultural History,* ed. Jan E.M. Houben and Karel R. Van Kooji, Leiden, 1999, pp. 117–24.
62. Anthropomorphic animals were a common element of nineteenth-century children's literature all across the world. In Europe, numerous nineteenth-century authors wrote stories anthropomorphizing animals, usually for juvenile audiences, including Anna Sewell, Jack London, Rudyard Kipling and Lewis Carroll. This current ran, as some scholars have examined, against the larger theme of exploitation and moderated it because teamsters and animal owners internalized some of these values, which led to anti-cruelty regulation. For a detailed discussion, see Clay McShane and Joel Tarr, *The Horse in the City: Living Machines in the Nineteenth Century,* Baltimore, 2007, p. 9.
63. *Krishi Lakshmi,* vol. 7, no. 1, Calcutta, April 1937.
64. Projit Mukherji, *Nationalizing the Body: The Medical Market, Print and Dakdari Medicine,* London, 2009, p. 139.

65. Meera Nanda has demonstrated how sanitation itself was equated to an ancient 'national' cultural practice that was only too often equated with 'Hindu national culture'. Every aspect of sanitation and hygiene was, therefore, found to be identical to 'ancient Hindu lifestyles'. See Meera Nanda, 'Hindu Nationalism and Vedic Science', *Prophets Facing Backwards: Postmodern Critiques of Science and Hindu Nationalism in India*, Piscataway, 2003, pp. 37–124.
66. Bankimchandra Chattopadhyay is perhaps a case in point who believed that the Indians were never lacking in physical prowess in the past. The Mauryas and Guptas could carve out an empire over a vast territory and ward off the Greek advance beyond the Sutlej. Bankimchandra Chattopadhyay, '*Bangalir Bahubal*', '*Bibidha Prabandha*', *Bankim Rachanabali*, vol. 2, Calcutta, 1991, pp. 209–12.
67. William Adam, *Third Report on the State of Education in Bengal*, Calcutta, 1838, p. 235.
68. Agricultural and Horticultural Society of India, *Transactions of the Agricultural and Horticultural Society of India*, vol. 1, 1937, p. 1.
69. *General Report on the Public Instruction in Bengal, 1865–66*, pp. 35–6.
70. K.C. Mitra, *Agriculture and Agricultural Exhibition in Bengal*, Calcutta, 1865, p. 17.
71. '*Banglay Krishi-shiksha*', *Sambad Prabhakar*, 9 December 1892, CrossAsia, http://www.ub.uniheidelberg.de/Englisch/fachinfo/suedasien/zeitschriften/bengali/overview.html (accessed on 19 December 2015).
72. Letter from Babu Joteendro Mohun Tagore, Honorary Secretary to the British Indian Association to F.R. Cockrell, Officiating Secretary to the Government of Bengal, General Department, March 1864, Proceeding no. 43, WBSA.
73. Ibid.
74. Letter from J. Beckwith, Secretary to the Landholders and Commercial Association to F.R. Cockrell, Officiating Secretary to the Government of Bengal, General Department, March 1864, Proceeding no. 43, WBSA.
75. Voelcker, *Report on the Improvement of Indian Agriculture*, pp. 1–19, 93–134, 191–438.
76. John A. Voelcker, 'Livestock and Dairying' (Chapter XI), *Report on the Improvement of Indian Agriculture*, p. 198, paragraph 245.
77. Ibid., paragraph 249.
78. '*Gorur Duraboshtha*' (The Miserable Plight of Cows), *Krishi Gazette*, 1913, p. 44, National Library of India, Kolkata (hereafter, NL).
79. '*Gojatir Upakarita*' (The Usefulness of Cow), *Krishak*, vol. 14, 1913, NL.
80. *Krishi Gazette*, 1913, p. 44, NL.
81. Ibid., p. 45, NL.
82. Ibid., p. 46, NL.
83. Shula Marks, 'What is Colonial about Colonial Medicine? And What has Happened to Imperialism and Health?' *Social History of Medicine*, vol. 10, 1997, pp. 205–19.
84. Some scholars have hinted at the Bengali *bhadralok*'s attempt to internalize modern science and medicine. David Arnold's *Colonizing the Body* had ended

with comments about how Indian middle classes had appropriated 'modern science' and integrated these into the latter's rhetoric of legitimation.
85. Mukherji, *Nationalizing the Body*, p. 9.
86. Ishita Pande makes a similar argument as she demonstrates that by the end of the nineteenth century, the Bengali *bhadralok* was no longer the passive object of medical gaze because he refused to be pathologized as the other. Ishita Pande, *Medicine, Race and Liberalism in British Bengal: Symptoms of Empire*, New York, 2010, p. 16. Also, see Samanta, 'Dealing with Disease'.

11

Cattle Breeding Policies in Colonial India

Himanshu Upadhyaya

This essay discusses the role played by science institutions on the front of animal husbandry with special reference to cattle breeding policies during the period 1905 to 1940. In this essay, I shall argue that cattle breeding policies that evolved in the first half of the twentieth century in colonial India gradually shifted focus to the aim of improving milking capabilities of Indian cattle, almost to an extent of draught qualities receiving scant attention, or peasants being asked to prepare themselves to tolerate deficiency on draught qualities, in the interest to see increase in milk yields. In those years, prior to the introduction of artificial insemination on a large scale, cattle breeding was being perceived as a long-range work. With reference to government cattle farms and institutional herds, frequent changes in cattle breeding policies were observed sometimes. However, these frequent changes were deprecated by experts, who opined that such changes, which accompanied change in leadership, or following financial stringency, had the potential to undo the advances already made.

This essay also engages with the impact of the introduction of perennial irrigation in regions known for excellent cattle breeding traditions. While earlier accounts by ecological historians who work on the colonial period have talked about the lack of empathy for native pastoral and nomadic tribes among colonial policymakers, I shall re-examine the question here. Engaging with contemporary sources, in the form of academic papers written on the eve of the extension of such canal projects, I shall argue that at least few agricultural scientists from colonial establishment voiced cautionary notes on

---

* This paper was first presented at a seminar at the National Bureau of Animal Genetics Research, Karnal, Haryana, and later at a faculty seminar at Azim Premji University, Bangalore, and the author has gained from comments made by the audience at both these seminars. The author also acknowledges Prof. Deepak Kumar, Prof. Purnendu Kavoori, Prof. Anil E. Nivsarkar, Dr Kamal Kishor, Prof. N. Kandasamy, Prof. Sasheej Hegde and Prof. Anil Sethi for their comments on different drafts of this paper.

the probable impacts of such policies with sedentary bias on some of the cattle breeds.

This essay reviews the issue of cross-breeding of Indian cattle with exotic bulls and argues that after the early efforts at introducing cross-breeding, an opinion had emerged among colonial experts that large-scale extension of cross-breeding with exotics will lead to adverse impact on the agrarian balance. On the contrary, even while keeping the focus on aiming at improving milk yields, the official discussions around cattle breeding policy repeatedly revolved around the idea of 'dual purpose' breed. The concept of 'dual purpose' breed was never more intense than at the time of Royal Commission on Agriculture proceedings. While we witness the Commission pronounce an opinion that draught and milk characteristics were physiologically incompatible, and hence, it shall be desirable to attempt one task at a time, imperial animal husbandry expert, Arthur Olver (who served in that capacity during the period 1930–8), refers to Indian peasant's preoccupation with 'general utility' cattle.

As Arthur Olver points out, within three decades, even some of the best-known draught breeds had come to show promising milk yields. I shall argue that such a focus on improving milking capabilities of Indian cattle was intricately linked with exigencies of colonial political economy, wherein the empire tried to thrust major transformations upon large-scale ecosystem and production system in the dominion, but tried to mend the agrarian landscape to suit the empire's interests in the colonies. The introduction of organized dairying, with a mission to solve the question of city milk supply, was intricately linked to an extractive policy, which linked milk producers in Kheda and other such dairying tracts, with the nutritional needs of troops fighting the two World Wars on behalf of the empire.

Norman C. Wright who travelled to India to review the veterinary side work of Imperial Council of Agricultural Research in 1937 had stressed upon the need to carefully chart out a cattle breeding policy for institutional herds and then not change it at the drop of the hat, or as soon as there is change in the organization:

It is, however also essential that, having decided what policy to adopt, this should be continued uninterrupted over a long period of years. There is no branch of animal husbandry in which changes of policy can do so much harm as in breeding: if the changes are frequent, no constructive breeding is possible. On the other hand if a policy which has been maintained over a long period is ultimately abandoned through some adventitious factor such as temporary financial stringency, many years of constructive work may be wasted.[1]

Wright also cited an example of an institution where such frequent changes (seven changes over a period of less than twenty-five years) in cattle breeding policy had produced poor results. Wright had not elaborated on it further, but in an oblique reference, he had called upon animal husbandry experts in India to reflect on the history of Scindi breed as the one that provides 'the worst instances of such short sighted policy'.

Early Discussions on Need to Introduce Scientific Breeding of Cattle

The colonial intervention to introduce the rational application of modern science to crop cultivation and animal husbandry primarily took the route by attempting to address the pathological aspects, and thereby, tried to understand diseases that plagued plants and cattle. Thus, in 1871, we find Col. J.H.B. Hallen, the first Inspector General of the Indian Civil Veterinary Department publish a popular handbook meant for the use of cattle owners titled *A Manual of the More Deadly Forms of Cattle Diseases in India*. The book underwent a second revision by Hallen in 1883 and then a third revision by Major G.K. Walker in 1903. In 1916, Major G.K. Walker undertook substantial rewriting, under instructions from the Government of India and published it as *Some Diseases of Cattle in India*. This was again revised in 1928 by J.T. Edwards, the then Director of Imperial Institute of Veterinary Research, Muktesar.[2]

Parallel to this, we also witness in 1883, the then Inspector of Cattle Diseases in Madras Presidency, James Mill, had published a book titled *Plain Hints on the Diseases of Cattle in India*. Writing an introduction to his book, Mill had no pretence that his was the first book on the subject of cattle disease which has been written in India, but strived to note:

With the changes which have taken place in our knowledge of the disease of cattle, and with the increased attention which Government is now paying to this subject, in the interests of the public, the want of a fresh work for the information of cattle owners and district officials, has been many times forced on my notice, in the performance of my duties as Inspector of Cattle Diseases.[3]

In 1894, when Mill published the second revised edition, he expressed that he had been thinking to work on the task for at least three years, but the duties that he had to perform as the Principal of the Government Veterinary College had not left him with enough time to undertake the task.

Similar to both these books that were authored by practising veterinary experts, there was another penned by a European lady called Isa Tweed, who, in 1890, felt the need to publish 'a simple and practical book on cattle and their care and treatment in India, after having kept cows for the last eighteen years and undertaken their medical treatment'. Tweed was of the opinion that 'there were many books on cattle and their care in Europe' and some books on Indian cattle, but went on to assert that the 'best of these is far from complete'. Tweed's book was well received and within nine years, in 1899, the second edition got published, and then within the next twelve years, the third edition appeared. In the preface, Tweed writes:

I have often thought, if more Europeans and well-to-do native families in India were to keep cows, there would be less disease, and less annoyance about procuring milk, butter and ghee; and there would also be a great improvement in the breed of cattle. I am convinced that very few persons in India know how to properly care for cattle when in health, or treat them when sick. Most people trust the care of their cattle entirely to their servants, and believe every word the stupid and dishonest gowallah says. There are a number of good Veterinary surgeons and gentlemen practicing medicine in every town and it would be more economical in the end to consult the best of them than to trust the stupid native servants.[4]

Such pronounced contempt for native ways of handling milk and an advocacy for introducing the English style dairying was a recurrent theme in writings right from the 1890s. A book titled *The Indian Amateur Dairy Farm* by Landolicus presented the native milk vendor in a similar condescending tone:

The few 'Dairy Farms' properly speaking that there are in India, are chiefly confined to towns and these are principally in the hands of natives, managed in a very unsatisfactory way, and their products much adulterated, so much so in such a way as not only to be repugnant, but they are also deleterious to health, and a frequent cause of the spread of diseases of most serious character, such as Typhoid fever, Cholera, Small Pox etc. . . . It is true that English modes of treating cattle are not entirely suitable to this country, but at the same time there is much in practice of treatment of cattle in India which should be avoided and the English or Continental system be followed as nearly as possible. The introduction of a new system to the goalah or cowherd caste in India seems one that he abominates and abjures; but let me ask, what new ways are ever pleasing to the natives of this country? Any innovation is always fought against and shirked, whether in gardening, the keeping of a poultry yard or cows. . . . It is well known that goalahs will always do as natives always will, and that is to follow same lines as their forefathers did from time immemorial.[5]

Both these quotes, from the preface of books that had started to get published and put in circulation with an aim to aid the colonial mission of 'livestock improvement', show a peculiar way of presenting the native livestock keeper as steadfastly traditional, who will display enormous amount of resistance to receive the modern scientific and rational knowledge, whether on cattle diseases or on cattle breeding. As far as the dairy development agenda was concerned, the European dairy animal was held up as a standard to emulate and institutional herds were established at agricultural colleges to demonstrate scientific principles in cattle breeding. Traditional modes of cattle keeping and breeding were now being confronted with scientific knowledge and advice that was communicated through written and printed words.

Thus, J.B. Knight, Special Assistant to Director of Agriculture in Bombay presidency wrote in a note prepared for the first meeting of the Board of Agriculture, what he planned to take up at the experimental dairy farm at Poona:

The farm is devoted to growing fodder for its animals and as far as possible trials are made with the various fodder plants available, especially fodder of the other leguminous. . . . The dairy farm consists of 250 animals, composed of the representatives of three breeds of cows, viz. Sind, Gir and Aden and three breeds of buffaloes, viz. Delhi, Jafrabadi and Gujerat. A record is kept for each animal and a comparative test of the profits of these six breeds is being made.[6]

Knight goes on to underline that the Poona experimental dairy farm was set up to demonstrate the value of scientific principles of breeding dairy cattle. The way he articulates the agenda of Poona experimental farm on cattle breeding clearly sets forth contradictions in the meanings assigned by a native cattle breeder and colonial experts when it came to undertake cattle breeding. Whereas for the native cattle breeder, cow or buffalo were primarily being bred for the role they played in agricultural pursuits, for a colonial expert, the introduction of scientific principles in cattle breeding was essential in the task to improve milk yields of the Indian dairy cow or buffalo to such a level as to make it compare favourably with the European dairy animal. Corollary to such a viewpoint was an eagerness to attempt description of such cattle breeds that had shown remarkable performance as milch breed primarily. Knight remarks,

Scientific breeding is almost unknown in the case of Indian cattle. In the breeding of these several breeds at Poona, it is proposed to demonstrate the value of scientific principles of breeding dairy cattle. It is believed that in a few generations the Indian dairy cow or buffalo could become so improved that she will compare favourably with the European dairy animal.[7]

However, it would be wrong to assume that the kind of condescending tone with which Landolicus, Tweed and Knight had written about the native cattle breeder was shared by all the colonial experts. There were others such as Albert Howard and Arthur Olver who looked at the agrarian landscape of India with a careful eye and spoke in an empathetic tone about the native, nomadic cattle breeders.

If we peruse through colonial records such as the Minutes of Board of Agriculture in India, Notes written by the then Imperial Dairy Expert, W. Smith and Imperial Animal Husbandry Expert, Sir Arthur Olver, etc., we get to see that there existed a school of thinking that laid great emphasis on the potential of indigenous breeds of cattle for dairy development in India. While there existed a few colonial experts who came with a typical imperial arrogance that looked down upon poor, illiterate cattle breeders, there were also some visionary thinkers amongst colonial agricultural experts, such as James Mollison, Albert Howard and others, who reflected on the potential contribution of native cattle breeders who had taken pains to preserve phenotypes in indigenous breeds. Writing in 1901, James Mollison, Director of Agriculture in the Bombay presidency had voiced his impression that 'the numerous pure Indian breeds are results of careful breeding through many generations'.[8] Mollison had also clearly pointed out that the chief purpose of cattle breeding in India was draught power requirement and buffalo was the prominent milch animal. Similarly, Arthur Olver appreciated the role played by native cattle breeder in his article, 'The Necessity for Authoritative Definition of Breed Characteristics and Unchanging Control of Breeding Policy in India' by stating:

To have established such well-marked breeds as, for instance, the Kankrej, Kangyam, and Ongole breeds, long continued and careful breeding to a definite type, with rigid elimination of every variation, must, however, have been practiced, and it is clear that this work must have been carried on from generation to generation. . . . It is evident however that the breeders who were capable of producing such even results must have thoroughly understood the fundamental principles which underlie all successful breeding of the larger domesticated animals, and judging from the care with which they still look for in the selection of their breeding stock, there is no doubt that this knowledge is still fairly widely applied by practical breeders in India, particularly amongst the professional cattle breeders, who maintain considerable herds in large grazing areas, and are able to keep them reasonably free from intrusion of alien blood.[9]

Direction in Cattle Breeding Work: Plough Bullock or Milch Cow?

During the second half of the nineteenth century, trotting bullocks had started to become less and less important and the native peasant slowly moved to concentrate on breeding cattle for ploughing on farm and carting, as well as aiming to improve milking capabilities of cows. Railways had now arrived, and Isa Tweed, while describing a few notable cattle breeds, commented thus on Nagourie bullocks, in her book, *Cow Keeping in India*:

Nagourie cattle are of the well-known trotting breeds, the bullocks are much prized, and used by native gentlemen for their carriages. Nearly half a century ago, they were extensively used in large cities by native gentlemen, and in those days were carefully bred for the purpose, but now they are not so well preserved and good cattle are scarce.[10]

At the eighth meeting of the Board of Agriculture held at Coimbatore on 8 December 1913, the subject of cattle breeding policy came under discussion by a committee headed by Hailey. The terms of reference set out for the committee had stated:

A survey of cattle breeding industry has been made or is in progress in most provinces and the time would now seem ripe for the Board to advise on the lines of future policy to be adopted. The measures to be taken for the preservation of fine breeds which already exist and the protection of cattle breeding industry, the preservation of grazing areas and the general question of fodder supply (together with the implied one of grazing versus stall-feeding) seem to call for further discussion. In view of the growing opinion in favour of stall-feeding, it is now desirable to start a systematic investigation of the relative food value of Indian cattle foods and to frame a working plan for this, on the lines of combined feeding and chemical tests.[11]

At the ninth meeting of the Board of Agriculture held at Pusa from 7 to 12 February 1916, the subject of cattle breeding and dairying in India was taken up for discussion. The Board resolved that in order to make satisfactory progress in the development of good milch cattle in India and in dairying, an officer should be appointed on the imperial staff under the title 'Imperial Dairy Expert', his duties being—the control of cattle breeding farms and dairy operations, the supervision of dairy instruction and the study and improvement of existing dairy methods and establishment of the dairy industry on commercial lines. The Board had also observed that the offer which was made by military authorities, who had expressed willingness to grant the herds of various breeds of pure-bred Indian cows and buffalo maintained by them to civil side so that these herds could be used by institutional farms, as well as their willingness to extend the facilities for conducting further cattle breeding operations on the military dairy farms, was of extreme value and should be gladly accepted. Other principal resolutions were with regard to establishment of dairy schools to fill the need for trained dairy managers, arrangements of immunization of cattle against diseases by increasing Muktesar staff, the advisability of instituting an investigation into the existing supply and demand for dairy products on the lines of the inquiry made by the Bombay department, and lastly, the legislation against adulteration in dairy produce.[12]

Around this time, the cattle breeding policy discussions had started to witness a marked shift towards focus on milch cows. E.W. Oliver, the then superintendent of Civil Veterinary Department of United Provinces read out a paper titled 'Cattle Breeding, with Special Reference to Milch Cow' at the Provincial Cooperative Conference held at Lucknow in February 1916.[13] This gradual shift towards introducing organized dairying in British India affected the aims of cattle breeding policy towards increasing milking capabilities of Indian cattle through selective breeding. Alongside this, departments of agriculture in the provinces were now seen attempting to upgrade the local buffalo stock by importing Murrah bulls. For example, Murrah bulls were maintained at the Agriculture College Dairy, Coimbatore from 1917 and a buffalo breeding centre was started at Lam, near Guntur in 1923 and the Murrah herd from Coimbatore College was transferred there.

Deterioration of Cattle Breeds in Irrigated Tracts

The colonial policies that sprang from the then prevailing notion of looking at sedentary cultivation as a more evolved form had already started to plan extending perennial irrigation to even those zones that were famous for their cattle breeding traditions of nomadic tribes. In a paper that Mollison wrote along with L. French, he had sounded a warning by stating:

The breeding of Montegomery cattle is likely to suffer unless special precautions are taken to maintain the purity of the breed, because the extraordinary prosperity of the Chenab colony across the Ravi, has diverted the attention of the nomadic 'Bar' tribes to the profits derivable from Agriculture when assisted by canal irrigation. It is also to be remembered that the whole of 'Bar' tracts of the Montegomery district are destined within few years to receive irrigation from the projected Lower Bari Doab Canal.[14]

Similar to Mollison's opinion, W. Smith wrote a note for a meeting of the Agriculture Committee in 1927, where he stated that despite all the efforts made in the direction of cattle improvement and scientific breeding, it had become 'impossible in the open market today, to procure in good quantity, no matter what the price may be, as good draught bullocks and milch cows, as were obtainable 16 years ago'.[15] Smith also held the opinion that the quality of milch cattle available in India, including Punjab, but excluding Sindh, had become much worse than those available when he had just arrived in India twenty-six years ago to work with military dairy farms. Detailing the probable reasons for decline in the quality of cattle, Smith underlined that the root reason was 'want of knowledge, accentuated by many circumstances, such as the spread of irrigation canals and conservation of forest lands with consequent diminution of grazing areas, the increased facilities for transport and consequent mixing of breeds or types, increased price for human foodstuffs, and the erroneous idea that the development of dairying or milk production would injure the draught qualities of working bullocks'.[16] However, the last reason that Smith lists on the inventory, by calling it an erroneous idea, was not merely the belief held by native peasants and cattle keepers, but even by the Royal Commission on Agriculture, which in its report had underlined that draught power and milking capabilities were physiologically incompatible and the best way to go about cattle breeding was to undertake 'one task at a time'.

Ten years after Smith voiced this opinion, Norman C. Wright also spoke about the deleterious impact of irrigation canals on cattle breeds:

At present there is a very general impression that the introduction of irrigation rapidly leads to deterioration and even to the virtual extermination of good breeds of cattle. This for example is true of Sindh and certain tracts of United Provinces. If full advantage is to be derived from irrigation, I am convinced that farming in irrigated areas will have to be modified to allow the inclusion of mixed farming system in which both crop and animal husbandry plays their part.[17]

With the introduction of perennial irrigation, more and more land was being taken for cultivation, leaving out the nomadic cattle breeders with shrinking access to pastures and grazing resources. In addition to diminution of grazing resources that were lying at walking distances from the rural habitat, the restrictive policies put in force by scientific forestry had pushed the nomadic cattle breeder into penury. The related problem that had started to surface in tracts that had shown extensive area under cultivation was progressive depletion of the fertility of soil and its humus content, nullifying efforts to improve yields by growing improved varieties of seeds and control of plant diseases, as a result of the tendency on the part of colonial agricultural experts to consider that 'the only practicable policy is to produce more and more crops'. An annual report of the Veterinary Department of Madras presidency for the year 1940–1 highlighted the implications of cultivators switching to commercial crops on the cattle breeding conditions in the home of the famous Ongole breed:

The breeding of cattle in recent years has suffered a check owing to progressive reduction in the supply of fodder consequent to replacement of cereal crops by other commercial crops such as cotton, tobacco, chillies, turmeric, etc., which yield no fodder. The decrease in the total cultivable area during the quinquennium ending 1953 is estimated at 0.8 million acres. The quality of cattle has also suffered owing to the great dearth of bulls in the breeding tracts . . . proportion of breeding bulls to number of cows in the Nellore and Guntur districts was 1:231 and 1:122 respectively, whereas the proportion for the whole province on an average was 1:20.[18]

The logic of productivity sans the care for soil health, despite the warning bells repeatedly sounded by experts such as Albert Howard, W. Smith and Arthur Olver, was leading to crop cultivation and animal husbandry charting two separate paths.

An inquiry carried out under the aegis of Imperial Council of Agricultural Research in seven prominent dairying tracts in British India in 1936–7 reported that it no longer paid to breed cattle, in the breeding tract of Hariana cattle. The reason being that with the extension of irrigation, cultivation of commercial crops such as cotton and wheat was found more profitable by the local zamindars! The report states the transformations that were brought about by late 1930s in the home of the famed Hariana breed:

The profits from cattle breeding are very small, and it is adopted as an industry by the small land-owner as a means of eking out a living for himself. No doubt the cattle are the zamindar's treasury here, and he puts all his savings into them; yet the melancholy fact remains that breeding does not pay or only pays poorly.[19]

In a presentation before the Royal Society of Arts on 29 May 1942 and recounting his experiences as Imperial Animal Husbandry Expert, Olver reminded about 'the grave danger in the official policy pursued in some provinces of devoting large areas to intensive cash-crop production, without due provision for livestock to preserve fertility and provide the necessary labour and a better diet for people'.[20] This tells that Olver had very clearly focused his work on cattle breeding to keep the crop-cattle interaction alive and within four years, after he vacated his office, he could see the grave danger posed by what he termed 'intensive cash crop production'.

The Role of Imported Bulls in Cattle Breeding Policy

At the first meeting of the Board of Agriculture in India, speaking about cross-breeding, Col. J.W.A. Morgan, the then I.G., Civil Veterinary Department, had 'deprecated the attaching of undue importance to the possibilities of cross-breeding in India'. He had said without mincing words that 'crossing with imported cattle, especially had been, in a general a failure'. At that point, it was considered so since cross-bred animals were 'extremely susceptible to rinderpest and usually die off.' Morgan mentioned only one case where some amount of success was claimed, namely Taylor breed in the vicinity of Patna, but this success had practically been confined to the region around Patna and had not proved to be thriving in other regions.[21] Adding his voice in corollary, W.C. Renouf, the then Director of Land Records and Agriculture in Punjab, voiced his opinion that 'at Hissar, cross-breeding had been a total failure'.[22]

Thus, after a short discussion on the consequences of the cross-breeding work so far, the Board of Agriculture at the first meeting pronounced a collective opinion that 'the most satisfactory line of work from a general standpoint for the improvement of work cattle for agricultural purposes lies *in the selection of indigenous cattle rather than in cross breeding*' [emphasis mine]. In some provinces, there had been experiments to acclimatize certain milch breeds, in an environment different from their native breeding tracts. Thus, for example, Benerjee's experiments had found that Montgomery (Sahiwal) remain superior compared to local breeds from Bengal; Lawrence's experiments in Bombay presidency had found that Sindi cattle turned out to be the best performers in Poona. The Board of Agriculture took note of these efforts at acclimatization of superior milch breeds of indigenous cattle and pronounced a collective opinion that 'it would be advisable to try experiments on a small scale in various parts of India for the acclimatization of superior breeds of Indian milch cattle, especially the Montgomery breed'.[23]

At the fourth meeting of the Board of Agriculture in India, the then president of the Board, James Mollison, once again articulated his opinion that 'introduction of foreign breeds of cattle was very unlikely to be of practical value owing to their susceptibility to rinderpest and other diseases that cause extremely high mortality among such cattle and their offspring in India'. Mollison also cited an example where an effort to introduce English cattle failed after all of them died from a rinderpest attack, while various native breeds had shown considerable immunity in the same outbreak.[24]

Review of work done by Imperial Council of Agricultural Research till 1929 was discussed at the fifteenth meeting of the Board of Agriculture in India, and at this meeting, the board passed a resolution on cattle breeding policy which read: 'This Board, as a result of evidence placed before the meeting supports the view that to effect general improvement in the cattle of India, attention should be concentrated on the indigenous breeds.'[25]

Two decades from those early days of policy discussions on cattle breeding policy in India, Lal Chand Sikka, who graduated from the then Imperial Institute of Animal Husbandry and Dairying, Bangalore, as a Punjab provincial government-sponsored student, talked about Indian cattle as 'the best tropical cattle in the "World", thanks to their power to withstand epizootic diseases and ability to stand a hot climate'.[26] Sikka had believed that 'the best way of solving the Indian cattle problem is to improve them without sacrificing these qualities'. He stressed upon a need to 'establish and carefully maintain and breed herds of the best Indian breeds and eliminate from them by a process of selection and rejection all cattle below a certain fixed standard'.[27] Sikka had hoped that such an evolutionary (but with a room for human intervention) process would gradually lead to profitable herds of indigenous cattle. Similarly, in a paper contributed to the journal, *Agriculture and Livestock in India*, K.R.P. Kartha reported about systematic investigation carried out by the Animal Husbandry Bureau of Imperial Council of Agricultural Research 'had shown that, by careful selection and proper feeding and management, herds of pure Indian dairy cattle have already been produced within twenty-five years, which can more than hold their own in India, with European cattle and with the best Indian buffaloes, in economy of milk and butter fat production'.[28]

Reflecting on his time in India as the Imperial Animal Husbandry Expert, Arthur Olver shared his opinion with regard to breeding practice and European cattle with his audience at the Royal Society of Arts on 13 March 1942:

Except amongst these expert breeders, who, as a class, are dying out, there is little tendency to pay attention to systematic mating to produce a definite type, and one of the greatest dangers of importing European cattle into India is that their male, crossbred, progeny are certain to be used indiscriminately as sires, with results that in the end only degenerate mongrels are produced, instead of useful indigenous stock. *Though they may not, as a rule, be high class, these indigenous cattle are far more suited to the climate and conditions under which they have to live, and it has now been shown that the better breeds can easily be improved to a relatively high standard by proper feeding and management.*[29] [emphasis mine]

Olver was speaking in 1942 at a time when the Council of Scientific and Industrial Research in Australia had published a research report comparing data of the comparative efficiency of European and Zebu cattle in north Australia. Citing from this report to bolster his argument, Olver stated that the data presented therein fully supported the view that European cattle do not thrive under tropical conditions. Olver also alluded to a photograph which accompanied the data and showed very clearly the obvious distress of the European beast in the heat, compared with the appearance of comfort and well-being of the Zebu cross-bred standing alongside.

Arthur Olver stayed in India as Imperial Animal Husbandry Expert for around seven years, but his struggles within the colonial power circles must have been arduous. This can be gauged by his relentless advocacy for demanding higher fund allocations for animal husbandry departments and an advocacy that had predicted that, 'Moreover, in the general absence in India of expert animal husbandry organizations, there is every possibility that livestock improvement, which is unavoidably long range work, may, as in the past, be sacrificed to the interests of cash-crop production, when the financial show pinches'.[30]

Debates around the Breeding Policy that Aimed at Dual Purpose Cattle

A review of minutes of the animal husbandry wing of Imperial Council of Agricultural Research, reveals that the term 'dual purpose' cattle occurs right from the first Cattle Conference (1924) to denote a type of cattle that several livestock experts of provincial governments were aiming to breed at the government cattle breeding farms. While indirectly suggesting that mere propaganda that aimed at urging the cultivator to 'grow more fodder', without understanding the resource constraints faced by the cultivator will not do, Smith stressed upon giving a clear direction to cattle breeding policy towards the aim of developing 'dual purpose' cattle breeds and laying a greater stress upon dairy education. In the same note on cattle breeding presented at the meeting of Agriculture Committee at Poona in 1927, Smith stated,

The solution of the whole matter lies in the dual purpose animal. No matter what class or type of male plough bullock is required, the dam must always be a good milker and all bulls issued for stud purposes must be got from heavy milkers as well as be of right size, class and type. . . . Any propaganda outside of dual purpose efficiency is only perpetuating a great economic evil. No other basis can be profitable. If these are the reasons for the present state of affairs then the first step to remedy matters is dairy education. In every civilized country in the world today dairying occupies a very prominent position in its Agricultural Department. The crying need of this country agriculturally is dairy education, both of the cultivator and the masses.[31]

However, the Royal Commission on Agriculture had clearly suggested a caution on a breeding policy that aimed at 'dual purpose' type, arguing that the draught and milk characteristics are physiologically incompatible to be found in the same breed. The Royal Commission on Agriculture's *Summary Report* (1928) had highlighted the effects successive droughts and famines had on the cattle in certain provinces:

There are some districts in which such animals (i.e. those capable of rearing a strong calf and of supplying in addition some 1000 to 1500 pounds of milk per lactation for household use) are already common. There are others where, by selection, they could be produced from the existing breeds, and if produced might be maintained; but there are many districts in which cows can with difficulty rear their calves, where the bullocks are of very poor quality, and where fodder is so scarce that cows capable of rearing good calves and providing any considerable surplus milk could not be expected to thrive. The improvement of cattle in such conditions is most difficult, and, in these circumstances, it seems to us that, desirable though it be to secure a surplus of milk for the cultivator himself, the first step should be the production of cows which are capable of rearing calves that will make useful draught bullocks.[32]

However, on a closer analysis of the mindset that gave rise to such thinking, a clear sedentary bias and a propensity to misunderstand a settled cultivator undertaking the task of breeding his draught bullocks locally appears to be responsible for such views. It also shows that the Royal Commission probably relied too heavily on the written submissions and oral evidences provided to its members, rather than going through the early colonial documentation of cattle breeding practices that prevailed in India at the turn of the century.

As early as 1879, Hewlett had discussed at length the cattle breeding practised by nomadic breeders and outlined how there existed network of cattle fairs and nomadic migratory routes which made it possible for settled cultivators of Kheda to obtain a year-old male offspring belonging to a good draught breed from north Gujarat.[33] Although Baldrey shows how in certain districts in Rajputana, most of the breeding stock had been wiped out in successive droughts and famines at the turn of the century, he also highlights the efforts to reorient the nomadic cattle breeding enterprise.[34]

However, having failed to understand the interdependence between nomadic cattle breeders and settled cultivators, the Commission went on to state its reservations about pursuing 'dual purpose' breeding policy in all the provinces:

We are of opinion therefore that the attempt to provide the dual purpose cattle, equally suitable for draught and for milking and ghee production, should only be made in those districts in which the prospects for successful milk production are markedly better than on the average they now are; and that even in such districts, the question whether it is expedient to develop high milk production in cows or to resort to buffaloes should always receive careful consideration. The condition of the cattle in many parts of the country is, as we have pointed out, deplorable. We are impressed with the difficulties confronting the breeder and we are anxious that dual aims should not complicate the task.[35] [emphasis mine]

Not only did the Royal Commission fail to understand the crucial interdependence between settled cultivators and nomadic cattle breeders, it also appeared to be pushing exclusive dairy farming and a model of city milk supply based on cattle stabled within urban areas. Since it persisted to argue that one must look at the settled cultivator and his choice of animals flowing from his preoccupation with crop cultivation, the milk seller must be regarded as a separate professional entity altogether.

This sort of separation was also rooted in the bias of the Commission that considered female buffalo as primary milch animal of India and not the cow.

In what appears to be a very laborious argument, the Commission went on to say:

> We agree that there are tracts of the country—northern Gujarat, the south-eastern Punjab, and parts of the United Provinces, for example—where a dual purpose breed would meet local requirements, and there are irrigated areas such as those of the Western Punjab and Sind, where the abundance of fodder should enable cultivators to keep heavy milking strains successfully; but in general, we believe that better progress will be made with livestock improvement, if the needs of the ordinary cultivator and the milk seller are considered separately. Above everything else, the cultivator wants a strong and active bullock of a breed that can forage for itself and endure hardship when seasons make hardship inevitable. He also wants a cow giving enough milk to rear a good calf and a surplus for his own use, but in the interest of his young stock, it is undesirable that the ordinary cultivator in tracts where fodder is scarce should be a milk seller. We do not wish to see the calves of improved breeds dying a 'natural death from starvation' like the male buffaloes of Gujarat; and although the process would not be as speedy for the progeny of the cow as for that of the buffalo, starvation if not death, would undoubtedly be the fate of many calves if a good market existed for fresh milk in districts in which fodder is difficult to provide.[36]

A careful analysis of the aforementioned opinion suggests that not only is it wrong to make a comparison between 'the calves of improved breeds' dying an unnatural death by starvation and the male buffalo of Gujarat, it makes a paradoxically hypothetical suggestion that such a fate would await the male calves of improved breed of cattle 'in tracts where fodder is scarce', should in the same tracts 'a good market exist for fresh milk'.

The rising milk prices and emergence of milk markets in the early decades of the twentieth century in India was predominantly an urban phenomenon. This had not only led to the emergence of exclusive dairy farming in the form of city milch stables, but also, along with railways facilitating the transport of fluid milk and milk products, it had influenced the bovine-holding patterns among cultivators in dairying tracts such as Kheda. Male buffaloes meeting 'natural death by starvation' resulted from such restructuring of bovine holding patterns, since Kheda cultivators did not find any gainful use for these male offsprings in their cultivation practices.

In the way this extra cautiousness is expressed, the Royal Commission appears to have overplayed the role of state and colonial cattle breeding policies as if all the cattle breeders everywhere in India had taken to breed nothing else but 'dual purpose' breeds. A careful perusal of archival documents and cattle breeding policies pursued by certain native states shows beyond doubt that this was not the case and that they continued with their long-established traditions to breed draught bullocks, as well as a certain desire to spread the milch breeds. Notwithstanding the rhetorical concern for draught characteristics of some of the well-known Indian breeds expressed by the Royal Commission, the colonial policy right from early 1930s sought to discard the native wisdom of the nomadic cattle breeder. By legislating the Livestock Improvement Act, 1933 in Bombay presidency, purportedly to carry out the castration of undesirable scrub bulls roaming around the countryside, the high colonial State had made cattle breeding a highly centralized, authorized and technologized affair.[37]

This legislative action paved way for the licensing of the approved bulls in the area, and all persons were prohibited from keeping unlicensed bulls in possession. The Act granted overarching powers to colonial officers belonging to the veterinary stream, since such an officer could enter the cattle shed in order to inspect the licensed bulls at any time and take down notes on their condition. The Act had also made provision that if a veterinary officer noticed deterioration, the license could be cancelled. In about a decade's time, the jurisdiction of this Act had expanded to cover 73 notified villages.[38]

In his report reviewing animal husbandry and dairying work of the Imperial Council of Agricultural Research thus far, Wright, in 1937, stated in a footnote that he did not feel it necessary to discuss the controversial questions involved in dual purpose breeding because he believed that the controversy was largely founded on the misconception of the genetic implications of such breeding, as well as of the needs of the cultivator. Wright did not find the caution expressed by the Royal Commission on Agriculture very convincing and urged that: 'This indicates the extreme urgency of taking immediate steps to extend the facilities for selecting and improving indigenous milking strains of Indian cattle. . . . I should, perhaps add that I do not consider that such efforts should be limited to those breeds which are recognized as predominantly milking breeds.'[39]

By the time Wright was writing his advisory report, thanks to scattered efforts at milk recordings, the figures on milking capacity of most breeds, except the purely draught type such as Amrit Mahal and Hissar, had surfaced. The lactation yield data could now put to rest the reservations expressed by Royal Commission on Agriculture and later by Olver that draught and milk characteristics are mutually incompatible.

Wright stated that the subject had been adequately dealt with by Olver in an article titled 'The Inadequacy of the Dual Purpose Animal as a Goal of Cattle Breeding in India'.[40] He also directs his readers to the discussion of the question of dual purpose cattle in an informative paper by Col. A. Matson titled 'Cattle in Relation to Agriculture in India'.[41] In his note prepared for the Cattle Conference held at Shimla, Olver further pointed out that the point that he made in his earlier article appeared to have been missed by many. Olver clarifies that:

in that note, it was not intended to deny that it is possible for an expert breeder to achieve the duality of purpose—up to a certain point—provided that he is at liberty to select freely and discard animals which do not show the desired combination of factors, however, the Indian peasant cannot hope to be in this position.

Olver observed that between the high-grade milch cattle and high-grade work-type draught cattle lie those that are preferred by the great majority of cultivators who keep one or two cows and produce less specialized 'general utility' cattle, which cannot, in view of their heterogeneous origin, be relied upon to breed true and, therefore, should not be confused with 'dual purpose' stock in the strict sense of the term. Olver describes the prevailing condition and breeding policy stating:

Though much is said of breeding of dual purpose cattle in this country, the method usually adopted appears to be to pay strict attention to milk recording and to retain the best milking strain until such time as definite signs of unsuitability of draught purpose appear in the progeny. When that time comes the breeder will be faced with a decision whether to retain any high yielding milk strains thus evolved, or to destroy the advance thus achieved by crossing back to a working type bull. What the answer must be in the interest of progress is not difficult to foresee and in the meantime more milk is being bred into the stock. Along these lines so long as promising dairy strains are not crossed back to a work type bull, there need be no objection to so called dual purpose breeding but high capacity for work and for milk production are physiologically incompatible, and instances are not wanting in India where attempts to retain these factors in equal degree, in one and the same strain, have led to marked deterioration of previous valuable stock.[42]

At the same, Olver emphasized that the breeding policy needs to consider 'the environment under which animals had to live and produce' and ruled out cross-breeding of indigenous stock with imported exotic bulls on a large scale. Olver was of the firm opinion that 'mating the domestic breed with exotic breeds will do more harm than good unless they were accompanied by measures to improve the environment of the stock feeding'.

On the eve of his retirement, Olver contributed an important article to *Agriculture and Livestock in India* titled 'Systematic Improvement of Livestock in India' in which he states that, 'Such a system, however, postulates the breeding of a reasonable amount of milk into the recognized working breeds, and I am sure from my own observations and from the observations of experienced breeders of Indian cattle that, up to milk yields considerably beyond what would be necessary, this can be done without damage to the stock as work-animals'. Olver also argued that the question 'whether cows should not be bred and as well fed and maintained as are she-buffalo' needed a careful study. His argument indicates that Olver had by that time revised his earlier reservations about the applicability of dual purpose breeding policy, as he states:

Where abundance of coarse fodder is available and where the production of ghee is a major consideration, or where milk is produced by unscrupulous and uncontrolled hawkers, the she-buffalo is at present commonly preferred. *But investigation has shown that pure bred cows of certain Indian breeds of cattle can in a comparatively few years be improved by feeding and proper management to a point where they can compete successfully with the buffalo in the economy of milk or butter-fat production.*[43] [emphasis mine]

In a paper presented at the Tenth World Dairy Congress at Rome (30 April–6 May 1934), Zal R. Kothawala summed up the cattle breeding policy and cattle improvement work of three decades:

The cow in this country in the past was looked upon as merely the mother of work animal required for tillage operations, and other kinds of draught work, whereas the buffalo was the principal milk producing animal. The changing conditions in the mode of agriculture in the country and other economic factors demanded, however, that the cow should produce a sufficiently large quantity of milk to maintain the farmer and his family, besides producing a work animal. This gave an impetus to the improvement of the cow as a milk animal in this country through pure selective breeding and efforts were also made to create a new breed of cattle by the crossing of the European breeds with the breeds of cattle of the Zebu type found in India, which besides yielding a high quality of milk would also be found suitable for tropical conditions.... Results of this experiment over 30 years, on the whole, are far from encouraging, and the experience gained indicates that although the crossbred progeny is better than its Indian dam in its milk yield, in the first one or two generations, it deteriorates in the subsequent generations and altogether loses the constitution, stamina and immunity to diseases possessed by its ancestral dam to such marked degree, till eventually it becomes altogether unsuitable for tropical conditions. The surest way of bringing about improvement in the Indian breeds of cattle is, therefore, breeding by pure line selection, and although great strides have been made in recent years in this direction, much remains to be done to produce a really good 'dairy cow' out of the Indian breeds of cattle.[44]

In corollary to the aforementioned view expressed by Olver and Smith, Dasgupta shows that 'at the instance of Victor Hope, Governor General of India, an enquiry was held in 1937 to find out conditions of the production and consumption of milk in typical breeding tracts of India in order to determine the future of cattle breeding policy of Government of India'. Dasgupta mentions that the report of this Enquiry Committee was considered by the Standing Cattle Breeding Committee of the Council and by the Advisory Board. It was recommended that 'the breeding policy should be to maintain and raise the milking capacity of the females of the draught breeds within reasonable limits in so far as this can be done without risking damage to the type.' This was followed by a report from Imperial Council of Agricultural Research in 1942 that recorded the council's decision to develop the milking capacity of the Kangayam cattle. Datar Singh summarizes the dual purpose cattle breeding policy discourse in a thoughtful piece he wrote for the nationalist magazine, *Harijan*, on 23 June 1946.

However, Dasgupta fails to outline the implications of laws, such as the Livestock Improvement Act, 1933 of Bombay presidency that aimed to make cattle breeding a highly centralized, authorized and technologized affair and the gradual marginalization of nomadic cattle breeders due to such a law. This shortfall probably remains in an otherwise brilliant work, *Cow in India*, which was authored by Dasgupta while undergoing a prison term at Alipur Jail, since he remains obsessively concerned with cows, rather than plough bullocks. It seems even elite nationalists could not escape the sedentary bias shaped by the then prevailing social evolutionary perspectives.[45]

Another Gandhian thinker, J.C. Kumarappa, stressed upon following the policy to breed different type of bullocks—small, medium and large—according to the needs of the agriculturists, probably as per agro-climatic and ecological conditions, and cautioned that 'a mere dual purpose cow will not do for all time'.[46] In this opinion, Kumarappa's grasp of social and economic stratification amongst the cultivating class as per landholding patterns stands out clearly. It is possible that Kumarappa was appealing to his audience to understand the social ecology of draught breeds of Indian cattle and appreciated even the so-called dwarf breeds in certain ecological conditions.

This essay has thus reviewed the changing contours of the meanings assigned to cattle by both the native cattle breeder or cultivator and the colonial experts. Along with the emerging milk markets, increasing demand for milk and milk products by British rulers and elites in the city, an altogether different form of dairying had emerged in the form of city-based milch stables. While colonial experts had voiced their caution against extending indiscriminate cross-breeding and rather laid stress on increasing milk yields through selective breeding, the cause of draught bullocks at best had received lip service. While we come across warning bells sounded by colonial experts such as Albert Howard, Arthur Olver, N.C. Wright, etc., the question of retaining the agriculture and livestock symbiosis did not receive the attention that it deserved in an environment fraught with the expediency of political economy of the times that witnessed two World Wars.

Notes and References

1. Norman C. Wright, *Report on the Development of the Cattle and Dairy Industries of India*, Government of India Press, 1937, p. 66.
2. J.T. Edwards, *Some Diseases of Cattle in India*, Government of India Press, 1928.
3. James Mill, *Plain Hints on the Diseases of Cattle in India*, Bombay, 1894.
4. Isa Tweed, *Cow Keeping in India: A Simple and Practical Book on their Care and Treatment, their Various Breeds and the Means of Rendering them Profitable*, Calcutta, 1911.
5. Landolicus, *The Indian Amateur Dairy Farm*, 1895.
6. J.B. Knight, in *Proceedings of the Board of Agriculture in India held at Pusa on 6th January 1905 and Following Days*, Government of India Press, 1905, pp. 41–2.
7. Ibid., p. 42.
8. James Mollison, *A Textbook in Indian Agriculture*, vol. II, Bombay, 1901.
9. Arthur Olver, 'The Necessity for Authoritative Definition of Breed Characteristics and Unchanging Control of Breeding Policy in India', *Agriculture and Livestock in India*, vol. I, part I, 1931, pp. 19–25.
10. Tweed, *Cow Keeping in India*.
11. *Proceedings of the Eighth Meeting of Board of Agriculture in India*, Government of India Press, p. 13.
12. See 'Report on the Ninth Meeting of Board of Agriculture in India', *The Agricultural Journal of India*, vol. XI, part II, 1916, pp. 93–107.
13. E.W. Oliver, 'Cattle Breeding, with Special Reference to Milch Cow', *The Agricultural Journal of India*, vol. XI, part II, 1916, pp. 168–73.
14. James Mollison and L. French, 'The Montgomery and Sind Breeds of Cattle', *The Agricultural Journal of India*, vol. II, part III, 1907, pp. 252–6.
15. W. Smith, 'Note on Cattle Breeding', presented at the meeting of Agriculture Committee at Poona, 1927.
16. Ibid.
17. Wright, *Report on the Development of the Cattle and Dairy Industries of India*.
18. *Annual Report of the Veterinary Department of Madras for the Year 1940-41*, Government of India Press, Delhi.
19. *Report of the Inquiry in Breeding Tracts of Seven Principal Milch Breeds*, Imperial Council of Agricultural Research, Delhi, 1939.
20. Arthur Olver, 'Animal Husbandry in India', *Journal of the Royal Society of the Arts*, vol. 90, no. 4614, 29 May 1942, pp. 433–51.
21. It is said to be among the first known experiments on cross-breeding in as early as 1875, around Patna when Taylor breed was evolved using Shorthorn bulls on native cows. See *Handbook of Animal Husbandry*, Imperial Council of Agricultural Research, New Delhi, 2011, p. 13.
22. See *Proceedings of the First Meeting of Board of Agriculture in India, held at Pusa on 6th January 1905 and Following Days*, Government of India Press, Simla, 1905.
23. Ibid.

24. See *Proceedings of the Board of Agriculture in India held at Pusa on the 17th February 1908 and Following Days*, Government of India Press, Calcutta, 1909.
25. See *Proceedings of the Board of Agriculture in India held at Pusa on the 9th December 1929 and Following Days*, Government of India Press, 1931, Calcutta.
26. Lal Chand Sikka, 'Standardisation of Lactation Period Milk Records', in *The Indian Journal of Veterinary Science and Animal Husbandry*, vol. I, part 2, 1931, pp. 63–98. The article had used milk records data on pure-bred Sahiwal herds and Sahiwal-Ayrshire cross-bred herds at military dairy farms and states that in 1931 such lactation data was available for the past 10-year-period (that is, 1920–30). This move towards institutionalizing and standardizing milk records and lactation yield data points at the early efforts by the colonial science and technology establishment to develop statistical records on milk yields.
27. Ibid.
28. Even the then Imperial Expert on Animal Husbandry had taken note of this paper by Kartha and added that such herds of pure Indian cattle already exceeded 'the average milk yields of commercial dairy herds in Europe and America and there is evidence that these results are likely to be steadily improved upon for years to come, while it has clearly been demonstrated that even with the best of care and under the best possible conditions, cattle of European origin always tend to degenerate in India'. See K.R.P. Kartha, *Agriculture and Livestock in India*, vol. IV, part II, 1934, p. 605.
29. Olver, 'Animal Husbandry in India'.
30. Ibid.
31. Smith, 'Note on Cattle Breeding'.
32. See *Royal Commission on Agriculture*, Summary Report, Government of India Press, Calcutta, p. 224.
33. K. Hewlett, *Breeds of India Cattle: Bombay Presidency*, Calcutta, 1879.
34. F.S.H. Baldrey, *The Indigenous Breeds of Cattle in Rajputana*, Simla, 1911.
35. *Royal Commission on Agriculture*, Summary Report, Government of India Press, Calcutta, 1928, p. 225.
36. Ibid.
37. A similar legislation, namely Livestock Improvement Act, 1940, was enacted in the Madras presidency.
38. The legacy of such colonial acts continued even during the post-independence period in a series of legislations such as Kerala Livestock Improvement Act, 1961. Under the provisions of these Acts, veterinary officers of State were authorized to penalize any livestock keeper who maintained a bull belonging to a native breed with a penalty worth Rs. 500 or a year in prison. Veterinary department officers indiscriminately promoted cross-breeding and carried out indiscriminate castration of indigenous bulls. Such a zeal to wield the power of law in order to turn cattle breeding into a highly centralized, authorized and technologized an affair has led to the extinction of even the well-known indigenous milch breeds in Kerala and Punjab.
39. Wright, *Report on the Development of the Cattle and Dairy Industries of India*.

40. Arthur Olver, 'The Inadequacy of the Dual Purpose Animal as a Goal of Cattle Breeding in India', in *Agriculture and Livestock in India*, vol. 6, no. 389, 1936.
41. A. Matson, 'Cattle in Relation to Agriculture in India', in *Journal of Central Bureau of Animal Husbandry and Dairying in India*, vol. 2, 1928, p. 185.
42. Olver was pointing to the impossibility for an expert cattle breeder working in Indian condition to discard unsuitable animals, given the taboo against beef eating and the then vigorous campaign against cattle slaughter.
43. Arthur Olver, 'Livestock Improvement in India', in *Agriculture and Livestock in India*, vol. VII, part IV, 1937, p. 464.
44. Zal R. Kothawala, 'The Indian Buffalo as a Milch Animal Suitable for Tropical Countries', *Agriculture and Livestock in India*, vol. V, part I, 1935, pp. 47–9.
45. However, to be fair to the Gandhian school of thought that Satishchandra Dasgupta was aligned with, there were other authors like Kumarappa and Mira Bahen who had kept the concern on draught bullock pronounced in their writings.
46. J.C. Kumarappa, 'The Cow and Peace', speech delivered to Go Sevaks at Pipri, Wardha in January 1953; in Kumarappa et al., *Cow in Our Economy*, Akhil Bharatiya Sarva Seva Sangh, 1957, p. 8.

12

Agrarian Distress:
The Political Economy of
British India in the 1930s

S.M. Mishra

The agriculturist has suffered from heavy assessment of land revenue by the policy of the alien government which has ruled India for so many years. . . . But the situation that was created by the (Currency) Ratio Bill of 1925, has brought great economic distress in the country.

—*Legislative Assembly Debates*, 10 September 1931

Never in the history of India was the agricultural population so distressed, nor did it suffer so much as it has been suffering during the last few years, and they are quite incapable of paying their rents to their landlords.

—*Legislative Assembly Debates*, 14 February 1934

By the first decade of the twentieth century, colonial India was an integral part of the world economy with increased demand for its cereals, jute and cotton. Its primary products became essential for the feeding of men and machines in the West, and India became a steady market for the finished goods of European industries. As a result, colonial India had a full share in the Depression which started in October 1929. Its public finance, as well as the private finances of its people, came to be seriously affected by the slump.[1] A deep-seated agricultural depression coincided with the world financial collapse, and the primary producing economies like India experienced a large slump in prices, following the instability of the 1920s. Within the space of five years after 1928–9, the value of India's foreign trade was very nearly halved and recovery thereafter was sluggish and restricted.[2] It is a truism that cumulative causation played an important role in economic affairs, but the perception of such processes is very often difficult for those who are involved in them, as information is at a discount and preconceived notions prevail.[3] Nevertheless, this essay is based on a study of the debates by the contemporary legislators in the Central Legislative Assembly of British India on the agricultural distress that overtook the country in the 1930s.

Inviting the attention of the government to the economic depression, Rai Bahadur Lala Brij Kishore (Lucknow Division, Non-Muhammadan Rural) suggested the appointment of a committee which might devise means to better the condition of the agriculturists in order to give them permanent relief. This was most poignantly reflected in the Resolution moved by him on 14 February 1934 and resumed on 6 April 1934:

That this Assembly recommends to the Governor General in Council to appoint a committee of enquiry consisting of officials, experts and Members of the Assembly to enquire into the causes of the present agricultural distress and to devise means for improving the conditions of landholders and peasants.[4]

Some significant questions arise from a study of the Proceedings on the Resolution. Did the cause of the agricultural distress merely mean the decrease of the prices of agricultural products? Did the existence of multiple land tenures across British India compound the problem? Did the provincial governments examine their own agricultural conditions to devise means of amelioration of the distress? Did the conservative nature of India's agriculturists add to their distress? Was the educated Indian apathetic to the problems faced by rural India? Was there a government bias for industry against agriculture so that the latter suffered? One may be able to pose as many questions as there might be causes, both natural and man-made, of agrarian distress, yet in this exposé, we seek answers to these questions in the deliberations on the Resolution in the Central Legislative Assembly.

Members agreed that the condition of the agriculturists was pitiable throughout the country. Their submissions provide an all-India survey, as it were, 'of the dimensions and patterns of agrarian distress and relief in the various provinces' as narrated further in the chapter. Identified communally, the members represented both rural and urban constituencies as far apart as the Lucknow Division, Tanjore-cum-Trichinopoly, Central Provinces, Madras City, West Punjab, Muzaffarpur-cum-Champaran, Bihar and Orissa, Bombay Central Division, United Provinces Southern Divisions, Bombay Northern Division, Bengal, Ganjam-cum-Vizagapatnam, West Coast and Nilgiris, Burdwan Division, Chittagong and Rajshahi Divisions, Punjab, Ambala Division, Lucknow and Fyzabad Divisions, Orissa Division, Rohilkhand and Kumaon Divisions, United Provinces, Madras-ceded districts and Chittoor, North Punjab, and Bombay Southern Division.

All-India Distress

Of all human conveniences, food is of primary importance. Traditionally, agriculture is the basic means of production of this indispensable convenience.

Therefore, tilling the land was obligatory to the filling of the stomach. And the agriculturist's distress began with the tilling, so thought Raja Bahadur G. Krishnamachariar:

The cause of the present agricultural distress does not merely mean the decrease of the prices of agricultural products. The distress starts from the time you begin to prepare your field after the harvest. Then, slowly, step by step, the sowing and the transplanting starts. The wages that you have got to pay increase at every step, and when all these troubles are over and we have done our weeding, we look up to the sky and we also look up to the flood in places like mine where we live, and when all that is escaped, you come to the harvest time. And then there is the thief who begins his operations from the threshing floor. And when all that is escaped and we gather in something, what is the result? We do not get any prices. And there are pious hopes and Resolutions and many things said about helping us with a marketing board, and, in the meanwhile, the railway companies go on merrily charging rates which are ruinous, and, by the time we settle with the middleman, the actual purchaser and the producer, there come the Government, the 10th of January is the last day, and if the money is not paid by the 20th, there comes the distress warrant. Things are really becoming very hard. . . .[5]

For industrial labour, there were provisions for saving them from indebtedness, provision of good houses, medical relief, and education for their children, and all sorts of amenities of life. They had fixed working hours in the factories.[6] Also, 'railways grant full pay on "Empire Day" and on the "King-Emperor's Birthday" on which days the shops are closed'.[7] But,

Unlike his fortunate brother, the zamindar must get up at three or four in the morning and work till midnight, there are no fixed hours for him, no one cares about his food or where he lives and whether his sons are properly educated. There is no water supply for him and he has got a very small share of what are called the necessaries of life. He deserves something to be done to ameliorate his lot.[8]

According to Khan Bahadur Mian Abdul Aziz, the phrase 'agricultural distress' ought to be properly understood because the land continued to yield practically all over the country as before. There was no diminution in the yield. The livestock continued to breed practically as before and the main difference was that 'the people who breed money have suddenly become barren'. They did not produce money at the same rate as they did before, and 'that is where the distress comes in'.[9] The driving force of the Depression was the contraction of credit which emanated from the financial centre and reached the remotest village at the periphery with amazing speed.[10] Also, for example, the amount of money orders sent home by the Indian labourers abroad diminished largely during this period.[11]

Low Prices of the Agrarian Produce

> At present it is the experience of the agriculturists that the money which is spent in raising a crop cannot be recovered from the sale of the produce of the land, because the prices of produce have gone down so considerably, and with it the purchasing power of the people.[12]
>
> —GAYA PRASAD SINGH, Muzaffarpur-cum-Champaran, Non-Muhammadan, LAD, 6 April 1934, p. 3331.

The most notable feature of the slump in India was the wide disparity between the prices of primary products and finished goods. While rice slumped 52 per cent, oil seeds 55 per cent, raw jute 53 per cent and raw cotton 51 per cent between September 1929 and March 1934, cotton manufactures slumped only 29 per cent, metals 22 per cent and sugar 26 per cent during the same period. Since India's exports were chiefly primary products, this disparity affected the barter terms of trade.[12] Members admitted that the greatest trouble of the agriculturist was the fall in prices. So that his condition was precarious and on that account, the condition of the landholders was equally bad and the relations between the landholders and tenants strained.[13] From 1929, the Indian ryot found his income dwindling and his burdens increasing; and in this respect, the troubles of Indian agriculturists had been much greater than those of the large farmers in the United States of America (USA), Canada, Australia and Argentina, who had a greater staying power and easier means of recouping themselves.[14]

N.M. Joshi felt that the low prices of agricultural goods could not rise unless those who were to purchase them had sufficient money in their pockets. The first thing necessary to be done, therefore, was to develop industries and to start public works, so that the workers of the country would have sufficient money to purchase agricultural goods.[15] It was a question of circulation of money; it was a question of infusion of purchasing power into the community by increase in the salaries and by programmes of reconstruction and large capital.[16] However, on 29 September 1929, the Legislative Assembly had ratified three proposals of the government for meeting the impending deficit. The proposals were to retrench, introduce an emergency cut in salaries and fresh taxation,[17] although the problem was the contraction and paucity of money. The big fall in the real wage in the 1930s was due to the Depression which sharply reduced the income from agriculture and consequently the level of employment of agricultural labour.[18]

Distress of Taxation

Rai Bahadur Kunwar Raghuvir Singh cited the book *India's Plight: Debts Doubled, Development Damned* by the Karachi industrialist Sir Montagu

Webb, who suggested that the steep fall in prices of commodities could be attributed to the exchange and currency policy of the government. The book gave figures of taxation which were 'extracted from failing agriculture and vanishing trade' and observed, 'It will be noticed that although the prices of agricultural produce have fallen by over 50 per cent, Government has nevertheless extracted from agriculturists in 1931-32 only about six per cent less land revenue than in 1923-24.' The author attributed the fall in prices to 'a prolonged policy of currency restriction, deliberately adopted at the close of the War by bankers of the West, enforced in India and elsewhere and throughout the world and still adhered to by the financial powers of London with tragic stubbornness and almost incredible lack of world vision.' Sir Montagu Webb quoted figures of production to prove that the calamitous fall in India was due not so much to overproduction in the country as to the chronic shortage of money in the hands of the people; that the author urged reduction of taxation in land revenue all round which, at the then prevailing price levels, was more than agriculture could bear, and to derate the rupee.[19] However, on the subject of taxation, Khan Bahadur Mian Abdul Aziz pointed out that the province of Punjab had taken measures to deal to the fullest possible extent with the situation that had arisen:

For instance, the richest district in the whole of India, Lyallpur, is now being resettled, not with the object of increasing the assessment, but of decreasing it, and the decrease will be not of one, or two or ten lakhs, but it will be much more. I cannot give it, but it will really be much more. The poorest district, which lies next door to Delhi, is Gurgaon. During the last six years we have remitted there nearly 50 lakhs of rupees. We have remitted taccavi of over 14 lakhs which we had given in cash, on account of their distressed condition. But that is not all. In the last kharif, in that one district alone, on account of floods, crops were ruined, and we remitted, not suspended, seven lakhs, practically the whole of the demand.[20]

Thus, the whole of the demand was remitted. Also, the Punjab Government, on account of the drop in prices, remitted over Rs.80 lakhs in water rates and land revenue. In many cases, the remission was 4*an.* in the rupee, in a number of cases it was 6*an.* in the rupee, and everything possible was done. From harvest to harvest, as occasion arose, remissions were given. Sitakanta Mahapatra held that the Government of India's financial and commercial policy was responsible for the plight of the agriculturists. The ratio policy of the government undermined the purchasing power of the people, and their protection policy had raised the prices of necessities to such an extent that they could not afford to purchase their vital needs.[21]

Distress of Rural Indebtedness: Budget Deficits

A perennial source of the agriculturist's distress was his indebtedness. The one important factor for consideration was the volume of rural indebtedness in the country. The Banking Inquiry Committee had calculated rural indebtedness in the neighbourhood of 900 crores as the debt in 1931 for the whole of India. Sir Darcy Lindsay also felt that the real cause of the trouble of the agriculturist was his heavy debt and any assistance that could be given in that direction was all to the good. He cited the example of 'one of my own community', Sir Daniel Hamilton, who, in the Sunderbans, reclaimed land and settled cultivators thereon and had a colony of about 12,000 persons. He took over all the debts of the people, abolished the moneylenders and would not allow any of his tenants to borrow in the open market, otherwise they had to repay what he had advanced them and they could no longer be his tenants. He charged them a fair rate of interest, half of which went towards a fund for the betterment of the people themselves in the shape of the establishment of dispensaries, schools, and so on. Sir Lindsay wished that there was a spread of that movement throughout the country, that would be all for the betterment of the people. He recommended a study of Sir Daniel Hamilton's scheme to the members.[22] B. Sitaramaraju in sympathy with the efforts of Sir Daniel Hamilton invited the Government of India to do the same.[23] The heavy load of agricultural debt, mostly inherited, compounded the distress of the agriculturist. Hony. Captain Rai Bahadur Chaudhuri Lal Chand quoted M.L. Darling, a Commissioner of the province of Punjab: 'The agriculturist is born in debt, he lives in debt, and he dies in debt',[24] a view supported by the Royal Commission on Agriculture that so far back as 1928 held that the agriculturists were heavily in debt and that most of the debt was irrecoverable.

Moneylenders and Landlords

Of the several factors which created a sort of permanent distress in Indian villages, N.M. Joshi gave the first place to India's system of land tenure. Creating private rights of property in land was the greatest mistake, because over time those rights in private property were amalgamated in the hands of a small number of people. Lands belonging to the small cultivators passed to the moneylenders or to the bigger landlords on account of the uncontrolled system of moneylending and usurious practices.

In India, even today, people have to cultivate their land from generation to generation at the will of the landlord. The Governments have passed legislation to give some kind of security to tenants, but still there is a very large class of tenants who are mere tenants at will and they have no security. Not only there is a very large class of men who are mere tenants at will, but there are classes of people even today who are tied to their fields as if they were slaves: when land is sold, these people, who are tied to the land, are, as it were, sold: they pass to the new landlord. . . . Even today in Madras, at least in some districts, a landlord, who has got the land to which some field workers are attached, can lease the services of these field workers to others as if they were his slaves. So long these practices exist in our country, what is the use of asking the Government to appoint committees to go into the question of improving the conditions of the agricultural classes?[25]

Recognizing this evil, the Government of India and the provincial governments took steps to prevent or restrict land passing into the hands of people who had no real interest in the cultivation of the land. In Punjab, they passed legislation restricting the passing of land into the hands of moneylenders; in Bombay and the United Provinces, there was a similar legislation. Unfortunately, although the government saw the evil, the remedial measures they took were not bold enough, so that the land passed into the hands of landlords and moneylenders belonging to what were called the agricultural classes; although, really speaking, they were not agricultural classes. It was in 1901 that the Punjab Land Alienation Act was passed and since then, though the land was saved to the agriculturist, the produce had all along been going to the moneylenders, and so they became, as it were, labourers for the moneylenders, fixed upon a particular piece of land. Nawab Major Malik Talib Mehdi Khan observed that whenever the provincial governments took measures to solve the problem of agricultural distress, they met with opposition from the moneylenders. Therefore, it was not wrong to say that most of them were half-hearted measures, and their actual working proved beyond doubt that they were so. The colonial state tried from time to time, all in vain, to restrict the rural moneylender's control over indebted peasants:[26]

Recently the Hindus held meetings in Khanewal (District Multan) and other places to oppose the new agriculturist debt law put up by the Punjab Government; and it is clear from the Resolutions passed there that one section of the public which represents the moneylenders would not like that the poor debtor should have breathing time or that he may be able to get rid of some of his debt. At the same time, we are very grateful to the Government for so kindly remitting lakhs and lakhs of rupees at the time of each harvest when they find that the people are not in a position to pay their dues. Government also advances lakhs of rupees by way of *taccavi* loans to the ryots; they not only advance these loans, but they also make huge remissions whenever there is necessity for it. . . .[27]

However, it is striking that in many places rural indebtedness, in money terms, did not appreciably increase during the Depression. Moneylenders themselves sharply reduced their debt operations, partly because of their inability to lend and partly of their reluctance to lend.[28]

Distress of Attachment

Uppi Saheb Bahadur informed the Assembly that in Malabar, the collection of rent and taxes and assessment was going on. As they were unable to pay the demands, the landholders and cultivators were running away and hiding from the village officers. Their properties were being attached and sold. There was nobody to defend them; they were not able to pay up the remaining one *kisht* of the assessment. They had sold all their property to pay up each instalment, so much so, for the last instalment, they were not able to find means to pay the demand of the government officials. The village officers were attaching standing crops and also household property. The situation in the region, which was mainly agricultural and depended mainly upon coconuts as its mainstay, was distressful. The price which ranged between 40 and 50 had come down to 12 and 15. During early 1930s, the price of pepper, another source of income in Malabar, had fallen from 600 to 120 and 130. Yet, the provincial government went on increasing the taxes. They came forward with a demand at the rate of 10, 11 and 12 *annas* per acre. The people could not pay but the government was unyielding. They attached and sold property, attached standing crops, without even allowing time to the people to collect their dues.[29] In 1928, the Congress launched a no-tax campaign in Bardoli. Following the revision of assessment, there was an increase in the land revenue demand by about 30 per cent. This revision had coincided with the Great Depression which rallied spontaneous discontent. The people of Bardoli resolved not to pay any assessment. The government intensified the coercive process and attached herds of buffalo, utensils, cots, sheets, etc.[30]

Distress of Disease/Want

Rao Bahadur B.L. Patil thought that for better living the agriculturist should have access to potable water. Since lakhs of people suffered from malaria, medical aid to combat the dreadful disease which generally affected the rural parts should be provided. Government ought to provide the agriculturist with better and selected varieties of seeds, like sugar cane, cotton, potatoes, and so on, along with marketing facilities. The government ought to take up legislation to check usurious loans, and undertake some legislation with regard to tenancies; and a speedy and radical revision of the system of giving suspensions and remissions in the provinces. Director General of the Indian Medical Service Major General Megaw stated that 50 per cent of the people were below par in health. The reason lay in the depression caused by their heavy debt.[31] The extent of suffering, however, varied between provinces and between different classes of the community. Of the provinces, Bengal, Bihar and Orissa were the worst hit. The principal crops of these two provinces fell in value by 61 per cent and 58 per cent respectively between 1929 and 1933, while those of Bombay, UP and the Punjab fell only by 30 per cent, 35 per cent, and 36 per cent respectively. Of the different agricultural classes, those who cultivated with hired labour were worse hit than the small holder who cultivated with his own labour and raised various kinds of produce. Landowners who got their rent in money were well off, but those who received in kind encountered losses. The labouring classes had comparatively gained where there was ample employment, but in several areas there had been an increase of unemployment and underemployment. This was particularly the case in the jute areas of Bengal. But all these agricultural classes suffered much more than the industrial population.[32]

Exemptions *de jure*

Various measures were needed to provide more credit, because the cooperative societies were not able to take the place of rural moneylenders entirely. It was proposed that arbitration tribunals should be formed with regard to loan transactions, that advances should be made by the government on a system of equated instalments for long terms and that there should be a regular debt redemption scheme. Another proposal made was that there should be a scheme of compulsory saving for tenants on the lines of insurance policies.[33] Quoting from paragraph 364 of the *Royal Commission on Agriculture Report*, Rai Bahadur Chaudhuri Lal Chand observed that the Commission's recommendations were not paid attention to:

The importance of the co-operative movement is accentuated by the comparative failure of legislative measures designed to deal with the problem of indebtedness to achieve their objects. We received evidence in Burma that the provisions of the Civil Procedure Code exempting the cattle, implements and produce of agriculturists from sale may be ignored by the courts. We have mentioned that the Kamiauti agreements Act in Bihar and Orissa has proved ineffective. The provisions of the Deccan Agriculturists' Relief Act are being evaded and the Usurious Loans Act is practically a dead letter in every province in India.[34]

He drew attention to those provisions of law that were there and that were being ignored by the courts. Section 60 of the Civil Procedure Code laid down certain exemptions, whereby implements of agriculture, seed, grain and cattle houses were exempt from attachment in execution of decrees:

As I have been practicing in mufassil Courts, I find that all these articles, every one of them, have been attached and are being attached by Courts and section 60 is being ignored. Then, there was section 61 of the Civil Procedure Code. At the time when this Code was enacted, this Legislature left it to the Provincial Governments to frame rules for the exemption of produce for the maintenance of the family of the agriculturist up to the next harvest. But up to this time no Government, except perhaps the United Provinces, have framed any rules under this section, and, therefore, it remains a dead letter.[35]

Thus, the procedure laid down by the Land Revenue Code regarding forfeiture and sale was often disregarded.[36]

Insolvency Laws

The Agricultural Commission recommended that irrecoverable debts should be wiped out. It was perfectly reasonable that irrecoverable debt should not be persisted in, and any instalment that might be fixed for that debt would only perpetuate it. Therefore, they recommended the passing of simple insolvency laws. From paragraph 364 to paragraph 367, they concluded that simple insolvency laws might be considered by provincial governments. But except the Bombay Government, no government had taken any action on those lines.

To tackle the problem of rural indebtedness, B. Sitaramaraju considered the establishment of a land mortgage bank for the specific purpose of alleviating the distress caused by rural indebtedness. The government could lend at a low interest and they might also take in statutory authority for making it compulsory for the moneylenders to take the bonds of the government for their money and the loan would, thus, be transferred to the land mortgage bank which would be able to control the debt and fix the period of its repayment on a graduated scale. Thus, if Rs.100 were to represent the debt, the instalment payable would be about Rs.4.60 or thereabouts for a period of fifty years. It would be easy for him, within a period of fifty years, to pay that Rs.100 in such small parts. A small sum of money might, however, be found necessary for the land mortgage banks to inaugurate the scheme and advance it.[37] T.N. Ramakrishna Reddi thought it was not enough to reduce the indebtedness of the ryots:

The Government should also take steps to see that the agriculturist does not again fall into the hands of his creditors and the Government should create facilities for easy credit and at the same time the credit must be made self-liquidating. This the

Government may do through the agency of co-operative societies that exist in the country and also through the land mortgage banks. No doubt the Co-operative Act has been working for some years, and societies do exist, but the societies have so far failed in the discharge of the work that has been expected of them.[38]

This failure was due to various causes. One chief reason was that they were purely lending societies. They did not take any interest in seeing how that credit was spent by the debtors. The cooperative societies were to be so improved as to make them function as agricultural societies also. They must combine the function of both agriculture and cooperation, and every loan that was given from that society must be only for agricultural purposes. The society should look to its utilization, and that society should also take upon itself the marketing of the product and get better prices for the produce of these debtors. Also, the cooperative societies should only advance short-term loans and it is only through the land mortgage banks that long-term loans should be given. Joshi felt that after the distribution of land, the agriculturists should be taught to start a cooperative movement, both for production and sale and also for credit. If they did that, the need for taking loans for agricultural purposes to a great extent would be minimized, and, as regards the loans themselves, he suggested that the Government of India and the provincial governments should take immediate steps to see that every practice of usury was discontinued immediately.[39]

Multiple Land Tenures

Did the existence of multiple land tenures across British India compound the problem? India was a vast country and the land tenures were of different kinds in different provinces of India: in Bengal, as well as in Bihar, there was Permanent Settlement. In the United Provinces, there was the talukdari system and large tracts of land were owned by talukdars who got their lands cultivated by tenants. In the presidency of Bombay, there was the ryotwari system and every individual cultivator dealt directly with or was dealt directly with by the Crown. The provinces under the Permanent Settlement were assessed very lightly,[40] whereas, the ryotwari areas had to bear a heavier burden. Thus, the conditions of the tenure and the conditions of land revenue and the relations of the landlord and the tenant were different in different parts of the country.[41] Therefore, a committee, however small or efficient or big, dealing with the indebtedness and the relations between the landlords and the tenants of various provinces was not likely to be of great use. Similarly, Muhammad Anwar-ul-Azim thought that the issues involved in the Resolution needed to be varied materially, according to the geographical position of the various provinces in this big country. The conditions necessary to alleviate the distress of the cultivator in eastern Bengal would not be applicable to relieve the distress of the frontier Pathan on the North-West Frontier Province.[42] Given the multiplicity of tenures and consequently the diversity of relations of productions across India, a few members were of the opinion that measures taken by the provincial governments would be more beneficial to improve the condition of the landlords as well as the tenants.

Did the provincial governments examine their own agricultural conditions to devise means of amelioration of the distress? It may be noted that countries that concentrated on the production of one or two commodities for the world market suffered both from the slump in prices and from the fall of demand from outside, for example, Brazil (coffee), Cuba (sugar), and certain part of Africa, e.g. Gambia and Senegal (groundnut). India chiefly suffered from the slump.

Fortunately for India, hardly any part of the country depends solely on one commodity. It must be admitted, however, that some parts of the country depend more than others on export trade; Bengal, with nearly all the jute area in the country, is therefore the worst exposed to the risks of world dependence, but even Bengal has only 6 per cent of her total cropped area under jute; and rice, which accounts for 80 per cent of the area, is nearly all consumed at home.[43]

Bengal was immediately affected by the Depression because the sudden drop in the prices of rice and jute had a deep impact. Punjab felt the impact of the Depression very early because it produced mainly cotton and wheat. Punjab was also the first province to adopt relief measures, as it was a major concern of the British to keep the Punjab peasant away from the national movement as most soldiers of the British-India army were recruited from the Punjab countryside. The Punjabi tenant mostly paid grain rents which were not affected by the fall in prices. The revenue burden was not very heavy and a major part of the peasant's dues consisted of water rates (*abiana*). The United Provinces were in a more precarious position due to their peculiar land system. A scheme for the fluctuating assessment of rent and revenue was implemented on an experimental basis. The United Provinces were divided into five zones for this purpose in order to adapt it to different crops, and the rent and revenue demands were linked to a price index. The Central Provinces, which included large cotton tracts, felt the impact of the Depression very early. The provincial government relied initially on the low pitch of revenue and granted taccavi loans liberally in order to tide over the difficulties of the most affected areas. In the field of legislation for debt adjustment, the provincial government played a pioneering role with its conciliation boards which were soon imitated by other provinces. There was a great variety of provincial legislation concerning the protection of indebted peasants and landlords and of tenants about to be evicted for arrears of rent: legislation with the objectives of defining interest rates, dealing with the control of moneylenders, creating institutions for debt management, and stopgap measures aimed at a temporary moratorium or a stay of execution of decrees. For example, Usurious Loans (United Provinces Amendment) Act, 1934; Usurious Loans (Central Provinces Amendment) Act, 1934; Punjab Regulation of Accounts Act, 1930; Bengal Moneylenders Act, 1933; Central Provinces Moneylenders Act, 1934; United Provinces Arrears of Rent Act, 1931; United Provinces Assistance of Tenants Act, 1932; Central Provinces Debt Conciliation Act, 1933; Punjab Relief of Indebtedness Act, 1934; United Provinces Encumbered Estates Act, 1934; Bengal Agricultural Debtors Act, 1935; United Provinces Temporary Regulation of Execution Act, 1934; etc.[44]

Agriculture and Nature

By the mere efflux of time, the fertility of the soil had a tendency of getting deteriorated—this needed to be restored. Bones of animals were great fertilizers for the fields but thousands and thousands of tons of bones, either crushed or otherwise, were being exported from India every year. This resulted in impairing the fertility of the soil to a great extent:

> There is another point. In ancient times we used to have a lot of trees and forests in and around the villages. That afforded the much needed moisture to the fields, and that was also an element which tended to increase the fertility of the land. Now, with the pressure on agricultural land, with the increase in population, de-forestation has been going on at a rapid pace, and that also, in my humble view, has contributed to a large extent in impairing the fertility of the soil.[45]

It was estimated that the cattle, which was the wealth of an agricultural country, was deteriorating both in terms of number and quality as thousands and thousands of cattle were being slaughtered, and the hides and skins exported. However, government institutions, and also private enterprises, were trying to improve the breed and quality of the cattle.

I will mention the case of my own province of Bihar which has recently suffered so terribly from the earthquake. The most serious problem in Bihar at present is the vast amount of sand that has been deposited as a result of the earthquake, from the bowels of the earth.... It is very difficult for the agriculturists to clear the sands and to make the lands as fertile as they were before.[46]

Science and Agriculture

> Scientific research is the life blood of the economic development of a country and the other countries have forged ahead of India. In India also, we have been

taking some interest of late in the advancement of scientific agriculture and we have established the Imperial Council of Agricultural Research [1929]. But beyond passing some grants for the research institutions, we take very little interest afterwards and generally we do know little of the activities of that institution.

—T.N. RAMAKRISHNA REDDI, LAD, 6 April 1934, p. 3319.

Gaya Prasad Singh believed that the question of agricultural produce was one that needed to receive very serious attention from the government. For instance, the question of supplying improved seeds and seedlings of wheat, rice and sugar cane had, in his opinion, not received as much practical importance as it should have received. Institutions at Coimbatore and Pusa, for instance, were making a lot of investigations into the matter, but as far as the actual agriculturists were concerned, they had not been afforded the full benefit of the results of the investigations carried on in those institutions. But the government was not entirely to blame. It was a notorious fact that the Indian agriculturist was very conservative and his habits and methods of cultivation were still the same as existed from time immemorial; there had not been much development in the field of agriculture.

New technology also contributed to the rural distress. For example, in 1931, 18 Bolpur rice mills produced 8.10 lakh maunds of rice. At the rate of two women per *dhenki* (old husking tool) cleaning one maund of paddy a day, 8.10 lakh maund of milled rice could have ensured full-time job for 4,438 women in a year. Bolpur mills had thus displaced 247 women per year. Also, unemployment created by the rice mill in Bengal in 1930, for example, amounted to the displacement of 1,01,745 huskers.[47] The sugar industry was given nearly 200 per cent protection, yet Java had been dumping its sugar in Indian markets over this high tariff wall. This was because Java was able to produce sugar at nearly one-third or one-fourth of the cost of production in India. Again, India was threatened with importation of rice from Siam and Japan. It was also due to the fact that they were able to produce nearly double or treble the quantity of rice which the Indian agriculturist produced with his conventional methods of production, and hence, much improvement had still to be made in the methods of agriculture. The conservatism of the agriculturists was one thing that prevented the people from taking to innovations of a novel kind to which they had not been accustomed before. India's agriculturists had been using, from time immemorial, those old implements to which they had been accustomed to from the time of their forefathers.[48] It was necessary, under the new conditions of things, to introduce new methods for the tilling of the soil and for other agricultural purposes. How far had the government been able to help the agriculturists in this matter? Sometimes, there were exhibitions in which agricultural implements were shown by way of demonstration to the people, but these were few and far between and had little effect on the agriculturists as far as their daily avocations were concerned.[49] The government might allay whatever feelings of suspicion there might be lurking in people's hearts—that the interests of the agriculturists do not receive as careful and earnest a consideration at the hands of the government as some other industries receive.[50] Nevertheless, the Royal Commission on Agriculture was sympathetic to 'the ordinary cultivator on his tiny plot [who] is still a man of small resources, with small means for meeting his small needs. He requires all the help which science can afford, and which organization, education and training can bring within his reach'.[51]

Village versus Town

Regarding the devising of means for improving the condition of the landholders and peasants,[52] Khan Bahadur Abdul Aziz said that all those who had the welfare of the village at heart demanded that the educated part of the community come to the solution of this problem:

The Agricultural Commission dealt with this question five years ago, and one of the complaints was that the educated people will not settle in a village. We have throughout tried our very best to get one educated man to make his home in village and to furnish a living guidance to the villager, but we cannot get the educated man. Even the teacher who earns his living there, as soon as he gets his pension, moves away.[53]

Educated Indians[54] were so wedded to the towns that they would not come and settle in the villages and devote their time to the villager. Even for carrying on propaganda, such as telling the villager that his house was a death trap and that all he needed was more light and air and a proper skylight, the educated man was not available. He was not available even for the simple work of telling the villagers to keep the roads clean and the cattle away from human dwellings.

There are in my division at present nearly 300 villages where for three months in the year the water they drink from the ponds is more contaminated mud than water and we cannot get rich people to supply them with wells. Formerly they did. Ever since the village has ceased to be a self contained unit, the people who earn the money there spend it elsewhere. It is no longer that the village sahucar will build a well. It is no longer that he will build a serai there. It is no longer that he will build a school or pathshala there. That is our trouble. We cannot get people to live there and to give guidance to these villagers. The very men who have grown fat on the income of the village will not stay in the village, but say to the villager 'You are not of us'.[55]

Drawing attention to the fact that educated Indians were moving from villages to towns, Rai Bahadur Chaudhuri Lal Chand explained that although 90 per cent of the population lived in villages, yet, in the matter of spending money for amenities, the government was spending only 10 per cent on villages, while 90 per cent went to the towns. For example, the town of Rohtak got 75 per cent of the expenditure on waterworks from the government through the Rohtak Sanitary Board:

We have got a school which is just outside the municipal limits—only two miles from there—which is in the rural areas. It is a full-fledged residential High School with four or five hundred boys in it and we asked that waterworks be extended to us also. The reply was that the Sanitary Board could not give us a grant, because we were not living in a town. The result was that although within two miles there is good water available on which 75 per cent of the money has been spent by Government, yet we do not get it.[56]

So up to that time, they were giving everything to the towns and not to the villages; so naturally, educated people began to move into towns. It was up to the government then to spend 90 per cent of their money, according to the population or taxation basis, on the villages and only 10 per cent on the towns 'and then you will see how the tide is turned and all of us will go back to the villages'.[57] That tendency was apparent in every walk of life.

For instance, there are about 11 very good colleges in Lahore. After having all those colleges, there was no need for a Government College in that area, yet they are spending huge sums on a Government College at Lahore although in the whole of the south-east of the Punjab, in the whole of the Ambala Division, there is not one college and they will not provide one. They can say that Lahore is an educational centre and the Government College should be located in the principal centre for education. But why should the Veterinary College be also there? That ought to be in the Hissar district which is the home of cattle. There has been a tendency on the part of the Government to spend money on towns, and, therefore, there is this move.[58]

Moreover, the demand for labour was determined by the growth of industry and requirements of the auxiliary services that caused movement of people from the village to the town. For instance, the growth of the cotton industry created a permanent demand for industrial labour in Bombay.

Agriculture versus Industry

K.P. Thampan held that the importance of agriculture was not adequately addressed by the Government of India, which took steps to safeguard and protect minor industries than agriculture. For instance, the money invested in the textile industry was only Rs.80 crores, and the number of people employed was only 7 lakhs, while in agriculture, the investment was Rs.20,000 crores and it engaged about 90 per cent of the population of the country. So, there was no comparison between the agricultural and textile industries. Then there were other industries such as steel, sugar and paper industries: all these had been given protection. There were only a few crores of rupees invested in those industries and the number of employees in them was infinitesimally small compared with that of the agricultural industry.[59] Owing to a vigorous policy of 'Discriminating Protection' followed since 1925, an inordinate increase of revenue tariff in past years and the Swadeshi spirit of the freedom struggle, the demand for products of Indian manufactures increased, and industrial production in the country greatly expanded. In 1928–9, Indian mills produced only 1,893 million yards of cotton goods (out of a total available quantity of 3,830 million yards), but by 1933–4, Indian mill production increased to 2,945 million yards—an increase of 55 per cent in five years. The progress of the sugar industry was even more striking. In 1925–6, there were only 41 sugar mills producing 91,399 tons of white sugar; in 1933–4, there were 128 mills producing 5,54,000 tons. Thus, sugar production increased six-fold in eight years. Large progress had also been made in iron and steel, cement, woollens and several small industries and, in particular, the provinces of Bombay and UP largely benefited from these developments. While such progress went on in India, industrial production had drastically curtailed in Europe and America, and no other country except Japan made such rapid strides in industrial production as India had done during the early 1930s.[60]

Government Response

The Secretary, Department of Education, Health and Lands, G.S. Bajpai, held that since 90 per cent of the population of India directly or indirectly subsisted on land and agricultural operations, it was inconceivable that the government could assign to agriculture a secondary place in its policy. As to the effectiveness of the proposed method, namely, investigation by a committee for carrying out the government's 'sympathetic ideas and intentions', it was to be recognized that there had been a Royal Commission on Agriculture, which took nearly two years to complete that work and cost more than Rs.14 lakhs. There had been a Banking Inquiry Committee since. Other investigations were afoot.

Two results followed from that: first, the agriculturist had a much smaller margin left for purchasing those necessities which he had to buy in the shape of manufactured goods; second, the burden on his fixed monetary charges, be that land revenue due to the government, or the interest and principal due to his creditors, the margin left for him was either practically non-existent or completely inadequate to his requirements. To provide a permanent remedy for the problem of the fall in agricultural prices, the United Provinces evolved a formula that provided for an automatic adjustment of the rent and revenue demand to fluctuations in prices. Obligation of the agriculturist to his creditor was a difficult problem. On one extreme there was the suggestion to repudiate all those debts. Since expropriation was not to be attempted, some remedy had to be found. The cooperative movement had not met with the measure of success that was expected.

The Government of India took steps under six heads to deal with the problem of indebtedness of the agriculturists: first, the policy of discriminating protection, for instance, to cotton, sugar cane and wheat; second, to find markets abroad for India's agricultural products; third, the improvement of the quality of the country's produce through research; fourth, the organization of commercial intelligence for the collection and the classification and presentation of statistics; fifth, on the question of freights, steps were taken by the railways with regard to wheat and rice; sixth, employment of a marketing expert to help the task of economic rehabilitation. Stressing that there was no conflict between agriculture and industry, Bajpai paraphrased Sir Syed Ahmad and described the two as the lotus eyes of a lovely maiden: 'You cannot hurt the one without marring the beauty of the face.'

Conclusion

The agriculturist's was a perennially distressed lot as his daily avocation was akin to hard labour and necessitated constant vigil against both elements and living beings. The income of the peasants fell while rent and revenue demands and debt service remained as before. The agriculturist was oppressed by his heavy debt. That was the real cause of his trouble. He did not really earn for himself, as a very large proportion of the money that he got for his crops went to pay the interest on the money he had borrowed. It was the usual charge against landlords and moneylenders that they deprived the agriculturist of his lands and contributed largely to his misery.

An examination of the position of the government as landlord undoubtedly showed that the policy that the government followed with regard to assessment and realization of revenue also contributed greatly to the indebtedness of the agriculturists. While the government as a landlord was a creditor of the agriculturist and also a guarantor of agricultural or taccavi loans, it held a prominent position of a creditor by the side of the moneylenders. These loans were granted on an undertaking taken from the borrowers that these would be repaid within a short period. When the debtors failed to pay up these debts within these stipulated periods, on account of failure of crops or for some other reason, it was usually the case that distress warrants were issued against defaulting debtors for seizing their movables for the payment of their debt to the government. It can, therefore, be easily imagined how the government as creditor contributed to, and even aggravated, the miserable plight of the agriculturists.

The government's deflationary policy to support the exchange rate intensified the impact of the Depression. The price movement in the 1930s, characterized by a sharp downward trend, had disastrous consequences for the peasantry.[61] Because the prosperity of industry and other trades depended upon agriculture, unless the condition of the agriculturists was improved, there was no hope for any betterment of the condition of industrialists, as well as of other trades. Machine technology over time caused all-round unemployment and the consequent distress; for example, in the rice mill, caused the dislocation of the extensive network of huskers, peddlers, carters and boatmen. Steam engines affected the oil and sugar industries, with the old ox-driven oil mill disappearing faster than the old sugar factories.[62] As ever, the burden of the distress of the economic depression was passed on to the rural poor.

Notes and References

1. J. Thomas Parakunnel, 'India in the World Depression', *The Economic Journal*, vol. 45, no. 179, September 1935, pp. 469–83.
2. Neil Charlesworth, 'The Peasant and the Depression: The Case of the Bombay Presidency, India', in *The Economies of Africa and Asia in the Inter-War Depression*, ed. Ian Brown, London, 1989, pp. 59–73.
3. Dietmar Rothermund, *India in the Great Depression 1929–1939*, Delhi, 1992, p. 8.
4. Legislative Assembly Debates (hereafter, LAD), Wednesday, 14 February 1934, pp. 841–6; resumed on Friday, 6 April 1934, pp. 3293–342.
5. Raja Bahadur G. Krishnamachariar, Tanjore-cum-Trichinopoly, Non-Muhammadan Rural, LAD, Wednesday, 14 February 1934, pp. 842–3. 'The nationalist critic did not deny the advantage of railways as a mode of transport, but they were more concerned about the boatman and the carter who lost their

Agrarian Distress 279

livelihood'. Quoted in Smritikumar Sarkar, *Technology and Rural Change in Eastern India, 1830–1980*, Delhi, 2014, p. 255.

6. Nawab Major Malik Talib Mehdi Khan, North Punjab, Muhammadan, LAD, Friday, 6 April 1934, p. 3327.
7. A.R. Desai and Sunil Dighe, eds., *Labour Movement in India 1928–1930*, vol. 7, New Delhi, 2003, p. 684.
8. Raja Bahadur G. Krishnamachariar, LAD, Wednesday, 14 February 1934, pp. 842–3.
9. Khan Bahadur Mian Abdul Aziz, Punjab, Nominated Official, LAD, 6 April 1934, p. 3306.
10. Rothermund, *India in the Great Depression 1929–1939*, Delhi, 1992, p. 4.
11. Parakunnel, 'India in the World Depression', pp. 469–83.
12. Ibid.
13. B.V. Jadhav, Bombay Central Division, Non-Muhammadan Rural, LAD, 6 April 1934, p. 3295.
14. Parakunnel, 'India in the World Depression', pp. 469–83.
15. N.M. Joshi, Nominated, Non-Official, LAD, 6 April 1934, p. 3314. He was essentially a leader of the industrial workers.
16. Diwan Bahadur A. Ramaswami Mudaliar, LAD, 14 February 1934, p. 846.
17. Parakunnel, 'India in the World Depression', pp. 469–83.
18. Dharma Kumar and Tapan Raychaudhuri, eds., *The Cambridge Economic History of India*, vol. II, Hyderabad, 1982, p. 173.
19. Rai Bahadur Kunwar Raghubir Singh, Agra Division, Non-Muhammadan Rural, LAD, 6 April 1934, pp. 3296–7.
20. Khan Bahadur Mian Abdul Aziz, LAD, 6 April 1934, p. 3307.
21. Sitakanta Mahapatra, Orissa Division, Non-Muhammadan, LAD, 6 April 1934, p. 3313.
22. Sir Darcy Lindsay, Bengal, European, LAD, 6 April 1934, p. 3298.
23. B. Sitaramaraju, Ganjam-cum-Vizagapatnam, Non-Muhammadan, LAD, 6 April 1934, p. 3299.
24. M.L. Darling, *The Punjab Peasant in Prosperity and Debt*, London, 1925. 'The book renders much detailed description unnecessary', *Royal Commission on Agriculture in India*, Government Press, Calcutta, 1927, p. 432.
25. N.M. Joshi, LAD, 6 April 1934, p. 3314.
26. Binay Bhushan Chaudhuri, 'Background to Land Reforms in Post Independence Bengal: Shifts in the Predominant Modes of Land Control and Appropriation in Colonial Bengal', General Presidential Address, 41st Session, *Punjab History Conference*, p. 2.
27. Nawab Major Malik Talib Mehdi Khan, LAD, 6 April 1934, p. 3325.
28. Kumar and Raychaudhuri, *The Cambridge Economic History of India*, vol. II, pp. 148–9.
29. Uppi Saheb Bahadur, West Coast and Nilgiris, Muhammadan, LAD, 6 April 1934, pp. 3327–8.
30. Brahma Nand, *Fields and Farmers in Western India 1850-1950*, Delhi, 2003, pp. 783–4.

31. Nawab Major Malik Talib Mehdi Khan, LAD, 6 April 1934, p. 3326.
32. Parakunnel, 'India in the World Depression', pp. 469–83.
33. J.H. Darwin, United Provinces, Nominated Official, LAD, 6 April 1934, p. 3319.
34. Hony. Captain Rai Bahadur Chaudhuri Lal Chand, Nominated, Non-Official, LAD, pp. 3324–5.
35. Ibid.
36. Nand, *Fields and Farmers in Western India 1850–1950*, pp. 783–4.
37. Muhammad Anwar-ul-Azim, Chittagong Division, Muhammadan Rural, LAD, 6 April 1934, p. 3311.
38. T.N. Ramakrishna Reddi, Madras ceded Districts and Chittoor, Non-Muhammadan Rural, 6 April 1934, pp. 3320–1.
39. N.M. Joshi, LAD, 6 April 1934, p. 3316.
40. 'Bihar is a permanently settled Province, and the revenue taken there is absolutely out of all proportion even to the reduced prices of agricultural commodities that prevail today.' G.S. Bajpai, Secretary, Department of Education, Health and Lands, LAD, 6 April 1934, p. 3335.
41. Rothermund, *India in the Great Depression 1929–1939*, p. 115.
42. Muhammad Anwar-ul-Azim, LAD, 6 April 1934, p. 3309.
43. Parakunnel, 'India in the World Depression', pp. 469–83.
44. Rothermund, *India in the Great Depression 1929–1939*, pp. 119–26.
45. Gaya Prasad Singh, LAD, 6 April 1934, p. 3330.
46. Ibid., p. 3331.
47. Sarkar, *Technology and Rural Change in Eastern India, 1830–1980*, p. 253.
48. Ibid.
49. 'What discouraged the use of the new technology was the shortage of capital, given the small-peasant system of farming, not social inhibition.' Ibid., p. 301.
50. Mr Gaya Prasad Singh, LAD, p. 3329.
51. *Royal Commission on Agriculture in India*, 1927, p. 14.
52. 'The 1931 census found only one Indian out of nine living in the towns, showing how little employment outside agriculture was available.' Irfan Habib, *The National Movement: Studies in Ideology and History*, New Delhi, 2011, p. 59.
53. Khan Bahadur Mian Abdul Aziz, LAD, 6 April 1934, p. 3307.
54. 'Only 9 per cent of the population was returned as literate in 1931, which reflected the acute cultural backwardness from which India suffered after some 150 years of British rule.' Irfan Habib, *The National Movement: Studies in Ideology and History*, op. cit., p. 59.
55. Khan Bahadur Mian Abdul Aziz, LAD, 6 April 1934, p. 3308.
56. Hony. Captain Rai Bahadur Chaudhuri Lal Chand, ibid., p. 3324.
57. With the gradual evolution of municipal administration and a greater sense of civic responsibilities, there was a growing sense of urgency to make urban life as comfortable and decent as possible.
58. Hony. Captain Rai Bahadur Chaudhuri Lal Chand, LAD, 6 April 1934, p. 3324.

59. K.P. Thampan, West Coast and Nilgiris, Non-Muhammadan Rural, AD, 6 April 1934, p. 3301.
60. Parakunnel, 'India in the World Depression', pp. 469–83.
61. Kumar and Raychaudhuri, *The Cambridge Economic History of India*, vol. II, pp. 148–9.
62. Sarkar, *Technology and Rural Change in Eastern India, 1830–1980*, p. xv.

13

In Quest of Plenty: Hunger and Agricultural Technology in India, 1955–67

Madhumita Saha

As reports came in of people starving amidst severe drought in the mid-1960s, it became clear that end of British rule did not end hunger in India.[1] Though the estimates of the incidence of undernutrition and malnutrition varied, the Famine Enquiry Commission on the eve of India's independence concluded that some 30 per cent of India's people were underfed even in normal times.[2] Eradication of hunger from everyday lives of Indians through 'scientific' planning was part of the core food policy agenda of Jawaharlal Nehru's government. The Grow More Food campaign, carried over from the last years of the British Raj, envisioned food self-sufficiency by the year 1951, yet undernutrition continued to be a lingering problem, to the exasperation of the members of the Planning Commission, the political and research establishment.[3] Touring the famine-stricken districts of northern India, in the early 1950s, a visibly distraught Nehru explicitly asked people not to shout slogans in his name. 'Why do you shout slogans in my praise,' he was quoted as asking the inhabitants of one village, 'when I cannot feed you to keep strong?'[4]

In sync with the changing notion of hunger from the middle of the nineteenth and through the twentieth centuries, the Indian government recognized hunger as a 'collective social problem'.[5] The worldwide discourse on hunger increasingly saw it not as a consequence of divine punishment or human laziness; instead, the new political order saw the hungry as 'innocent victims of failing political and economic systems over which they had no control'.[6] Contrary to the arguments of Adam Smith and John Malthus who saw free markets as capable of generating sufficient wealth to end poverty and hence hunger, nation states gradually took upon the responsibility to protect its citizens from economic downturn. The deep involvement of the state machinery in the eradication of hunger was especially true for India; the effort of the political establishment to modernize the country underlined the necessity of having a well-fed citizenry; thus, the story of modernity in India as elsewhere 'became partially organized around the conquest of hunger'.[7] The national leadership, as well as the planners, was confident that hunger had to be banished if India has to come out of the 'waiting room' of development.

Government of India's strategy to eradicate hunger among its rural population gradually changed over the period since independence through the introduction of Green Revolution technology in the country. From being conceptualized largely as a major structural problem, indicating chronic social and economic inequity, it would shift to being understood as a technical problem of failing production, requiring technological intervention. Right after independence, the understanding of the central government, as well as the Planning Commission, was that as poverty was indubitably associated with rampant hunger, the task of eliminating it and improving agricultural production had to be integral to a larger goal of rural development. Researchers at the Indian Agricultural Statistical Institute statistically corroborated the role of poverty as one of the important contributory causes of undernutrition as well. They observed that though majority of Indians suffered from undernutrition, 'the deficiency clearly falls heavily on the poorer sections.'[8] To eliminate hunger, therefore, the First Five-Year Plan emphasized on achieving a higher production target as well as on the necessity of having a proper food distribution system. If 'food for all' was to be the effective basis of policy, the planners reasoned that it would be imperative to keep down the price of food through state procurement and rationing, until a substantial and enduring improvement in domestic production could be achieved.[9] Such measures, however, generated much controversy since its earliest days of inception as many considered it to be against free market economy and detrimental to accomplishing higher production. Prioritizing institutional reforms and continuing with state control in the food sector was especially not to the liking of many state leaders of the Congress party who were sceptical that any such measures might disappoint the big landlords and negatively impact the party's electoral performance.[10] Nehru's political stature, especially his indispensability in winning elections, helped the Planning Commission to override most objections and implement, with mixed successes, institutional reforms as well as retain a government-regulated food distribution system.

The belief that a modernized countryside based on social and economic equality would have a better chance of avoiding hunger was very integral to the plans of those who believed in the socialist reconstruction of India. Thus, the Prime Minister and the planners argued that if the people were to be fed adequately, any plan for improving the country's agricultural production could not possibly ignore the necessity of 'institutional reforms'.[11] It was through redistribution of land among cultivators that additional labour could be mobilized to push up production; instituting farmers' cooperatives would give farmers of all economic means an access to improved inputs; whereas, decentralizing power in the hands of village panchayats would help to weaken the semi-feudal nature of the Indian countryside, strengthening thereby the forces of modernization and democracy, as the national leadership reasoned.

Social equity as a goal was not only at the heart of institutional reforms; the Planning Commission documents reveal that the central government wanted the agricultural scientists to keep the same in consideration while conducting research or recommending biochemical inputs towards augmenting production. This was especially evident in the plans for the widest coverage, including small and big farmers, as well as irrigated and unirrigated tracts, rather than concentrating the inputs on selective areas. If encouraging the use of cheap, locally available organic inputs in improving agricultural production was an important way to save precious foreign exchange, commitment to the goals of social equity was no less a crucial consideration.[12] Faced with increasing criticism at home and abroad, the Indian government would, however, gradually shift from its commitment to institutional approach to a more technocratic strategy that emphasized price incentive to farmers and uses of capital-intensive technology to improve food production.[13]

Scholars of South Asian studies such as Francine Frankel and Ashutosh Varshney have written extensively on the debate over institutional reforms and its eventual replacement by a policy of favouring capitalization of Indian agriculture.[14] In much of the existing work, however, little had been said about how, since independence, the Indian government and other actors constantly grappled over the policies to tackle hunger.[15] Little is known about how India's own experience with recurrent famine, its feeling of helplessness at the need for repeated food aid and the constant pressure to win the race against population explosion shaped its understanding of hunger and framed a nation-wide debate as to how it could be managed. An exploration into the history of the debate over hunger in India, therefore, brings out clearly how the country's identity as a modern nation, a sovereign nation, a healthy nation came to be constructed around the food question.

As technical experts backed by the political establishment increasingly championed high yield as the way to overcome hunger, this study looks into how the shift impacted their research on agricultural issues. In looking into the world of scientific and technological research, along with that of the planners and political establishment, the study bridges the realm of practice and policy, revealing how they interacted and constituted each other. This essay explores how the policy of privileging the technocratic over the social, and in disassociating the two, led to stereotyping, abstraction and reductionist approaches towards the farming communities among the practitioners of Green Revolution technology. Focused on getting high yield, the advocates of Green Revolution technology sidestepped issues for which technological intervention was not possible, for instance, questions of equity and farmers' access to resources were ignored, as were complexities of rural situations, agro-ecological variance, and the gap between farmers' needs and priorities of the experts. Similarly, the technological success in making the dwarf seeds produce 'spectacular' yield provided a context and helped to justify the shift towards a technocratic policy in Indian agriculture over the years.

Debate over Institutional Reforms and the Food Crisis

Economist David Hopper's 'strategy for the conquest of hunger' best enunciated the views of those who believed in a strict separation of the goals of agricultural production from that of social equity. An employee of the Rockefeller Foundation, Hopper was categorical in his pronouncement that 'to conquer hunger is a large task. To ensure social equity and opportunity is another large task. Each aim must be held separately and pursued so'.[16] The Indian government had its own share of doubt over land reform measures. Though the First Five-Year Plan confidently justified land ceiling on the grounds of giving the tiller his 'rightful' place in the agrarian system, to provide him with fuller incentives for increasing agricultural production and to put an end to the 'tenant–landlord nexus' as 'essential' steps in the establishment of a 'stable' rural economy, the planners were more sceptical while drafting it. They were initially nervous about 'land-ceiling measures bringing down the production in the larger farms and of having a serious effect on the well-being and stability of rural society as a whole.'[17] Over the years, critics of land ceiling act would persistently protest that the economic use of machinery and the adoption of modern methods of agriculture would not be easy to obtain for smaller farms; tiny parcels would lower the standards of efficiency and manoeuvrability in management.[18] It was the 'economic' logic behind land reform, especially understanding how sizes of farms would contribute to technological modernization, leading ultimately to higher yields, which became the crux of the debate, rather than differences over its radical and conservative nature.

The surplus food production of the mid-1950s, however, kept alive the relevance of institutional reforms in public policy. The government believed that institutional reforms, along with organic and cheap input-based research work of the scientific establishment, would bring more production, better lifestyle, and energize the peasantry. The economic planning of the country, however, received a serious setback soon after the Second Five-Year Plan was launched; acute shortfall in the foreign currency reserve and drought in eastern parts of India severely impacted the economic stability of the country.

With monsoon playing truant, foodgrain output declined by 10 per cent, pushing up the price by 50 per cent from October 1955 to August 1957.[19] In its annual meeting, the All India Congress Committee (AICC) underlined that it was urgently necessary to regard self-sufficiency in food as an integral part of national self defence.[20] American observers, keeping a close eye on the performance of Indian food sector, expressed deep concern at the growing 'gap' between rapid growth of population in India and stagnating food crop production.[21] By late 1950s, many in the United States (US) administration openly doubted if land reform could actually be an antidote to a slacking agrarian sector of the developing world.

In the midst of this discontent, doubt and confusion that marked the institutional reforms, the Ford Foundation team visited the country in 1959 to inspect the food situation and recommend measures towards its 'development'. The Foundation's work in India, involving Community Development Projects, had by then displayed no tangible results. The meagre gains in living standards and food production failed to justify the huge cost of the programme, which at the end of eight years of work was running to ten million dollars. Forced to evaluate its approach, the Ford Foundation reinterpreted existence of hunger in India as a failing race between food production and alarming growth in population.[22] In the next seven years, the population of India would be 480 million, the Foundation experts calculated, and to provide 15oz. of food for daily consumption to this growing number, India needed to increase its food production at the rate of 8.20 per cent per year, for the same number of years. Thus, a hunger-free India could only be achieved through increasing production by 10 million tons to 110 million tons.[23] The Foundation's changing approach was part of a global trend to see the population growth dangerously running out global food production.[24]

With food supply calculated as out of balance with the rising population, participants in a symposium of the US National Academy of Science would gloomily portend that the world was 'living at the edge of the knife'. It was mainly through increasing food supplies, the conference participants concluded, that the possibility of a communist revolution could be avoided in Asia.[25] At the outbreak of successive droughts in India, the US President's Science Advisory Committee (PSAC) insisted on countries adopting massive, long-range, innovative efforts unprecedented in human history. The Advisory body urged the US government to especially use its food aid to make countries such as India reconsider their ways of boosting agricultural production. The debate, therefore, not only succeeded in framing the issue of hunger in terms of a binary relationship between food and population, it also created a strong case for technology as the solution.[26]

In its report on the 'food crisis', the Ford Foundation advocated 'an all-out emergency' programme build around large-scale adoption of expensive biochemical resources, credit supply, water conservation and price support measures to farmers. They wanted adequate number of personnel trained and assigned to the job of increasing production.[27] With a plan to use more chemical fertilizers and hybrid seeds to push up the yield, the Foundation experts discouraged land reforms as a way of improving the Indian agrarian sector. They considered 'insecurity of tenure' brought about by land ceiling measures as having a 'retarding' effect on food production. The Foundation representatives reasoned that if the well-off farmers could be assured that no further reforms would risk their possession, they would be encouraged to make the requisite investment in acquiring the new inputs and thus push up the production towards diffusing the food crisis.[28] It specifically pointed out that 'care should be exercised' so as not to break up farms that were 'efficiently and productively operated'. As little or no emphasis was put on questions of social inequity, income, food distribution and access to resources as possible factors causing the food crisis, the report proposed no structural measures. Instead, their sole emphasis was on making sure that Indian farmers adopt the latest technology and thereby diffuse the 'food crisis'.

Many agricultural officers of the Indian government displayed a similar approach to land reforms, ignoring it in favour of technological improvements. During her field work in the southern state of Tamil Nadu, one of the states to benefit most from Green Revolution technology, cultural anthropologist Joan P. Mencher wrote that the officers she interviewed did not consider land reform to be 'necessary'. They failed to see land reforms as being related to production, except negatively. They reportedly commented that if the government takes away land from the better-off farmers, it could only decrease production because the poor would not have the facilities to cultivate properly. In such a scheme of things, the idea of cooperative farming to help small holders apparently took a backseat, compared to plans for technological development or price support to encourage larger farmers to grow the high yielding varieties.[29]

In urging technology as a solution to hunger in India, the Ford Foundation ignored the most obvious problems, namely that India till then had a very limited capacity of producing sufficient amounts of chemical fertilizers and most farmers lacked the means to purchase expensive inputs. The report tried to overcome the issue of limited resources by advising the government not to dribble fertilizers across the entire country, rather concentrate these 'scarce' resources in areas that were adequately equipped to utilize them.[30] Following this strategy meant that the resources should be pumped into areas already enjoying a relative advantage on account of good irrigation facilities, fertile soil, active credit infrastructure, etc. The Ford experts' recommendation, however, conflicted with the long-standing position of the Indian government against any 'selective' approach, which the Planning Commission and the left-to-the-centre members of the Congress party had been considering as economically imprudent and against the principle of social equity.

Faced with a growing disapproval about the food situation, however, the Indian government decided to implement some of the recommendations of the Foundation report in launching the Intensive Agricultural Development Programme (IADP) in 1961. The Union Ministry of Agriculture released improved varieties of seeds and chemical fertilizers to saturate areas under cultivation in selected districts. The planners recommended an additional flow of credit to meet the demands of cultivators participating in the programme. The Planning Commission wanted these districts not only to be centres of higher production, but also to act as models to motivate other villages to adopt the new agricultural methods and innovations.[31]

While use of chemical fertilizers and 'improved' varieties of seeds did go up in certain districts under the IADP, there was little tangible improvement in the overall food scenario.[32] Agricultural production showed little progress in the first two years of the IADP and then started declining after 1962. Food prices rose rapidly, unchecked even by grain imports from the US. Nehru was jittery about the food situation. Not sure what exactly went wrong, he said to the Indian Parliament, 'I am . . . naturally disappointed at many things, more especially our performance in agriculture. . . . You may of course apportion blame between the Planning Commission, the Government of India, myself and the state governments.' At various places and times, he admitted that all the elements of the institutional strategy of land reforms, cooperatives and panchayats had failed.[33]

China's attack on India in 1962 came as a sudden jolt to the political leadership of the country. Nehru, already distraught by the food crisis, linked the drive for national defence to the quest for agricultural prosperity. He pointed out that the 'fact that we produce enough in our agricultural sector is as important as guns. . . . Real war is governed by scientific advance. Today you want scientific knowledge; you want scientific processes for production'.[34] Thus, under this pressure situation, the national leadership showed signs of lacking confidence in slow-paced and gradual institutional reforms as a way of improving agricultural production. Techno-scientific means, in moments of crisis, appeared as a more reliable and quicker deliverer.

By fall 1964, S.K. Patil, a Union Minister, openly declared that the country was passing through a 'national calamity' in terms of availability of food.[35] The rising food prices shook confidence in the national government among all communities of citizens. The opposition parties passed a no-confidence motion against the Congress government in the Lower House of the Indian Parliament.[36] Demonstrators appeared in thousands, and the police soon made about 1,300 arrests to keep the situation from spiralling out of control.[37] The Union Minister of Food and Agriculture C. Subramaniam emphasized that the real answer to the spiralling prices was to increase production. He urged the government to adopt the market approach of providing price incentives to private investment, while substantially increasing outlays on yield-enhancing inputs, especially chemical fertilisers.[38]

In January 1965, Subramaniam presented the framework of the new agricultural policy to the annual session of the Congress party at Durgapur. In the meeting, he vociferously argued that even though the 'immediate' concern apparently was food distribution, 'the food problem cannot be solved unless the agricultural production programme is attended to . . . and produce more to increase the availability of food to meet the needs of the people'. He defined the 'future programme' of Indian agriculture entirely in terms of 'need for scientific agriculture' that would be composed of extensive fertilizer use, bigger seed farms, plant protection, improved elements, etc. He planned the growth of entire administrative, research and extension sectors to facilitate the use of the aforementioned components of what he called 'scientific agriculture'.[39]

But it did not prove easy for Subramaniam to push through his plans for scientific agriculture and price incentive at the expense of mechanisms of strong government control and institutional reforms. There were a number of departures from the practices of the past; Nehru never was against using price as the sole incentive to higher production. The Planning reports repeatedly emphasized on maintaining a balance between profit of the producers and the affordability of the consumers that a price-driven policy stood to challenge. The main bone of contention was, however, whether the new emphasis on production through capital-intensive means in Subramaniam's plans amounted to abandoning socialist principles and the goal of social equity. The chief ministers of the state governments supported the Food Minister's policy, whereas, the more left-inclined members of the Congress party demanded a return to what they called 'Nehru's ideals'. The session, however, ended on a compromising note. A resolution was passed that reaffirmed the goal of making progress towards an ideal socialist society, but at the same time recognizing the need for rapidly stimulating the base of farm production.[40]

Troubled Time and the New Seeds

Subramaniam wrote in his memoir that in the spring of 1965, Dr Ralph Cummings of the Rockefeller Foundation came to visit him, as it became known that the minister was prepared to consider new approaches to agriculture based on the new advances in science and technology.[41] The Foundation had already been testing the high-yielding wheat and maize seed varieties in various experimental farms of the IARI, following their success in the Mexican programme where George Harrar and Norman Borlaug worked with a group of Mexican scientists in breeding wheat varieties that had higher yield potential and were resistant to stem rusts disease.[42]

The success of the new technology in significantly bringing down Mexico's import of corn and wheat went a long way to substantiate the claim that technology on being effectively used could modernize agricultural production and thereby fight hunger. In 1967, for example, the American scientists E.C. Stakman, Richard Bradfield and Paul C. Mangelsdorf, who were involved with the Foundation's Mexican programme, wrote a book, *Campaigns Against Hunger*, crediting science and technology with creating an 'agricultural revolution' that succeeded in conquering hunger in that country in a short span of two decades. Their account read like a fairy tale, with the good (in this case, techno-science) winning over evil (hunger), thanks to the valiant struggle of the heroes, the scientists. Their moral-laden account linked modern science to efficiency and contrasted hard work versus laziness in producing striking contrasts of wealth and poverty. The book promised that modern research could ensure a bright future for all, except the traditionalists who were too non-enterprising or too sceptical to embrace change.[43]

In proclaiming the gains achieved through a scientific revolution in farming, Stakman, Bradfield and Mangelsdorf made assertions that played into a larger context of international political arguments over agriculture. The scientist trio suggested that while land redistribution might satisfy the hunger of the landless for land, such measures did not automatically satisfy their hunger for food. They argued that the developing world would cease to be hungry only when the people learn how to master nature. It was science and technology that would empower nations to fight drought, soil infertility and crop loss. It is, therefore, understandable why the new technology appeared so promising to a significant number of agricultural scientists from the developing countries; for them, the prospect of taming nature to make it more productive appeared too alluring to be ignored.

To the agricultural scientists in India, who were struggling to come up with a solution to the country's food problem, the new dwarf seeds looked promising. In 1963, therefore, an invitation was send to Norman Borlaug from the Government of India to visit the wheat-growing areas in India and assess the possibilities of introducing the new varieties for cultivation in the country. After testing several Mexican varieties, the scientists at IARI agreed on releasing Sonora 64 and Lerma Rojo for cultivation in about 2,900 ha. (about 7,100 acres) in the 1965–6 cropping season.[44]

On April 1965, as IARI celebrated its sixtieth anniversary, Pakistani incursions occurred in the Rann of Kutch area at the western part of India. The volatile situation added to the national anxiety about the food situation. To make matters worse, it was apparent by late July that the year's main monsoon had failed over northern India, and the four-year agreement that Indian leaders had made with President Kennedy on P.L. 480 was about to come to an end too. A week following the declaration of war by the President of Pakistan, General Ayub Khan, the US government suspended all military and economic aid to both belligerents. In the context of the Cold War, the suspension of aid did not go down well with the members of India's political establishment. H.N. Mukherjee, the communist Member of Parliament from West Bengal, considered the suspension of aid as the American way of 'arm twisting' the Indian government. He argued that the US attitude towards the Indo-Pakistan conflict over Kashmir had shown 'where we are likely to stand if we depend on aid'.[45]

In October 1965, in a speech to the nation, the Prime Minister correlated the need for greater food production with the preservation of India's freedom. He told listeners that just as the *jawan* (soldier) is 'staking his life for the country', similarly, the *kisan* (farmer) should 'give their toil and their sweat'. Shastri's speech brought out the desperation of India's situation. He urged that every bit of land should be cultivated, 'a well-kept kitchen garden should be a matter of pride to every household . . .'. He asked ordinary consumers to practice self-restraint and not to hold parties, dinners and lunches, because these 'are not in tune with the time at all'. He wanted public opinion to 'encourage austerity'. He called upon women to economize on consumption of wheat and instead use maize, barley or gram.[46]

Shastri anticipated that the winter of 1965 would serve as a crucial turning point. He warned that whatever India would be able to do in the agricultural front during the coming three or four weeks would determine the fortunes of the country in the coming year. He wanted agricultural work to be undertaken on a war footing. The Prime Minister proposed a scheme in which groups of villages would be entrusted to officials whose responsibility would be to keep in close and direct touch with the farmers and to do everything possible to resolve their difficulties. Shastri visualized the district officer as equivalent to a military commander, who had to organize the drive and achieve the production target.[47]

Subramaniam stood by Shastri in this time of great national crisis, expressing anguish at how the 'inadequacy' of India's agricultural production has thrown a 'grave' challenge to the nation, especially as it was facing increasing responsibilities of its defence. He saw lagging productivity as a challenge to India's 'will to live in prosperity and freedom' and gave a call to the 'men of science' to provide the ideas and leadership 'for bringing into the field methods and techniques which will effect a breakthrough in our agriculture and sustain its dynamic growth'.[48] To Subramaniam, the state of national emergency warranted, even more than ever, that Indian agricultural scientists conducted research on the new seed varieties from Mexico. He confidently declared that given the right type of political and administrative support and academic freedom, India's research workers could convert theoretical knowledge into tangible advances in the food sector.[49] A close confidant of Subramaniam and a crucial actor in the introduction of the dwarf varieties in India, M.S. Swaminathan pointed out how rapid advances in agronomic practices, especially in fertilizer application and water management, was making it possible to obtain very high yields in the range of 5,000lb. per acre from the Mexican dwarf varieties of wheat. In the successful trials conducted at farms in Delhi, Uttar Pradesh, Punjab, and Bihar, Swaminathan saw a great possibility of solving the food situation.[50]

By the middle of 1965, the swirling debate around the food situation had thrown open several important questions as to the fundamental causes of crisis. People were divided over whether to blame stagnating production (the result of unsophisticated technology) or to blame the government for its lack of political determination and administrative failure. In the pages of *The Times of India*, the government's slack procurement policy was under vicious attack. Consideration of 'political expediency', critics pointed out, barred the states from taking the procurement programme with any seriousness. The fear of doing anything that might alienate the middle-class and rich peasants was paramount among the state leadership; the apprehension was that alienating propertied cultivators could cost the ruling party a large number of votes at the upcoming general elections.[51] The critics were relentless in protesting that government's negligence at building up a buffer stock through efficient procurement had made the people of India vulnerable to vagaries of weather. Given that droughts and floods were not uncommon in India, the unpreparedness of the national government became the butt of much public criticism.

Classical liberal economist B.R. Shenoy as well as the former food minister, A.P. Jain, however, came out in support of C. Subramaniam in locating the locus of the crisis not in an administrative failure but in what they saw as the long-term technical backwardness of Indian agriculture. They argued that the Indian government must seek long-term remedies for food scarcity by investing in research for greater agricultural production. They wanted measures that would help farmers get financial resources, essential supplies and cheaper fertilizers, which could help bring a reasonable return on their produce.[52] The Food and Agricultural Ministry under Subramaniam's leadership proposed a comprehensive outline for the new agricultural strategy in the approach paper prepared for the Fourth Five-Year Plan, 1966–71. It specified that fertilizers and pesticides should be used only in regions that had assured rainfall and sufficient irrigation, where such investments could return the maximum production.[53] The policy, therefore, 'favoured' areas which were already comparatively well developed with regard to irrigation. As debate raged among scientists, economists, bureaucrats, politicians and others over the incumbent changes in India's agricultural policy, the food situation worsened further. As winter set in, the spectre of having to further curtail food rations from 12oz. to 10oz. or even 8oz. per day was turning into a reality.

It was apparently only with one positive note that the year came to an end: the war between India and Pakistan reached an inconclusive end. Lal Bahadur Shastri flew to Tashkent to sign a treaty with Pakistan under Soviet mediation. The treaty was successfully concluded, but Shastri very suddenly died the very next day on 11 January 1966. What followed Shastri's death was a long and intense power struggle between several factions and the subsequent succession of Indira Gandhi as the Prime Minister of India.

M.S. Swaminathan later recounted that Indira Gandhi gave the scientific community the full support necessary to implement the Green Revolution technological package in India.[54] Under tremendous pressure at home and from the developed countries to show perceptible increases in production in the agrarian sector, the Prime Minister evidently grabbed onto the promises made by the new technological package. To her, the immediate crisis made it a dire necessity to use science and technology for more specific, time-bound utilitarian projects. In her speech to the nation on 12 June 1966, she characterized the moment for India as one of 'frustration, agitation, [and] uncertainty'. She constantly reminded fellow politicians and scientists that the need of the hour was to increase production, especially food production. 'Unless we increase agricultural production rapidly, control our population, and thus achieve self-sufficiency in the next few years, we will have forfeited our right to call ourselves a free country, let alone a great country. We must become self-reliant. Aid and help should be a temporary phase,' she insisted.[55]

Hybrid Seeds, Promise of High Yield and Indian Ryots

Amidst food crisis, famine and worldwide speculations over human fertility overtaking the productivity of the soil, the Green Revolution brought with it a promise of 'spectacular' boom in cereal production. Deeply optimistic of the yield capacity of the technological package, Norman Borlaug wrote to J.A. Pelissier, the Head of the Product Research Division of ESSO Research and Engineering Company, that he and his researchers were at the verge of 'trigger(ing)' a 'real revolution' in wheat production.[56] The new ways of production, Borlaug was confident, would introduce 'dynamic new methods in one stroke . . . [and] kill old ideas and methods'.[57]

Borlaug's brief letter to Pelissier helps to bring out two most important features associated with the new technology that would determine the course of agricultural knowledge production in the years to come. Its advocates would untiringly urge that the concept of higher productivity or yield be given a pivotal place in the discourse of agricultural modernization as it was only through producing more that India and other countries suffering from food crisis can emerge out of it; their enthusiasm for the new technology would be fuelled further by their conviction that the farmers of the developing world, in witnessing the benefits of increased fertilizer use, would inevitably start 'clamouring for fertilizers', which, along with the hybrid seeds, irrigation and pesticides, was the core component of the Green Revolution's technological package. Greater yield, therefore, carried the combined potential of solving the question of hunger and bringing material benefits to the farmers. Since the yield factor could be maximized through the application of dwarf seeds and chemical fertilizers purchased from the market and the use of industrial technology, it arguably was the shortest path that Indian agriculture needed to travel from being traditional to achieving modernity. Thus, the researchers involved with the Green Revolution technology worked towards establishing a simple linear relation between use of chemical fertilizers, high yield and strong chances of profit to convince anyone doubting the utility of the technological package.

Putting yield at the heart of the agricultural discourse, however, deeply impacted the ways agricultural scientists especially economists, breeders and agronomists designed their experiments. Research projects and experiments were subsumed under the overarching need to produce more per unit. Ryots, regions and plant varieties as objects of study were categorized and labelled depending on their relevance to the productionist practices. Consequently, researchers' focus mainly narrowed down to three priorities: breeding shorter plants, testing these dwarf varieties in a high-quality environment defined by regulated water supply and fertile soil and identifying the 'risk taking' farmers who were ipso facto the progressive ones as the users of the new technology.

For the sake of raising the cereal yield, it became imperative for the agricultural scientists to identify the kind of farmers, regional features and types of seeds conducive to the goal, rather than analyse and address factors that were not. As the experts went about distinguishing the 'modern' seeds from the 'traditional', 'entrepreneur' farmers from the 'laggards' and the 'irrigated' fertile parts of the country from those that were not, such acts of 'labelling' rarely gave a comprehensive picture;[58] social, economic and agro-ecological challenges rarely featured in the scientific experts' understanding and formulation of the research problems. It kept farmers' experience fragmented and interpreted it from the perspective of the need of the technological package.

Agricultural economists employed in Indian research institutes collaborated with practitioners of other branches of agricultural science to develop a new framework of persuasion based on economic profitability. Under the new framework, using high doses of chemical fertilizers to push up the crop yield appeared the most profitable thing for the farmers to do. The economists undertook a series of studies centred on high doses of fertilizer application. The goal of these studies, Deborah Fitzgerald observed, was to marshal evidences by which ordinary people would be 'inclined to change the way they did things . . .'.[59] The agricultural economists in India were part of a transnational 'rationalizing elite' who worked to convince the cultivators about the economic benefits of the new technology.

Agricultural economists, Bill C. Wright of the Rockefeller Foundation and R.B.L. Bharadwaj was one such professional duo working on the promotion of chemical fertilizers among farmers. Using the data gathered by agronomists of the All-India Coordinated Wheat Improvement Project, they calculated that with the price of wheat at Rs.75 per quintal and nitrogen at Rs.2 per kg., farmers would gain a net profit of Rs.825 per ha. for an application of 40 kg. nitrogen per ha.; this profit would continue to soar with every additional 40 kg. of fertilizer use, until it starts sending a negative return with an application of 120 kg. nitrogen per ha. Experiments evidently revealed that the farmers, cultivating traditional varieties would be making on an average Rs.600–Rs.700 less per ha.[60] The nature of the experiments indicate that the efforts of the agricultural economists were primarily motivated by the idea of bringing numerical and quantitative insight into the agricultural questions to overcome 'romantic and impractical ideas' of existing farm practices.[61] It was thus, necessary to discard the ways of subsistence agriculture for the promises of high yield and possibility of high economic return; it was here that lay the deliverance of the country and its people from hunger and poverty.

Instances of unprecedented yield achieved by any Indian farmer were widely reported by the print media throughout this period. Scientists, such as M.S. Swaminathan, made repeated references to these accomplishments to justify what the new technology could do to transform the productivity of Indian farms. To win the war against hunger, the experts reminded that farmers needed to be the heroes in adopting the improved varieties and by producing more in their fields. Such a farming hero was Kanwar Mohinderpal Singh. A farmer from the Delhi region, Singh became a familiar name to many. His claim to fame rested in the fact that he harvested nearly 8.40 tons of crops per ha., which was the highest yield so far recorded anywhere in the world in 1967 for a crop of 150-days duration. Mohinderpal Singh came to represent all those 'progressive farmers [who] adopt[ed] the high-yielding varieties for improving their financial conditions'.[62] The vision of a hunger-free India came to revolve around these progressive farmers who, the development experts believed, in discarding traditional ways of farming showed the way to overcome backwardness.

Equating progressivism with the adoption of the new technology, however, amounted to reducing the whole reality of a person's life to a single feature or trait.[63] Analysing local situation was not part of the discourse on hunger or in understanding what kept some farmers away from adopting the dwarf seeds. Bent on providing a professional or experts' view on the subject, rather than dwelling on the peasants' perspective, Escobar observed in his critique of development, that 'one never finds in these accounts consideration given to peasants' struggle and oppression, nor accounts of how the peasants world may contain a different way of seeing problems and life'.[64] 'Risk-averse' farmers were as much a problem as were the 'malnourished' and often the two categories coincided. The only solution was to fix them through 'effective' development.

The agenda of effective development involved stereotyping farmers into risk-taking and risk-averse categories. It involved fragmenting their experiences and labelling them based on the experts' understanding whether farmers were willing to take 'risks' or at least approach the question of investing in the new technology with an extent of 'risk neutrality'. A 'risk-taking' farmer was good for capitalist development in agriculture, because such a farmer would try to maximize average or expected net returns, resulting, according to the agricultural economists, in the highest returns over the long run. In contrast, a 'risk-averse' farmer would not be keen to invest if there were some possibilities of loss; he would rather be willing to forego some expected returns, if that meant reducing the variability of his income stream. As instilling higher production through investment in the new technology was slowly gaining ground among professional circles of agricultural scientists and economists, the image of a 'risk-taking' farmer gained significance. This essentialism of the farmers' approach to the use of new technology to define their identity helps explain the importance that scientists accorded to farmers like Mahinderpal. He, and many others like him, who decided to invest in the dwarf seeds and chemical fertilizers to receive high yield became representative examples of risk-taking farmers.[65]

Much of the labelling, therefore, depended on the farmer's response to the new technology. The strategies of agricultural growth that the experts recommended did not take into account the interests, needs, concerns, as well as dreams of the farmers, who chose not to or were unable to adopt the new technological package. Economists, such as Sudhir Sen, who saw the promises of a 'richer' harvest as opening up new horizons for developing countries, urged the farmers to take risks, to dispense with the much longer process of initial trials and experiments under local conditions and even to run the risks of diseases to which the new varieties might have been susceptible. He considered it 'perfectly sensible' to 'trade-off traditional safety' for the time India and, especially, its farmers could salvage in the 'fateful race between population growth and economic development'.[66]

No matter how good the economic prospect of the new technology might have been, the possibility of crop failure was very serious in the early days of Green Revolution in India. Moreover, inadequate measures of crop insurance accentuated the economic cost of crop loss. This threat was not always obvious to agricultural scientists fascinated by the yield capacity of the new seeds over the traditional cultivars. 'Below the white mountains was a stretch of terraced fields covered with dark green and pale green strips of wheat. The dark green wheat had been fertilized with calcium ammonium nitrate and super-phosphate and gave promise of a bumper crop,' wrote M.S. Randhawa, the Director General of the country's Intensive Agriculture Area Programme. An impressed Randhawa surmised that nitrogen was changing the landscape of the Punjab Himalaya region and found this change very gratifying.[67] The fields sown with the new rice and wheat varieties uniquely displayed to many, just like Randhawa, 'the arrival of modernity'.[68] The dark-shoot, stubby-looking plants, aside the thin pale green strips, were a visible metaphoric representation of modernity and tradition. What remained a less appealing feature and perhaps, therefore, not mentioned in the writings of Randhawa was that the new varieties with dark green leaves produced an ecological condition in the fields that was more favourable to diseases and pests, requiring cultivators who decide to grow these varieties to always have the resources to counter such possibilities of crop loss.

It became a matter of great concern when cultivators discovered to their dismay that a large number of dwarf wheat varieties, such as Kalyansona, PV 18 and Sonalika, though highly resistant to rusts under field conditions at the time of their release, soon became susceptible to a new virulence of the rust pathogen. With little or no infrastructure to fight widespread pest or disease incidence, many were alarmed that such attacks carried the risk of spiralling out of control and causing widespread crop loss. A deeply anguished A.H. Moseman, the assistant administrator for the Technical Cooperation and Research, observed that although important people in the Indian administration might think that availability of the Taiwan rice and Mexican wheat varieties had largely resolved India's food problems, the country was really facing a potentially 'unhealthy situation' of practising large-scale 'one variety' agriculture. Moseman wanted still wider dissemination of the dwarf varieties, but what 'seriously concerned' him was India's lack of technological capability to meet the challenges of a series of destructive new diseases that might attack the crops.[69]

In case of cultivating the dwarf varieties, a farmer's susceptibility to losing crop in pest attacks was, therefore, undeniably more, especially if he lacked the wherewithal to purchase chemical pesticides on which the crop protection measures of the new seeds was primarily dependent. Being largely imported, most farmers found chemical fertilizers to be an expensive input as was equipment for aerial spraying.[70] Moreover, non-availability of dusting and spraying equipment also discouraged the uses of pesticides. Although the development block officials made frequent arrangements to keep some equipment in their offices for farmers' use, the frequent complaint was that their number had been inadequate, their breakdown rate was high and that the users did not return them in time.[71] Farmers' use of pesticide, therefore, remained far from satisfactory in India. By March 1974, the level of pesticide consumption was only 45,000 tons.[72] In distinguishing the entrepreneurs from others who displayed a lack of confidence towards the 'seeds of plenty', the experts evidently discounted material factors, including infrastructural shortcomings that guided farmers' decision towards adoption of technology.

The disassociation of the 'experts' from the world of farmers who could not adopt most of the components of a capital- and chemical-intensive agricultural model was, therefore, a consistent feature in the heyday of Green Revolution technology in India. As the experts persuasively argued in favour of a technical solution of the food problem, rather than exploring its social and economic dimensions, the juggernaut of the Green Revolution technological package could roll over—ignoring the arguments of its critics, needs of small farmers, and any possible advantages of a research model that considered social and the material aspects of technology to be important. In reducing hunger as primarily a problem of low yield, the advocates of the capital-intensive agricultural model could set the direction of agricultural modernization in India along the technocratic lines, away from its broader social and political underpinning.

Notes and References

1. The term 'hunger' is synonymous with 'undernutrition'. Undernutrition means an insufficient per caput calorie intake due to inadequate quantity of food available to the individual. Malnutrition refers to the quality of the diet. While

the quantity may be sufficient, it may be unbalanced, that is, it may be composed heavily of cereals and starchy foods, such as potatoes and cassava, with too few of the protective foods, such as meat, fruit, and vegetables. Food and Agricultural Organization, 'Six Billions to Feed', World Food Problems, no. 4, RomeFeed, United Nations, 1962.
2. *Report on Bengal*, Famine Enquiry Commission, Government of India, New Delhi, 1945. Quoted from P.V. Sukhatme, 'The Food and Nutrition Situation in India', in *Indian Journal of Agricultural Economic (1940-1964): Selected Readings*, Bombay, 1965, p. 320.
3. The planners were especially worried at the experts' recommendation of 3,000 calories as the 'minimum' requirement because from the reports of the Nutrition Advisory Committee, they concluded that Indians were consuming on an average 2,200 calories or less. Second Five-Year Plan, Chapter 13, http://planningcommission.nic.in/plans/planrel/fiveyr/index9.html (accessed on 2 April 2012). V.G. Panse of the Indian Agricultural Statistics Research Institute, however, calculated that India had on the average a calorie supply of some 1,950 to 2,000 calories per person per day during 1956–9. The supply was 300 calories short of the requirement. Sukhatme, 'The Food and Nutrition Situation in India', pp. 326, 339.
4. Dennis Merrill, *Bread and Ballot: The United States and India's Economic Development, 1947-1963*, Chapel Hill, 1990, p. 61.
5. James Vernon, *Hunger: A Modern History*, Cambridge/London, 2007, p. 2.
6. Ibid., p. 3.
7. Ibid., p. 4.
8. Sukhatme, 'The Food and Nutrition Situation in India', p. 339.
9. First Five-Year Plan, *Food Policy*, http://planningcommission.nic.in/plans/planrel/fiveyr/index9.html (accessed on 8 November 2014).
10. Ashutosh Varshney, *Democracy, Development, and the Countryside*, Cambridge, 1998, p. 49.
11. Varshney had identified three constitutive elements in Nehru's institutional strategy: land reforms, farm and service cooperatives, and local self-government at the village level. See Varshney, *Democracy, Development, and the Countryside*, p. 34.
12. Francine Frankel, *India's Political Economy: 1947-2004*, Oxford University Press, 2008, p. 175; Varshney, *Democracy, Development, and the Countryside*, pp. 29–35.
13. Varshney, *Democracy, Development, and the Countryside*, p. 30.
14. Frankel, *India's Political Economy*, pp. 201–92; Varshney, *Democracy, Development, and the Countryside*, pp. 48–80.
15. Nick Cullather's *The Hungry World: America's Cold War Battle Against Poverty in Asia*, Cambridge, Massachusetts, 2010, provides the US government's perspective to the need of building a hunger-free world as part of the Cold War strategy and includes India in the narrative as part of the global scope of the War; similarly, John Vernon's work on the history of hunger as part of the colonial history of

Great Britain only occasionally mentions India as part of Britain's test case of fighting malnutrition.
16. Quoted in Andrew Pearse, *Seeds of Plenty, Seeds of Want: Social and Economic Implications of the Green Revolution*, New York, 1980, p. 79.
17. S.K. Ray, 'Land Reforms in Post-Independent India', in *Indian Agriculture in the Changing Environment*, vol. 2, 2002, pp. 194–5.
18. For a discussion on the economic logic of the land reform, see Ronald Herring, *Land to the Tiller: The Political Economy of Agrarian Reform in South Asia*, New Haven, 1983, Chapter 9.
19. Frankel, *India's Political Economy*, p. 142.
20. Resolutions on Economic Policy, Programme and Allied Matters (1924–69), Indian National Congress, New Delhi, 1969, p. 105.
21. Cullather, *The Hungry World*, pp. 145–6.
22. Ibid., p. 90.
23. The Foundation derived the figure of 110 million tons as follows: Consumption requirements for cereals and pulses—88 million tons; seed, feed and wastage—12.60 million tons and stock requirements and safety margins—9.40 million tons. Agricultural Production Team (sponsored by the Ford Foundation), *Report on India's Food Crisis and Steps to Meet It*, Government of India, 1959, p. 12.
24. For a list of publications during this period on food and population discourses, see Sterling Wortman and Ralph W. Cummings Jr., *To Feed This World: The Challenge and the Strategy*, Baltimore, 1978, pp. 85–7.
25. John H. Perkins, *Geopolitics and the Green Revolution: Wheat, Genes, and the Cold War*, New York and Oxford, 1997, pp. 181–2.
26. The titles of the papers presented at the symposium make clear how the experts were looking for a technological solution to the food and population conflict. N.S. Scrimshaw presented a paper titled 'Application of Nutritional and Food Science to meeting World Food Needs'; Paul C. Mangelsdorf's paper was 'Genetic Potentials for Increasing Yields of Food Crops and Animals'; E.C. Stakman wrote on how increased production could be achieved through pest, pathogen and weed control; Frank W. Parker and Lewis B. Nelson argued a clear correlation between fertilizer use and food production in their paper, 'More Fertilizers for More Food'.
27. Agricultural Production Team, *Report on India's Food Crisis*, pp. 6–7.
28. Ibid., p. 6.
29. Joan P. Mencher, *Agriculture and Social Structure in Tamil Nadu: Past Origins, Present Transformations and Future Prospects*, Durham, 1978, p. 240.
30. Agricultural Production Team, *Report on India's Food Crisis*, p. 19.
31. Third Five-Year Plan, Planning Commission, Government of India, Chapter 19.
32. In the Ludhiana district of the state of Punjab, the application of nitrogenous and phosphatic fertilizers went up by 36 and 79 per cent respectively in the first two years of IADP.
33. Debate in the Lower House, 11 December 1963, Chapter 2. Quoted from Varshney, *Democracy, Development, and the Countryside*, p. 43.

34. Jawaharlal Nehru, *Development is Defence*, February 1963.
35. *The Times of India*, 23 August 1964, p. 4.
36. As the Communist Party of India was the main initiator of the no-confidence motion, the Swatantra Party, a rightist group was unwilling to be a part of it.
37. Frankel, *India's Political Economy*, p. 249.
38. Ibid., p. 257.
39. *Agricultural Development: Problems and Perspectives*, Ministry of Food and Agriculture, Department of Agriculture, April 1965, Appendix I.
40. C. Subramaniam, *Hands of Destiny: The Green Revolution*, New Delhi, 1995, pp. 115–16.
41. Ibid., p. 132.
42. For a detailed history of the Mexican Agricultural Programme and the role of the Rockefeller Foundation and American science, see Joseph Cotter, *Troubled Harvest: Agronomy and Revolution in Mexico, 1880–2002*, London, 2003.
43. Elvin Charles Stakman, Richard Bradfield and Paul Christoph Mangelsdorf, *Campaigns Against Hunger*, Cambridge, 1967, Chapter 5.
44. Perkins, *Geopolitics and the Green Revolution*, p. 241.
45. 'Imports of Foodgrains', Lok Sabha Debates, Indian Parliament, New Delhi, 12 November 1965, pp. 1725–8.
46. Lal Bahadur Shastri, *Produce More Food and Preserve our Freedom*, Broadcast over AIR, Delhi, 10 October 1965, published in *Indian Farming*, vol. 15, no. 7, pp. 3–4.
47. Ibid., p. 4.
48. C. Subramaniam, 'Message', *Indian Farming*, vol. 15, no. 7, 1965, p. 2.
49. C. Subramaniam, '60 Years of Agricultural Research', *Indian Farming*, vol. 15, April 1965, pp. 2–7.
50. M.S. Swaminathan, 'Plant Breeding Opens New Vistas in Crop Production', *Indian Farming*, vol. 15, issue 1, 8–10 April 1965.
51. *The Times of India*, 22 July 1965, p. 8.
52. *The Times of India*, 19 August 1964, p. 8 and 3 September 1964, p. 5.
53. *Approach to Agricultural Development in the Fourth Five Year Plan*, Ministry of Food and Agriculture, Government of India, 1964, pp. 15–18.
54. M.S. Swaminathan, *Indira Gandhi and Freedom from Hunger*, McDougall Memorial Lecture of the United Nation on 'Eradication of Hunger', Food and Agricultural Organization, November 1981.
55. From a speech at a conference of state chief ministers and agriculture ministers, New Delhi, 9 April 1966. *The Years of Challenge: Selected Speeches of Indira Gandhi, January 1966-August 1969*, Director, Publications Division, Ministry of Information and Broadcasting, Government of India, New Delhi, 1971.
56. Though not elaborated in so many words, Borlaug's insistence on portraying the technological breakthrough as the 'real' revolution can be read as implying its greater revolutionary potential over the political revolutions that the communists groups had been working for. William Gaud was more explicit about it when he coined the term 'Green Revolution' as opposed to the ongoing 'Red Revolution'.

Govindan Parayil, 'The Green Revolution in India: A Case Study of Technological Change', *Technology and Culture*, vol. 33, no. 4, October 1992, p. 737.
57. Borlaug to Mr J.A. Pelissier, 26 July 1965, *Norman Borlaug Papers,* Iowa State University, Special Collections, Box 5, Folder 20.
58. Arturo Escobar, *Encountering Development: The Making and Unmaking of the Third World*, Princeton, 1985, p. 160.
59. Deborah Fitzgerald, 'Accounting for Change: Farmers and the Modernising State', in *The Countryside in the Age of the Modern State*, ed. Catherine McNicol Stock and Robert D. Johnston, Ithaca, 2001, p. 192.
60. Bill C. Wright and R.B.L. Bharadwaj, 'Fertiliser Needs of Wheat', *Indian Farming*, March 1969, pp. 15–20.
61. Fitzgerald, 'Accounting for Change', p. 194.
62. K.K. Das and D.R. Sarkar, 'Attitude of Native Farmers toward "Taichung Native I", A High-Yielding Variety of Rice', *Indian Journal of Agricultural Science*, vol. 40, no. 1, January 1970, pp. 63–4.
63. Escobar, *Encountering Development*, p. 110.
64. Ibid., p. 111.
65. Agricultural economists in a study conducted towards the later years of 1970s, however, strongly argued that to encourage greater investment in agriculture there should be an effective crop insurance system, which would enable the farmers to shift the risks to the insurance system as a whole rather keep it on their own shoulder. Hans P. Binswanger, 'Risk Attitudes of Rural Households in Semi-Arid Tropical India', *Economic and Political Weekly*, vol. 13, no. 25, 24 June 1978, pp. A49–A62.
66. Sudhir Sen, *A Richer Harvest: New Horizons for Developing Countries*, New Delhi, 1974, p. 72.
67. M.S. Randhawa, 'The Miracle of Nitrogen', *Indian Farming*, June 1965.
68. Cullather, *The Hungry World*, p. 159.
69. A.H. Moseman to Norman Borlaug, 29 September 1966, *Norman Borlaug Papers,* Iowa State University, Special Collections, Box 5/34.
70. In 1970/2, only 2.73 million acres of cultivated land was covered by aerial spraying, a quarter of it by the state-owned Agro-Aviation Corporation. *Organisation, Evaluation Study of the High-Yielding Varieties Programme, 1968-69*, Planning Commission, Programme Evaluation, Government of India, Delhi, 1970.
71. Agricultural Economics Research Centre, *A Study of the High-Yielding Varieties Programme*, Rabi 1968–9, Waltair, December 1969.
72. In a sample in West Godavari, Andhra Pradesh, only 30 per cent of the total stock of fungicides and insecticides was sold, and in another sample from Tikamgarh, Madhya Pradesh, the corresponding figure was 20 per cent. Quoted from Biplab Dasgupta, *Agrarian Change and the New Technology in India*, Geneva, 1977, p. 67.

Select Bibliography

Contemporary Tracts and Publications

Adam, William, *Third Report on the State of Education in Bengal*, Calcutta, 1838.
Allen, B.C., *Assam District Gazetteer*, vol. X, 1906.
Allen, D.B., 'Demonstration Farms', *Indian Agricultural Gazette*, 31 July 1885.
Allen, W.J., *Report on the Administration of the Cossyah and Jynteah Hill Territory*, Calcutta, 1858.
Anonymous, *Undeveloped Wealth in India: The Ways to Prevent Famines and Advance the Material Progress of India*, London, 1875.
Anstey, Vera, *The Economic Development of India*, London, 1929.
Bagchi, Kanangopal, *The Ganges Delta*, Calcutta, 1944.
Baldrey, F.S.H., *The Indigenous Breeds of Cattle in Rajputana*, Simla, 1911.
Beveridge, H., *The District of Bakarganj: Its History and Statistics*, London, 1876.
Brayne, Frank, *Village Uplift in India*, Allahabad, 1927.
Campbell, A.C., *Glimpses of Bengal: A Comprehensive, Archaeological, Biographical and Pictorial History of Bengal, Behar and Orissa*, Calcutta, 1907.
Clouston, D., *Lessons on Indian Agriculture*, London, 1920.
———, 'The Development of Agriculture in India', *Agricultural Journal of India*, vol. xix, 1929, pp. 164–5.
Colebrooke, Henry Thomas, *Remarks on the Husbandry and Internal Commerce of Bengal*, Calcutta, 1804; repr., Calcutta, 1884.
Darling, M.L., *The Punjab Peasant in Prosperity and Debt*, London, 1925.
Das Gupta, Sugata, *A Poet and a Plan: Tagore's Experiments in Rural Reconstruction*, Calcutta, 1933.
Davidson, C.J.C., *Diary of Travels and Adventures in Upper India*, vol. 2, London, 1843.
Dutt, Rajani Palme, *India Today*, Delhi, 1940.
Dutt, Romesh Chandra, *The Economic History of India During Early British Rule*, London, 1903.
———, *The Economic History of India in the Victorian Age*, London, 1903.
Edwards, J.T., *Some Diseases of Cattle in India*, Simla, 1928.
Elmhirst, L.K., *Poet and Plowman*, Calcutta, 1975.
Ganguli, Birendra Nath, *Trends of Agriculture and Population in the Ganges Valley: A Study in Agricultural Economics*, London, 1938.

304 Select Bibliography

Gastrell, Col. J.E., *Geographical and Statistical Report of the Districts of Jessore, Fureedpore and Backergunge*, Calcutta, 1868.

Hallen, J.H.B., *Manual of the More Deadly Forms of Cattle Disease in India*, Calcutta, 1885.

Hamilton, Walter, *A Geographical, Statistical and Historical Description of Hindoostan and the Adjacent Countries*, vol. I, 1st edn, London, 1820; repr. Delhi, 1971.

Hatch, D. Spencer, *Up from Poverty in Rural India*, Bombay, 1933.

Hewlett, K., *Breeds of India Cattle: Bombay Presidency*, Calcutta, 1879.

Higginbottom, Sam, *The Gospel and the Plow or The Old Gospel and Modern Farming in Ancient India*, New York, 1929.

Hill, K.A. and P.K. Banerjee, *Final Report on the Survey and Settlement Operations in the District of Burdwan, 1927-34*, Calcutta, 1940.

Howard, Albert, 'Agriculture and Science', *Agricultural Journal of India*, vol. xxi, 1935, pp. 171–82.

———, *An Agricultural Testament*, London, 1940; repr. Goa, 1956.

Howard, A. and G.L.C. Howard, *The Application of Science to Crop Production*, London, 1929.

———, *The Development of Indian Agriculture*, vol. VIII, London, 1929.

Hunter, W.W., *A Statistical Account of Assam*, vol. II, Calcutta, 1880.

———, *A Statistical Account of Bengal, Districts of the 24 Parganas and Sundarbans*, vol. I, 1st edn, London, 1875; repr. Delhi, 1973.

Imperial Council of Agricultural Research, *Handbook of Animal Husbandry*, 3rd edn, New Delhi, 2011.

Ivanow, Wladimir, *Concise Descriptive Catalogue of the Persian Manuscripts in the Curzon Collection of the Asiatic Society of Bengal*, Calcutta, 1926.

Khan, M.N., *Post-war Agricultural Development in Bengal*, Calcutta, 1945.

Kothawala, Zal R., 'The Indian Buffalo as a Milch Animal Suitable for Tropical Countries', *Agriculture and Livestock in India*, vol. V, part I, 1935.

Lal, P.C., *Reconstruction and Education in Rural India*, London, 1932.

Landolicus, *The Indian Amateur Dairy Farm*, 1895.

Leake, H. Martin, *The Foundations of Indian Agriculture*, Cambridge, 1923.

Lees, W.N., *Tea Cultivation and other Agricultural Experiments in India*, Calcutta, 1863.

Lindsay, L., *Lives of Lindsay or Memoir of House of Crawford and Balcarres*, vol. III, London, 1849.

Long, J., *Introduction to Adam's Report*, Calcutta, 1868.

MacKenna, James, *Agriculture in India*, Calcutta, 1915.

Maconochy, G.C., *Report on the Protective Irrigation Works in Bengal*, Calcutta, 1902.

Majumdar, G.P., *Vanaspati: Plants and Plant Life as in Indian Treatises and Traditions*, Calcutta, 1927.

———, *Upavana Vinoda: A Sanskrit Treatise on Arbori and Horticulture*, Calcutta, 1935.

Mann, H.H., 'The Introduction of Improvements into Indian Agriculture', *Agricultural Journal of India*, vol. II, 1909, p. 7.

Matson, A., 'Cattle in Relation to Agriculture in India', *Journal of Central Bureau of Animal Husbandry and Dairying in India*, vol. 2, 1928.

Mills, A.J.M., *Report on the Khasi and Jaintia Hills*, Calcutta, 1853.

Mill, James, *Plain Hints on the Diseases of Cattle in India*, Bombay, 1894.

Mills, J.P., *Some Recent Contact Problems in the Khasi Hills: Essays in Anthropology, Presented to Rai Bahadur Sarat Chandra Roy*, Lucknow, n.d.

Misra, Babu Ram, *Economic Survey of a Village in Cawnpore District*, Allahabad, 1932.

Mitra, K.C., *Agriculture and Agricultural Exhibition in Bengal*, Calcutta, 1865.

Mitra, Sarat Chandra, *On Some Curious Cults of Southern and Western Bengal*, Calcutta, 1918.

Mitra, Satish, *Jassahaur Khulnar Itihas*, vol. I, Calcutta, 1914.

Mollison, James, *A Textbook in Indian Agriculture*, vol. II, Bombay, 1901.

Mollison, James and L. French, 'The Montgomery and Sind Breeds of Cattle', *The Agricultural Journal of India*, vol. II, part III, 1907, pp. 252–6.

Mukerjee, Radhakamal, 'Introduction', in *The Pressure of Population: Its Effects on the Rural Economy in Gorakhpur District*, Jai Krishna Mathur, Allahabad, 1931.

Oliver, E.W., 'Cattle Breeding, With Special Reference to Milch Cow', *The Agricultural Journal of India*, vol. XI, part II, 1916, pp. 168–73.

———, 'Animal Husbandry in India', *Journal of the Royal Society of the Arts*, vol. 90, no. 4614, 29 May 1942, pp. 433–51.

Olver, Arthur, 'The Necessity for Authoritative Definition of Breed Characteristics and Unchanging Control of Breeding Policy in India', *Agriculture and Livestock in India*, vol. I, part I, 1931, pp. 19–25.

———, 'Livestock Improvement in India', *Agriculture and Livestock in India*, vol. VII, part IV, 1937.

O'Malley, L.S.S., *Bengal District Gazetteers, Khulna*, Calcutta, 1908.

———, *Bengal District Gazetteers, 24 Parganas*, Calcutta, 1914.

———, *Bengal District Gazetteers, Rajshahi*, 1916.

———, *Bengal District Gazetteers, Birbhum*; repr. Calcutta, 1996.

Parakunnel, J. Thomas, 'India in the World Depression', *The Economic Journal*, vol. 45, no. 179, September 1935, pp. 469–83.

Peterson, J.C.K., *Bengal District Gazetteers: Burdwan*, 1st Indian repr., New Delhi, 1985.

Piddington, H., *On the Scientific Principles of Agriculture*, Calcutta, 1839.

Playne, Somerset, *Bengal and Assam, Bihar and Orissa: Their History, People, Commerce and Industrial Resources*, London, 1917.

Raju, A. Sarada, *Economic Conditions in Madras*, Madras, 1941.

Rao, C.S.R., ed., *Notable Speeches of Lord Curzon*, Madras, 1905.

Schrottky, E.C., *The Principles of Rational Agriculture Applied to India and its Staple Products*, Bombay, 1876.

Selections from the Records of the Bengal Government Relating to the Damodar Canal Project from May 1867 to January 1870 and July 1903 to July 1925, Calcutta, 1927.

Sen, A.C., *Report on the Agricultural Experiments and Enquiries in Burdwan Division*, Calcutta, 1897.
Seth, Santosh Nath, *Bange Chaltattwa* (All about Rice in Bengal), Chandennagar, 1925.
Sikka, Lal Chand, 'Standardisation of Lactation Period Milk Records', *The Indian Journal of Veterinary Science and Animal Husbandry*, vol. I, part 2, 1931, pp. 63–98.
Smyth, Ralph, *Statistical and Geographical Report of the 24 Pergunnahs District*, Calcutta, 1857.
Tagore, Rabindranath, *On the Edges of Time*, Calcutta, 1958.
———, 'City and Village', *Towards Universal Man*, Calcutta, 1961.
———, *Pitrismriti*, Calcutta, 1968.
———, *Letters from Russia*, Calcutta, 1984.
The Tea Cyclopaedia, Calcutta, 1881.
Tweed, Isa, *Cow-keeping in India: A Simple and Practical Book on their Care and Treatment, their Various Breeds, and the Means of Rendering them Profitable*, Calcutta, 1900.
Vaugh, Mason, *A Review of the Work done by the Agricultural Engineers in India*, Allahabad, 1941.
Voelcker, J.A., *Report on the Improvement of Indian Agriculture*, London, 1893.
———, *Report on the Improvement of Indian Agriculture*, 2nd edn, Calcutta, 1897.
Voorduin, W.L., *Preliminary Memorandum on the Unified Development of the Damodar River*, Calcutta, 1945.
Wallace, R.H., 'Agricultural Education in Greater Britain', *Journal of the Society of Arts*, vol. XLVIII, no. 2468, 9 March 1900.
Watt, George, *Memorandum on the Resources of British India*, Calcutta, 1894.
Westland, J., *A Report on the District of Jessore: Its Antiquities, Its History, and Its Commerce*, Calcutta, 1871.
White, A., *Memoir of David Scott*, London, 1832.
Willcocks, William, *Lectures on the Ancient System of Irrigation in Bengal and its Application to Modern Problems*, Calcutta, 1930.
Wissett, R., *On the Cultivation and Preparation of Hemp*, London, 1804.
Wright, Norman C., *Report on the Development of the Cattle and Dairy Industries of India*, Simla, 1937.

Secondary Sources

Articles

Alavi, Hamza, 'India and the Colonial Mode of Production', *Economic and Political Weekly*, vol. X, August 1975, pp. 1235–62.
———, 'Colonial State and Agrarian Society', in *Situating Indian History*, ed. Sabyasachi Bhattacharya and Romila Thapar, Delhi, 1986.

Select Bibliography 307

Binswanger, Hans P., 'Risk Attitudes of Rural Households in Semi-arid Tropical India', *Economic and Political Weekly*, vol. 13, no. 25, 24 June 1978, pp. A49–A62.

Biswas, Arabinda and Swapan Bardhan, 'Agrarian Crisis in Damodar-Bhagirathi Region 1850-1925', *Geographical Review in India*, vol. 37, no. 2, 1975, pp. 132–50.

Chandra, Bharat, 'Satyapirer Katha', in *Bharat Chandra Rachanabali*, Calcutta, 1963.

Charlesworth, Neil, 'The Peasant and the Depression: The Case of the Bombay Presidency, India', in *The Economies of Africa and Asia in the Inter-war Depression*, ed. Ian Brown, London, 1989, pp. 59–73.

Chaudhuri, B.B., 'Some Problems of the Peasantry of Bengal Before the Permanent Settlement', *Bengal Past and Present*, July–December 1956.

———, 'Some Problems of the Peasantry of Bengal After the Permanent Settlement', *Bengal Past and Present*, Jubilee Number, 1957.

———, 'A Chapter of Peasant Resistance in Bengal After the Permanent Settlement', *Enquiry*, vol. 3, New Delhi, 1960.

———, 'Agrarian Economy and Agrarian Relations in Bengal: 1859-1885', in *The History of Bengal: 1757-1905*, ed. N.K. Sinha, Calcutta, 1967.

———, 'Growth of Commercial Agriculture in Bengal, 1859-1885', *Indian Economic and Social History Review*, vol. 7, no. 2, 1970, pp. 211–51.

———, 'Peasant Movements in Bengal 1850-1900', *Nineteenth Century Studies*, vol. 3, July 1973.

———, 'Agriculture: Eastern India, 1757-1857', *Cambridge Economic History of India*, vol. 2, Cambridge, 1983.

———, 'Rural Power Structure and Agricultural Productivity', in *Agrarian Power and Agricultural Productivity in South Asia*, ed. Meghnad Desai, Susane Howber Rudolph and Ashok Rudra, Delhi, 1984.

Chaudhuri, Dulal, 'Folk Religion', in *A Focus on Sundarban*, ed. Amal Kumar Das, Sankaranda Mukherji and Manas Kamal Chowdhuri, Calcutta, 1981.

Das, K.K. and D.R. Sarkar, 'Attitude of Native Farmers toward "Taichung Native I", A High-yielding Variety of Rice', *Indian Journal of Agricultural Science*, vol. 40, no. 1, January 1970.

Dutta, Achintya Kumar, 'Rice Trade in the "Rice Bowl of Bengal": Burdwan 1880-1947', *Indian Economic and Social History Review*, vol. 49, no. 1, 2012, pp. 73–104.

Fitzgerald, Deborah, 'Accounting for Change: Farmers and the Modernising State', in *The Countryside in the Age of the Modern State*, ed. Catherine McNicol Stock and Robert D. Johnston, London, 2001.

Gough, Kathleen, 'Indian Peasant Uprising', *Economic and Political Weekly*, vol. 9, Special Number, August 1974, pp. 32–4.

Guha, Ranajit, 'Neel Darpan: The Image of a Peasant Revolt in a Liberal Mirror', *Journal of Peasant Studies*, vol. 2, no. 1, 1974, pp. 1–46.

Habib, Irfan, 'Colonization of the Indian Economy, 1757-1900', *Social Scientist*, vol. 3, no. 8, March 1975, pp. 23–53.

Select Bibliography

Harding, Christopher, 'The Christian Village Experiment in Punjab: School and Religious Reformation', *South Asia: Journal of South Asian Studies*, n.s., vol. XXXI, no. 3, December 2008.

Henry, R.J., 'Technology Transfer and Its Constraints: Early Writings from Agricultural Development in Colonial India', in *Technology and the Raj: Western Technology and Technical Transfer to India 1700-1947*, ed. Roy Macleod and Deepak Kumar, New Delhi, 1995, pp. 51–77.

Houben, Jan E.M., 'To Kill or Not to Kill the Sacrificial Animal (Yajna-Pasu): Arguments and Perspectives in Brahminical Ethical Philosophy', in *Violence Denied: Violence, Non-Violence and the Rationalization of Violence in South Asian Cultural History*, ed. Jan E.M. Houben and Karel R. Van Kooji, Leiden, 1999, pp. 117–24.

Inskter, Ian, 'Prometheus Bound: Technology and Industrialization in Japan, China and India prior to 1914: A Political Economy Approach', *Annals of Science*, vol. 45, 1988, pp. 399–426.

Kumar, Deepak, 'Patterns of Colonial Science in India', *Indian Journal of History of Science*, vol. 15, no. 1, 1980, pp. 105–13.

Kumar, Prakash, 'Scientific Experiments in British India: Scientists, Indigo Planters and the State, 1890-1930', *Indian Economic and Social History Review*, vol. 38, 2001, pp. 249–70.

Kurosaki, T., 'Agriculture in India and Pakistan, 1900-95: Productivity and Crop Mix', *Economic and Political Weekly*, vol. 34, no. 52, 1999, pp. A160–A168.

Mishra, Saurabh, 'Beasts, Murrains and the British Raj: Reassessing Colonial Medicine in India from the Veterinary Perspective, 1860-1900', *Bulletin of History of Medicine*, vol. 85, no. 4, Winter 2011, pp. 587–619.

———, 'Cattle, Dearth, and the Colonial State: Famines and Livestock in Colonial India, 1896-1900', *Journal of Social History*, vol. 46, no. 4, Summer 2013, pp. 989–1012.

Mohapatra, Prabhu Prasad, 'Class Conflict and Agrarian Regimes in Chota Nagpur, 1860–1950', *Indian Economic and Social History Review*, vol. 28, no. 1, 1991.

Mukherjee, Aditya, 'Agrarian Conditions in Assam, 1880-90, A Case Study of Five Districts of the Brahmaputra Valley', *Indian Economic and Social History Review*, April–June 1999.

Mukherjee, Mridula, 'Commercialization and Agrarian Change in Pre- Independence Punjab', in *Essays on the Commercialization of Indian Agriculture*, ed. K.N. Raj, Neeladri Bhattacharya, Sumit Guha and Sakti Padhi, Delhi, 1985.

Mukherjee, Saugata, 'Some Aspects of Commercialization of Agriculture in Eastern India 1891–1938', in *Perspectives in Social Sciences, vol. 2, Three Studies on its Agrarian Structure in Bengal Before Independence, 1850-1947*, ed. A. Sen, P. Chatterjee and S. Mukherjee, Calcutta, 1981.

Nelson, Lance, 'Cows, Elephants, Dogs, and Other Lesser Embodiments of Atman: Reflections on Hindu Attitudes Toward Nonhuman Animals', in *A Communion of Subjects: Animals in Religion, Science, and Ethics*, ed. Paul Wadu and Kimberly Patton, New York, 2006, pp. 179–93.

Laldinpui, Audrey and Laithangpuii, 'Rice Economy and Gender Concerns in Mizoram', *North East India Studies*, vol. 1, no. 2, January 2006, pp. 107–27.
Mody, A., 'Population Growth and Commercialization of Agriculture: India, 1890–1940', *Indian Economic and Social History Review*, vol. 19, nos. 3–4, 1982, pp. 237–66.
Parayil, Govindan, 'The Green Revolution in India: A Case Study of Technological Change', *Technology and Culture,* vol. 33, no. 4, October 1992.
Pray, C.E., 'The Impact of Agricultural Research in British India', *The Journal of Economic History*, vol. 44, no. 2, 1984, pp. 429–40.
Raha, Bipasha, 'Economic Thought of Rabindranath Tagore', *The Calcutta Historical Journal*, vol. XXV, no. 1, January–June 2005.
Richards, John F. and Elizabeth P. Flint, 'Long-term Transformations in the Sundarbans Wetlands Forests of Bengal', *Agriculture and Human Values*, vol. VII, no. 2, Spring 1990.
Roy, Aniruddha, 'Case Study of a Revolt in Medieval Bengal: Raja Pratapditya Guha Roy of Jessore', in *Essays in Honour of Professor S.C. Sarkar*, ed. Barun De et al., New Delhi, 1976.
Samanta, Arabinda, 'Plague and Prophylactics: Ecological Construction of an Epidemic in Colonial Eastern India', in *Situating Environmental History*, ed. Ranjan Chakrabarti, New Delhi, 2007, pp. 221–42.
Samanta, Samiparna, 'Dealing with Disease: Epizootics, Veterinarians, and Public Health in Colonial Bengal, 1850-1920', in *Medicine and Colonialism: Historical Perspectives in India and South Africa*, ed. Poonam Bala, London, 2014.
Sangwan, Satpal, 'Level of Agricultural Technology in India (1757-1857)', *Asian Agri-History*, vol. 11, no. 1, 2007, pp. 5–25.
Sen Gupta, Kalyan Kumar, 'Bengali Intelligentsia and the Politics of Rent, 1873-1885', *Social Scientist*, vol. 3, no. 2, September 1974, pp. 27–34.
Shukla, P.K., *Indigo and the Raj: Peasant Protests in Bihar, 1780–1980*, Delhi, 1993.
Singh, U.N., *Some Aspects of Rural Life in Bihar: An Economic Study, 1793–1833*, Patna, 1980.
Sihna, C.P.N., *From Decline to Destruction: Agriculture in Bihar During the Early British Rule, 1765–1813*, New Delhi, 1997.
Sihna-Kerkhoff, Kathinka, *Colonising Plants in Bihar (1760–1950)*, New Delhi, 2014.
Subramaniam, C., 'Message', *Indian Farming,* vol. 15, no. 7, 1965.
Sukhatme, P.V., 'The Food and Nutrition Situation in India', *Indian Journal of Agricultural Economics (1940-1964): Selected Readings*, Bombay, 1965.
Tuteja, K.L., 'Agricultural Technology in Gujarat: A Study of Exotic Seeds and Saw Gins, 1800-1850', *Indian Historical Review*, July 1990 and January 1991, pp. 136–51.
Washbrook, David, 'The Commercialization of Agriculture in Colonial India: Production, Subsistence and Reproduction in the "Dry South", *c.* 1870-1930', *Modern Asian Studies*, vol. 28, no. 1, 1994, pp. 129–64.

Zito, Angela, 'Secularizing the Pain of Foot Binding in China: Missionary and Medical Stagings of the Universal Body', *Journal of the American Academy of Religion*, vol. 75, no. 1, 2007, pp. 1–24.

Books

Ali, Imran, *The Punjab Under Imperialism, 1885-1947*, Princeton, 1988, Delhi and Bombay, 1989.
Ambirajan, S., 'Small Peasant Commodity Production and Rural Indebtedness: The Culture of Sugarcane in Eastern U.P., c. 1880-1920,' in *Subaltern Studies I: Writings on South Asian History and Society*, ed. Ranajit Guha, Delhi, 1984.
Amin, Shahid, *Sugarcane and Sugar in Gorakhpur: An Enquiry into Peasant Production for Capitalist Enterprise in Colonial India*, Delhi, 1984.
Alan, Sir Pim, *Colonial Agricultural Production: The Contribution Made By the Native Peasant and the Foreign Enterprise*, London, 1946.
Andrews, Birdie and Andrew Cunningham, eds., *Western Medicine as Contested Knowledge*, Manchester, 1997.
Arnold, David, *Colonizing the Body: State Medicine and Epidemic Disease in Nineteenth-century India*, London, 1993.
Baber, Zaheer, *The Science of Empire: Scientific Knowledge, Civilization and Colonial Rule in India*, New York, 1996.
Blyn, George, *Agricultural Trends in India 1891-1947: Output Availability, and Productivity*, Philadelphia, P.A., 1966.
Bandyopadhyay, Arun, *The Agrarian Economy of Tamilnadu, 1820–55*, Calcutta, 1992.
Banerjee, Himadri, *Agrarian Society of the Punjab, 1849–1901*, New Delhi, 1982.
Banga, Indu, *Five Punjabi Centuries: Policy, Economy, Society, and Culture, c.1500-1990*, New Delhi, 1997.
Bates, Crispin, 'Regional Dependence and Rural Development in Central India: The Pivotal Role of Migrant Labour', in *Agricultural Production and Indian History*, ed. David Ludden, Delhi, 1994.
Banerjee, S.B., *Evaluation of Damodar Canal (1959–60): A Study in the Benefits of Irrigation in the Damodar Region*, Bombay, 1965.
Bhattacharya, Ashutosh, ed., *Manasa Mangal*, Calcutta, 1954.
Bose, Sugata, *Agrarian Bengal: Economy, Social Structure and Politics, 1919–1947*, Cambridge, 1986.
———, *Peasant Labour and Colonial Capital: Rural Bengal Since 1770*, in The New Cambridge History of India, III.2, Cambridge, 1993.
Boyce, James K., *Agrarian Impasse in Bengal: Institutional Constraints to Technological Change*, New York, 1987.
Bremen, Jan, *Patronage and Exploitation: Changing Agrarian Relations in South Gujarat*, 1974; repr. New Delhi, 1979.
Brown, Karen and Daniel Gilfoyle, eds., *Healing the Herds: Disease, Livestock Economies and the Globalization of Veterinary Medicine*, Athens, 2010.

Catanach, I.J., *Rural Credit in Western India: Rural Credit and the Co-operative Movement in the Bombay Presidency, 1875–1930*, Berkeley, California, 1970.

Chandra, Bipan, *The Rise and Growth of Economic Nationalism in India: Economic Policies of Indian Nationalist Leadership, 1880–1905*, New Delhi, 1966.

Charlesworth, Neil, *Peasants and Imperial Rule: Agriculture and Agrarian Society in the Bombay Presidency, 1850–1935*, Cambridge, 1985.

Chatterjee, Partha, *Bengal, 1920–47: The Land Question*, Calcutta, 1984.

Chattopadhyay, Goutam, ed., *Bengal: Early Nineteenth Century, Selected Documents*, Calcutta, 1978.

Cotter, Joseph, *Troubled Harvest: Agronomy and Revolution in Mexico, 1880-2002*, Praeger, 2003.

Davis, Mike, *Late Victorian Holocausts: El Nino Famines and the Making of the Third World*, London, 2001.

De, Gourishankar and Shubhradip De, *A Lost Civilization*, Kolkata, 2004.

De, Rathindranath, *The Sundarbans*, Calcutta, 1990.

Dutta, Achintya Kuma, *Economy and Ecology in a Bengal District, Burdwan 1880-1947*, Kolkata, 2002.

Dutta, P.N., *Impact of the West on Khasi and Jaintias: A Survey of Political Economic and Social Change*, New Delhi, 1982.

Guha, Ranajit, *Elementary Aspects of Peasant Insurgency in Colonial India*, Delhi, 1983.

Habib, Irfan, *The Agrarian System of Mughal India 1556-1707*, 3rd edn, Delhi, 2014.

Harnetty, Peter, *Imperialism and Free Trade*, Manchester, 1972.

Harrison, Mark, *Public Health in British India: Anglo-Indian Preventive Medicine, 1859–1914*, Cambridge, 1994.

Headrick, Daniel R., *The Tools of Empire: Technology and European Imperialism in the Nineteenth Century*, New York, 1981.

Hill, A.V., *The Ethical Dilemma of Science and Other Essays*, New York, 1960.

Islam, Sirajul, *The Peasant Settlement in Bengal: A Study of its Operation, 1790–1819*, Dacca, 1979.

Jail, A.F.M. Abdul, *Sundarbaner Itihas*, 2nd edn, Dacca, 1986.

Jain, P.K., *The Indian Agricultural Service: A Study of Technical Personnel, 1906–1924*, New Delhi, 1978.

Kabiraj, Narahari, *A Peasant Uprising in Bengal*, New Delhi, 1972.

Kling, Blair B., *The Blue Mutiny: The Indigo Disturbances in Bengal 1859–1962*, Philadelphia, 1966.

Kumar, Deepak, *Science and the Raj: A Study of British India*, 2nd edn, New Delhi, 2006.

Kumar, Dharma, *Land and Caste in South India*, Cambridge, 1965.

Kumar, Dharma and Tapan Raychaudhuri, eds., *The Cambridge Economic History of India*, vol. II, Hyderabad, 1982.

Kumar, Ravinder, *Western India in the Nineteenth Century: A Study in the Social History of Maharashtra*, London, 1968.

Kumar, Prakash, *Indigo Plantations and Science in Colonial India*, Cambridge, 2012.

Lodrick, Deryck, *Sacred Cows, Sacred Places: Origins and Survivals of Animal Homes in India*, Berkeley, 1981.
Ludden, David, ed., *Agricultural Production and Indian History*, Delhi, 1994.
Macleod, Roy and Deepak Kumar, eds., *Technology and the Raj: Technical Transfers to India 1700–1947*, New Delhi, 1995.
Metcalf, Thomas R., *Land, Landlords and the British Raj: Northern India in the Nineteenth Century*, Delhi, 1979.
Mangamma, J., *Technical and Agricultural Education in Madras Presidency (1854-1921)*, Delhi, 1990.
Mandal, A.K. and R.K. Ghosh, *Sundarban, A Socio Bio-Ecological Study*, Calcutta, 1989.
Mayadas, Cedric, *Between Us and Hunger*, London, 1954.
Mukherjee, Mridula, *Colonizing Agriculture: The Myth of Punjab Exceptionalism*, New Delhi, 2005.
Mukherjee, Nilmani, *The Ryotwari Settlement in Madras, 1792–1827*, Calcutta, 1962.
———, *A Bengal Zamindar: Joykrishna Mukherjee of Uttarpara and His Times, 1808–88*, Calcutta, 1975.
Mukherjee, R.K., *The Changing Face of Bengal: A Study of Riverine Economy*, Calcutta, 1938.
Naik, K.C., *Agricultural Education in India: Institutes and Organizations*, New Delhi, 1961.
Nand, Brahma, *Fields and Farmers in Western India, 1850–1950*, Delhi, 2003.
Palit, Chittabrata, *Tensions in Bengal Rural Society: Landlords, Planters and Colonial Rule, 1830–1860*, Calcutta, 1975.
Prakash, G., *Another Reason: Science and the Imagination of Modern India*, Princeton, 1999.
Pratap, Ajay, *An Ethnology of Shifting Cultivation in Eastern India*, Delhi, 2000.
Raha, Bipasha, *The Plough and the Pen: Peasantry, Agriculture and the Literati in Colonial Bengal*, New Delhi, 2012.
———, *Living a Dream, Rabindranath Tagore and Rural Resuscitation*, New Delhi, 2014.
Randhawa, M.S., *A History of Agriculture in India*, vol. III, New Delhi, 1983.
Ratnam, R., *Agricultural Development in Madras State Prior to 1900*, Madras, 1966.
Ray, Kabita, *History of Public Health in Colonial Bengal 1921–1947*, Calcutta, 1998.
Ray, Ratnalekha, *Change in Bengal Agrarian Society, c.1760–1850*, Delhi, 1979.
Rothermund, Dietmar, *India in the Great Depression 1929–1939*, Delhi, 1992.
Sangwan, Satpal, *Science, Technology and Colonization: An Indian Experience, 1757–1857*, Delhi, 1991.
Sarkar, Smritikumar, *Technology and Rural Change in Eastern India, 1830–1980*, New Delhi, 2014.
Sarkar, Sutapa Chatterjee, *The Sundarbans: Folk Deities, Monsters and Mortals*, New Delhi, 2010.
Satya, Laxman D., *Ecology, Colonialism, and Cattle: Central India in the Nineteenth Century*, New Delhi, 2004.
Sen, Sunil, *Agrarian Struggle in Bengal, 1946–47*, New Delhi, 1972.

Sen Gupta, Kalyan Kumar, *Pabna Disturbances and the Politics of Rent, 1873–1885*, Calcutta, 1974.
Sen, Sudhir, *Rabindranath Tagore on Rural Reconstruction*, Calcutta, 1978.
Sengupta, Jayashree, *A Nation in Transition: Understanding the Indian Economy*, New Delhi, 2007.
Siddiqui, Asiya, *Agrarian Change in a Northern Indian State: Uttar Pradesh, 1819–1833*, Oxford, 1976.
Signage, Clive A., *Cattle Plague: A History*, New York, 2003.
Singh, S.B., *Second World War as Catalyst for Social Changes in India*, Delhi, 1998.
Sinha, Sasadhar, *Social Thinking of Rabindranath Tagore*, Calcutta, 1962.
Subramaniam, C., *Hands of Destiny: The Green Revolution*, New Delhi, 1995.
Thirumalai, S., *Post-war Agricultural Problems and Policies in India*, New York, 1954.
Thorner, Daniel and Alice Thorner, *Land and Labour in India*, Bombay, 1962.
Tomlinson, B.R., *The Economy of Modern India, 1860–1970*, New Delhi, 1998.
Waddington, Keir, *The Bovine Scourge: Meat, Tuberculosis and Public Health, 1850-1914*, Woodbridge, 2006.
Worboys, Michael, *Spreading Germs: Disease Theories and Medical Practice in Britain, 1865–1900*, Cambridge, 2000.

Editors and Contributors

DEEPAK KUMAR is Professor of History of Science and Education, Zakir Hussain Centre for Educational Studies, School of Social Sciences; and Concurrent Professor, Centre for Media Studies, Jawaharlal Nehru University, New Delhi, India. His published titles include *The Trishanku Nation: Memory, Self and Society in Contemporary India* (2016) and *Science and the Raj: A Study of British India* (2006).

BIPASHA RAHA is Professor and Head, Department of History, Visva-Bharati, Santiniketan, India. She is the author of *Living a Dream: Rabindranath Tagore and Rural Resuscitation* (2014) and *The Plough and the Pen: Peasantry, Agriculture and the Literati in Colonial Bengal* (2012).

SANDIPAN BAKSI is a doctoral research scholar at the Tata Institute of Social Sciences (TISS), Mumbai, India. He is trying to trace the modernization of agriculture in the United Provinces during the colonial period. His research interests include history and philosophy of science, economic history, and issues related to technological evolution and diffusion.

RAJSEKHAR BASU is Associate Professor, Department of History, University of Calcutta, Kolkata, India. He has specialized in different facets of social history and has written exclusively on the marginalized communities of south India. His publications include two monographs, *Nandanar's Children: The Paraiyans' Tryst with Destiny, Tamil Nadu 1850–1956* (2010) and *Dalit Experiences in Colonial and Post-Colonial India* (2010).

SUTAPA CHATTERJEE is Professor of History, West Bengal State University, Kolkata, India. She has published widely in reputed journals. She is the author of *The Sundarbans: Folk Deities, Monsters and Mortals* (2010).

ACHINTYA KUMAR DUTTA is Professor of History, University of Burdwan, India. His publications include *History of Medicine in India: The Medical Encounter* (2005, co-edited with Chittabrata Palit); *Economy and Ecology in a*

Bengal District: Burdwan, 1880–1947 (2002) and *Intellectual History: A New Undertaking* (1999, edited). He is currently working on the social history of the kala-azar epidemic in north-east India.

S.M. MISHRA recently retired as Deputy Director of the Indian Council of Historical Research and has helped researchers for almost four decades. He is the author of *Agriculture and Environment: Debates in the Central Legislature of India 1937–1957* (2015).

SAJAL NAG is Professor of History, Assam University, Silchar. He has been working extensively on the history of the north-east and has published a great deal on politics, marginality and ethnicity issues.

ARNAB ROY is a research scholar at the Zakir Hussain Centre for Educational Studies, Jawaharlal Nehru University, School of Social Sciences, New Delhi, India. He is working on agricultural science in colonial India and his research interests include history of science, economic history, environmental history, peasant movements and the history of the communist movement.

MADHUMITA SAHA is Assistant Teaching Professor at Drexel University, Philadelphia, United States, where she teaches history of science, technology and environment as well as courses on South Asian history. Currently, she is collaborating with scholars from Europe, North America and Asia to come up with a special volume on twentieth-century history of science in China and India in the *British Journal for the History of Science*.

SAMIPARNA SAMANTA is Assistant Professor of History, Georgia College and State University, Milledgeville, Georgia, United States. She works predominantly in the areas of history of science and medicine, socio-cultural and environmental history, colonialism, British Empire and human–animal interactions. She is currently working on a book that uses the lens of animal cruelty/protection to write a social history of Bengal from 1850 to 1920.

HIMANSHU UPADHYAY is Assistant Professor at School of Development, Azim Premji University, Bengaluru, India. His doctoral research at the Centre for Studies in Science Policy, Jawaharlal Nehru University, New Delhi, India, explored the interrelations between crops and cattle, and transformations brought about during the late colonial and postcolonial period (1890–1980) in western India.

Index

Agnihotri, Gangaprasad 77
agrarian distress 14–17, 75–6, 260–1
  all-India 261–2
  of attachment 267
  of disease and want 267–8
  disparity between the prices of primary products and finished goods 263
  in execution of decrees 268–9
  fertility of the soil, impairment of 272
  government's 'sympathetic ideas and intentions' 276–7
  importance to industries and 275–6
  insolvency laws and 269–70
  moneylenders and landlords, role of 265–6
  movement from villages to towns 274–5
  multiple land tenures and 270–2
  new technology and 272–4
  of rural indebtedness 264–5
  of taxation 263–4
Agricultural and Horticultural Society of India (AHSI) 5, 25, 27–9, 57
agricultural college 31
agricultural depression 14, 260
agricultural education 30–7, 69n44, 97n65, 115
  colleges, establishment of 31–4, 57
  framing of an agricultural education curriculum 59
  model farms 30–1
agricultural implements 4, 27, 37, 39, 47n168, 60, 83, 103, 119n33, 134, 149, 153, 161, 179, 185–6, 273
agricultural improvement 61
agricultural policy 27–30
agricultural production 71, 75–6, 78–9, 285

improvement of land 80–5
modern scientific knowledge and 55–7, 63–4
perennial deficit in 76
agricultural science and technology, in socio-economic context 86–91
agricultural technology in India, 1955–67 282
  cultivation of dwarf varieties 297–8
  debate over institutional reforms and the food crisis 285–9
  Green Revolution technology 283–5, 293–4, 298
  hybrid seeds 293–8
  Intensive Agricultural Development Programme (IADP) 288
  Nehru's ideals 288–9
  new approaches and new advances 289–93
  risk-taking and risk-averse strategies 296
  scientific revolution in farming 290
agricultural writings of Bengali literati 61–3
Agriculture Research Institute 7
ahimsa 13
Allahabad Agricultural Institute 102, 104–7, 115
The Allahabad Farmer 107
Allen, D.B. 20–1
Allen, W.J. 147–8
All-India Coordinated Wheat Improvement Project 295
Amarkosa 4
animal wealth (*pashu dhan*) 12–14
Arnold, David 219
Arthasastra 3–4
Aziz, Khan Bahadur Mian Abdul 262, 264,

274
Baber, Zaheer 56
Bajpai, G.S. 276
Bakla 198, 211n48
Baksi, Sandipan 8–9
Banabibi Jahuranama 201
Basu, Rajsekhar 9
Beckwith, J. 228
Belgachia Veterinary College 222
Bengal Agricultural Debtors Act of 1935 272
Bengali *bhadralok* 13, 39, 44n107, 179
Bengali literati 68n43, 71
   on agricultural modernization 50, 64
   agricultural writings of 61–3
Bengal Moneylenders Act of 1933 272
Bengal-Nagpur Cotton Mills Company 24
Bengal Silk Committee 37
Bengal Tenancy Act 27
Benson, C. 38
Bentinck, William 1
Bernier, Francois 193–4
Beveridge, H. 196–9
Bhagalpore Agriculture College 57
Bharadwaj, R.B.L. 295
Bhesajavidya 3
bigyan 13
Borlaug, Norman 290, 293–4
botanical knowledge 3–4
Bradfield, Richard 290
Brahmanas 3
Brayne, Frank 108
breeding policies 13
Buck, E. 20, 32–3
Burdwan, agricultural knowledge and practices of 11–12, 168
   changes in crop pattern 175–7
   constraints of irrigation in 187
   critique of government's role 183–6
   debate on growth rate of food crop output 169–72
   experimental works of Burdwan Agricultural Farm 186
   implements used in irrigation 179
   irrigational knowledge and practice 180–3
   melan and *sech* methods of irrigation 180–1
   method of cultivation and agricultural practices 177–80
   productive character 172
   raiyats of 178–9, 184–5
   resource endowments for agricultural production 172–4
   saja or *dhanthika* system of cultivation 180
   tank irrigation 182–3
Burdwan Raj family 172

Calcutta *gowkhana* (cow sheds) 221–2
Calcutta Society for Prevention of Cruelty to Animals (CSPCA) 221, 223
Carey, William 227
Carmichael, C.P. 30
Carter, Ray 109
cash crop economy 2, 21–6, 53
cattle breeding policies in colonial India 28, 238–40
   breeding cattle for ploughing on farm and carting 243–5
   cross-breeding of Indian cattle with exotic bulls 239, 247–9
   debates around breeding policy 250–6
   'dual purpose' breeding policy 250–6
   efforts in direction of cattle improvement and scientific breeding 245–7
   introduction of scientific breeding 240–3
cattle ecology in rural Bengal
   animal disease 216–18
   'anticattle ideology' of the British, impact of 220
   cattle murrains and mortality 215–16
   cattle plague 215
   colonial veterinary policy and 217, 219–21
   cruelty to animals 222–4
   death and distress 218–19
   epizootics 215–21
   go-dagas, case of 224–6
   health problems 227–30
   knowledge of veterinary science 226–7
   sentiment of compassion towards animals 221–2
   slaughterhouse regulations 218
Cattle Plague Commission in 1870 220
Central Provinces Debt Conciliation Act of 1933 272
Central Provinces Moneylenders Act of

1934 272
Chakrabarti, Pratik 56
Chand, Rai Bahadur Chaudhuri Lal 268, 275
Chandabhandas 198
Chatterjee, Sutapa 12
Cheetham, John 24
Christian communities 9
Christian communities in agriculture and cattle breeding. *see* missionary involvement in agriculture and cattle breeding
Civil Procedure Code 268–9
Clogstoun, H.F. 33
Clouston, D. 16
Colebrooke, H.T. 172
colonial Bengal, modernization of agriculture in
  application of modern science in agricultural production 55–7
  colonial interventions 51–5
  historiography 50
  improvement of Indian agricultural production 51–2, 54
  location and timeframe of study 50–1
  role of the Bengali population 59–60
  setting up of experimental farms 57–9
colonial economy 2
colonial experiments in agriculture 4–6
  cotton 5, 23–5
  dissemination of scientific agriculture, issues with 6
  hill slopes 147–8
  horticultural experiments 147–8
  Indian response to modern scientific tradition of agriculture 6–10
  indigo 25–6, 53
  intervention in tribal agriculture 147–8
  introduction of plants 5
  jute 53
  official attitudes to agricultural problems 36–40
  phases in the attitude of colonial government 5–6, 36–7
  shifting cultivation 146–7
  tea 23, 25, 53
  tobacco 26
Contagious Diseases (Animals) Act in 1869 220

Cooch Behar, agricultural farm of 58, 68n39
Cotton Supply Association of Manchester 23
Coventery, B. 35
crop rotation and mechanization 123
cross-breeding of Indian cattle 14
Curzon, Lord 26, 37–8, 56

Dar Fann-i-Falahat 4
Das, Chittaranjan 61
Dehra Dun Forest School 100
Dentith, A.W. 153
Department of Land Records and Agriculture 30
DeValois, J.J. 111
dual purpose breed 14
Dubey, Dayashankar 76
Durand, Henry 28

East India Company 2, 4–6, 25, 50
  botanical garden 5
  issues on dissemination of scientific agriculture 6
educational curriculum 7, 9
Elmhirst, L.K. 128
experimental farms 57–9, 67n32, 68n34
  dissemination project of 59

Famine Commissions of 1866, 1880 and 1901 28, 32, 52, 54–5, 282
famines 21, 27–8, 53–5, 76–7, 91, 116n6, 250–1
Fitzgerald, Deborah 295
food crisis 285–9, 293–4
foodgrain yield per acre, 1891–1947 17
Ford Foundation 286–7
Frankel, Francine 284
Fuller, J.B. 34–5

Gangopadhyay, Nagendranath 123
Gastrell, J.E. 196–7
Ghazi-Kalu-Champavati-Kanyar-Punthi 201
Ghosh, Hridoy Nath 222–3
Ghosh, Pratap Chandra 198
Gijang reserve 159
Go-Bandhab (Friend of Cow) 229
Goheen, John 9, 110–11
Gramer Dak 61

320 *Index*

Great Famine of 1942 15
Green Revolution technologies 16–17, 283–5, 293–4, 298
Groff, George Weissman 102
Grow More Food campaign 15, 282

'Halakarshan' 138
Hamilton, Sir Daniel 265
Harishchandra, Bharatendu 74
Harper, Arthur 110
Harper, Irene 110
Hatch, D. Spencer 112–13
Hayes, Brewster 109
Henry, R.J. 39
Higginbottom, Sam 9, 102, 104–5, 108–9, 118n20–4
Hindi literati 85, 92–3
Hindu revivalism 61
Hopper, David 285
Howard, Albert 14, 16

The Indian Agricultural Gazzette 39
The Indian Agriculturist 39
Indian Association of Cultivation of Science 51
Indian Jute Mill Association 116n4
Indian Tea Association 6, 37
Indigo Improvement Syndicate 25–6
Indus Valley Civilization 3
infirmary for animal protection 222
insolvency laws 269–70
Intensive Agricultural Development Programme (IADP) 288

Jackson, C.G. 23
Jain, A.P. 292
Jnandạyini Sabha 222–6
Joshi, N.M. 263, 265

Kerala Livestock Improvement Act in 1961 258n38
Khan, Nawab Major Malik Talib Mehdi 266
Khasi and Jaintia hills, agricultural practice of 145
  colonial interventions 147–8
  land tenure system 148
  potato cultivation 148

  practice of *jhumming* 149–50
  settled rice agriculture, introduction of 150–3
  Tejpat, pan or betel leaves and wild cinnamons, cultivation of 148
Kothawala, Zal R. 255
Krishak (Agriculturist) 229
Krishi Gazette 229–30
Krishi Lakshmi 229
Krishnamachariar, Raja Bahadur G. 262
Krisi Sampada 63
Kumar, Deepak 56, 71
  Science and the Raj: A Study of British India 71
Kumarappa, J.C. 256
Kyd, Robert 5, 25

Lahiri, Ashutosh 62
Lahore Veterinary College 221
Lamb, R.A. 40
Leake, Martin 17
Lindsay, Sir Darcy 265
Liverpool East India Association 23
Livestock Improvement Act of 1933 253, 256

machine technology 15
Madhuri 74, 79
Madras Presidency 103
Mahabharata 196
Majumdar, Santosh 127
Manchester Cotton Supply Association 100
Mangelsdorf, Paul C. 290
Mann, Harold 16
Maryada 74, 79
masterly inactivity phase 5, 36
Mauryans 3
Mayo, Lord 28, 55, 100
Mazumdar, Santosh Chandra 123
McKee, William 110
meteorological knowledge 4
Methodist missionaries 111. *see also* missionary involvement in agriculture and cattle breeding
Mikirs 155
milk markets 14
Mills, J.P. 148
Minto, Lord 56

Mishra, Saurabh 217
missionary involvement in agriculture and cattle breeding 99–101
　agricultural improvement in Travancore 111–12
　beginnings 101–4
　colleges, establishment of 102, 106, 117n10
　converts, training of 112, 116n6
　formation of agricultural societies 110
　goat breeding and research centre 114
　Gwalior project 108
　introduction of steel plough 109
　Katpadi experiment 111–12
　model farming 108
　model villages 111
　Poor Man's Cow project 112
　poultry farming 114
　poultry project 114
　in Punjab 109–11
　Rothampstead agricultural experiments 106
　rural concerns and the Allahabad Experiment 104–9, 119n31
　rural development at Etah in northern India 113–14, 121n53, 121n55
　1930s 103, 115
　in south India 111–12
　teenagers at Martandapuram, training of 113
　Ushagram experiment 111
Mitra, Satish 199
modernization of agriculture 6–10, 71, 77, 91–2
　role of science and technology in 79, 96n60
Moseman, A.H. 297
Mukerjee, Radhakamal 16
Mukherjee, H.N. 291
Mukherjee, N.G. 22
Mukherjee, R. K. 195
multiple land tenures 270–2

Nakhl-bandiya 4
napier grass cultivation 13
Nightingale, F. 54
North-East, agricultural practices in 10–11. see also Khasi and Jaintia hills, agricultural practice of; tribal agriculture of Khasis
Nuskha-i-Kukh-bad 4

Olver, Arthur 14, 239, 253–6

Pal, Babu Manmatha Nath 185
Pasteur Institute 35
Patil, Rao Bahadur B.L. 267
Patil, S.K. 288
Pearly, J.H. 23
Permanent Settlement 50–2, 186, 228
Phipps, Henry 35
plantations 2, 21
Poona Civil Engineering College 31, 35, 38–9
Prakash, Gyan
　Another Reason: Science and the Imagination of Modern India 71
pre-colonial economy 2
Premia System 111
Prevention of Cruelty to Animals Act
　of 1861 224
　of 1891 225
price movement of 1930s 15
Protestant agricultural missions 102–3. see also missionary involvement in agriculture and cattle breeding
Punjab Land Alienation Act in 1901 266
Punjab Regulation of Accounts Act of 1930 272
Punjab Relief of Indebtedness Act of 1934 272
punthi literature 201
Purana 61
Pusa agriculture college 35–6

Raha, Bipasha 10, 71
raiyats 26–7
Randhawa, M.S. 297
Reid, W.J. 153
risk-taking farmer 296
Risley, H.H. 35
Robertson, W.R. 54
Roxburgh, William 5
Roy, Arnab 7
Roy, M.N. 52
Royal Commission of Agriculture 14, 51, 114, 119n38, 239, 246, 250, 253, 265, 274, 276

rural indebtedness 264–5

Saha, Madhumita 15–16
Saidapet Agricultural College 32–3
sal forest 153–4, 158–60
Samanta, Samiparna 13
Samhitas 3
Santals 206, 213n77
Saraswati 74, 79
Sarkar, Smritikumar 71
Sarsudhanidhi 74
Schrottky, Eugene C. 55
scientific agriculture 6–7, 9–10, 33, 39, 82, 87, 89–91, 96n60
Scott, David 147
Sen, Sudhir 297
sericulture 26–7
settled rice agriculture 150–3, 163
Settlement Rules of the Assam Land and Revenue Manual 152
settlements in the Sundarbans
  aspects of settlers' lives in 200–2
  colonial ideas in 202–7
  lowlands 194–6
  mangroves 192–4
  marginalized people of 195
  nature of 196
  refugees in 207
  settlement debates 196–200
  topography of 195
  total area of Sundarbans 193
Shastri, Lal Bahadur 291–3
Shenoy, B.R. 292
Shibpur Engineering College 57, 100
shifting cultivation (*jhumming* cultivation) 146–7, 149–51, 153–62
  argument for 156–8
  as destroyer of forests 158–61
  use of fire in 147
Simla Conference on Agricultural Education 38
Singh, Datar 256
Singh, Gaya Prasad 273
Singh, Kanwar Mohinderpal 295
Singh, Rai Bahadur Kunwar Raghuvir 263
Slater, Arthur H. 113–14
Sly, F.G. 23
Smith, W.M.H. 22
speculation 1

Sriniketan 10
Stakman, E.C. 290
The Statesman 39
Stephens, Revd C.L. 150
Stevenson-Moore, C.J. 206
Subramaniam, C. 289, 292
superactivity phase 5, 37
Swaminathan, M.S. 291–3, 295
swidden agriculture 145–6

Tagore, Debendranath 122
Tagore, Joteendro Mohun 227–8
Tagore, Nagendranath 127, 136
Tagore, Rabindranath 10, 122, 138–40
  agrarian festivals, fairs and exhibitions 138
  Brati Balaks 130
  classification of land and varieties of soil 130–2
  cropping pattern 130, 132
  crossing of Leghorn cocks 137
  cultivation of rabi crops without irrigation 132
  dairy farming 136
  district agricultural association, establishment of 134
  enduring phase at Sriniketan 127–38
  experiment with Nainital potatoes 142n21
  'extension' programme 135
  idea of cooperative farming 135
  impartment of training 134–5
  maintenance of a seed bank 129–30
  paddy experimentation 129
  Patisar agrarian experiments 126–7
  poultry industry 137
  principal crops cultivated 133
  Shilaidaha agrarian experiments 123–6, 142n25
  silo method of fodder preservation 130
  Suri Cattle and Produce Show 134, 136
  system of crop rotation 129
  tank irrigation 134, 143n58
  well irrigation 132
Tagore, Rathindranath 127
Tamaskar, Gopal Damodar 76
Tebhaga movement of 1946 203
Temple, Richard 6, 28, 31, 37
Thampan, K.P. 275

Index 323

Tomlinson, B.R. 169
Travers, Robert 203
Treaty of Yandabo 147
tribal agriculture of Khasis 145–6
   practice of *jhumming* 146, 149–51, 153–61
   Sikkim variety of rice, experiment with 151
   terrace cultivation 151–2

United Provinces Arrears of Rent Act of 1931 272
United Provinces Assistance of Tenants Act of 1932 272
United Provinces Encumbered Estates Act of 1934 272
United Provinces Temporary Regulation of Execution Act of 1934 272
Upadhyay, Himanshu 13
Upavanvinoda 3
Usurious Loans (Central Provinces Amendment) Act of 1934 272
Usurious Loans (United Provinces Amendment) Act of 1934 272

Vana Mahatsova 138
Varsha Mangal festival 138
Varshney, Ashutosh 284
Vaugh, Mason 109
Vayu Puran 196
Vedic farmers 3

vernacular press 74
veterinary education as curriculum 217
Vigyan 74, 79
village improvement (*gram sudhar*) 9, 93
   land improvement 80–5
Voelcker, J.A. 5, 24, 29–30, 32–3, 37–8, 55, 100, 171, 221, 228
Vrksayurveda 3

Wardle, T. 22
Warner, James 109
Watt, George 26
Wellesley, Lord 30
Wilson, W. 32
Wright, Bill C. 295
Wright, N.C. 14, 239–40
writings on agriculture
   concern of declining fertility 75–6
   development of agriculture 78
   distress in agriculture 75
   exploitation of land 80
   general writings 75
   in Hindi periodicals 75–6, 91
   improvement of land 80–5, 93
   modernization of agriculture 79
   references to British rule as cause for agriculture decline 78

Zamindari School 35
zamindari system 38, 50, 123